Standard Methods
of
GEOPHYSICAL
FORMATION
EVALUATION

James K. Hallenburg

CRC Press
Taylor & Francis Group
Boca Raton London New York

CRC Press is an imprint of the
Taylor & Francis Group, an **informa** business

First published 1998 by Lewis Publishers

Published 2019 by CRC Press
Taylor & Francis Group
6000 Broken Sound Parkway NW, Suite 300
Boca Raton, FL 33487-2742

First issued in paperback 2020

ISBN-13: 978-0-367-57940-1 (pbk)
ISBN-13: 978-1-56670-261-4 (hbk)

Visit the Taylor & Francis Web site at
http://www.taylorandfrancis.com

and the CRC Press Web site at
http://www.crcpress.com

Library of Congress Cataloging-in-Publication Data

Hallenburg, James K.
Standard methods of geophysical formation evaluation/James K. Hallenburg.
      p.   cm.
    Includes bibliographical references and index.
    ISBN 1-56670-261-5
    1. Engineering geology—instruments.   2. Geophysical instruments.
  3. Prospecting—geophysical methods.   I. Hallenburg, James K. II. Title.
TA826.H34   1997
551'.028—dc21

                                 97-29872
                                     CIP

Library of Congress Card Number 97-29872

Contents

Preface

This text, *Standard Methods of Geophysical Formation Evaluation*, examines the theory, practice, and instrumentation of some of the standard, older (but still of primary importance) geophysical and logging methods. It examines electrical and porosity determination methods. While none of these are examined in exhaustive detail, the author has included references for the reader who wishes to examine any of the areas more closely. This text is not intended to replace any of the many available books on log analysis. Some of them are good, some are poor, some are highly specialized. Almost all of them restrict their discussion to borehole geophysics (logging). Most treat only the subject of data reduction: reading logs. This book is intended as a reference text, for teaching classes or individuals. Also, I have tried, in addition, to cover subjects which are not always covered in logging and geophysical volumes and explain why these things occur and are important. Finally, I have covered other uses than petroleum well logging.

The question has appeared many times: "Why do we need another text about petroleum logging?" The facts are that the geophysical field, including borehole geophysics (logging) has been dominated and largely financed by the petroleum industry. Most of the basic research and engineering in geophysics has been done by the petroleum industry. The basic ideas of the geophysical methods are mostly the same of both petroleum and non-hydrocarbon geophysics; these must be understood in order to do any effective work in geophysics. Where do we do to get information about these methods? We go first to the petroleum methods, the petroleum research and the petroleum engineering. Thus, this text about the standard methods in geophysics is largely about borehole geophysics and oil-field systems, but it does dare to suggest that logging is not the only geophysical method available. I have also attempted to draw the connections between petroleum usage and non-hydrocarbon usage.

It is a truism that logging people think in logging terms, seismic people think in surface seismic terms, and geologists consider mostly geology. This is natural. The fact is, however, that a good explorationist (or any kind of an analyst) must consider *all* of the information he can garner, whether it is in his field of expertise or not. Time and again we see projects repeating work that was already done, because the researcher had not looked at other, similar fields.

Many of the older logging systems have been improved tremendously since they were first introduced. In fact, some of them are barely recognizable as similar to the older methods. Many of the older logging systems, however, are quite suitable for some forms of analytical work. And frequently, they are faster and less expensive. A digital system is a marvel and wonderful to work with. If one is back-packing into a remote location, however, a faulty replacement circuit board is hard to find and even more difficult to replace. A pocket-full of common components is easy to carry and an analog system can be easily repaired.

Any geophysical method can be expensive. Thus, it (or they) should not be used without some idea of *how to use the information.* If one is informed in this manner, geophysical methods can be extremely cost effective and instructive. I have seen projects run everything available simply because the manager did not know what was needed. This is a terrible waste of a budget.

Please note that the nomenclatures and notations used in the drawings and tables of this volume are not always identical to those of the contractor from whom much of this information originated. This was done to preserve uniformity. Each contractor, of course, has his own nomenclature. This same comment is true of the trade names of the systems. I have tried to use standard notations, as listed in the CRC *Handbook of Chemistry and Physics.* Where possible, generic names are used. Of course, if a specific system or device of a particular contractor is mentioned, the trade name is used and credited.

Many tables, diagrams, figures, and charts have been used in this volume and are from other volumes of mine. Quite a few of them originated from the several contractor's chart books and literature. Some came from existing literature. *Many of these illustrations have been copied with a scanner and may have been redrawn or cleaned up.* The purpose was to emphasize a particular point. It was also to put them into a form which could be handled by the computer. It is not always possible to achieve the degree of accuracy as in the original. In *every* case, a great effort was made to realize as great an accuracy as possible. Nevertheless, when one uses one of these diagrams, figures, or charts, he should go to the original. This will ensure the best results and avoid any possible inaccuracies which may inadvertently be present in these copies. The same caution applies to the numerical tables. These values are under constant investigation and sometimes change from day to day. Always use the latest figures available.

The several contractors and other sources who kindly allowed me to use their drawings are due a sincere vote of thanks. The series would be much poorer without their contributions. I have tried to give credit wherever it was due. I would especially like to thank Schlumberger Well Services, Inc., Western Atlas Logging Services Division of Western Atlas International, Inc., Baroid Well Fluids, and Century Geophysical Corporation, BPB Instruments. This does not mean that I do not appreciate the many other corporations and individuals who allowed me to use their data and illustrations.

James K Hallenburg
Tulsa, Oklahoma

The Author

James K. Hallenburg is currently a retired geophysicist/petrophysicist living with his wife, Jaquelyn, in Tulsa, Oklahoma. Jim was born in Chicago in 1921, and has traveled to and worked in many corners of the world. During World War II, Jim was a bomber pilot and weather observer in the U.S. Army Air Corps. and still enjoys flying.

Following his stint in the military, Jim returned to his studies, graduating with a B.A. in physics from Northwestern University. Throughout his career, he has attended a number of schools to further his education and has taught courses at several of them. Jim spent 18 years as an engineer, designing geophysical systems for Schlumberger Well Services He was a Senior Engineer for the Mohole Project and Chief Engineer for the Western Company of North America. He also operated and owned Data Line Logging Company in Casper, Wyoming. Later, he was Manager of Applications Engineering for Century Geophysical in Tulsa. Jim has also been a consultant for the International Atomic Energy Commission.

Subsequently, Jim spent several years giving seminars in geophysics and formation evaluation. He is the author, co-author, and editor of several books and computer programs and has held office in numerous technical societies.

1

Resistivity Methods

1.1 Introduction

Most of the discussion in this book will involve downhole methods. However, the techniques and instrumentation descriptions will generally apply to laboratory, oceanographic, and surface applications, also. We find three general classes of downhole, geophysical, resistivity measuring instruments: (1) standard or unfocused systems, (2) focused systems, and (3) electromagnetic systems. The unfocused methods include the single point, normal, lateral, micro-resistivities, older dipmeters, and many of the surface methods (except for the electromagnetic methods). The focused methods include the guard logs, micro-guard logs, high-resolution dipmeters, and the spherically focused logs. Focused methods are also used occasionally on the surface. These methods all use electrodes in the mud column (occasionally against the wall of the borehole) to introduce current into the formation material and to read the resulting voltage drops as signals.

The electromagnetic methods (EM) are the induction logs, the electromagnetic propagation time logs (EPT), the EM logs, and the EM surface methods. The electromagnetic methods generally use a coil or an antenna to induce an electromagnetic field into the formation and usually to read it, also.

Array and scanning methods are new categories. They will be covered in a separate chapter. Focused and unfocused resistivity methods and the electromagnetic methods are widely used in oceanographic studies and laboratory methods.

Electrodes may be combined if the circuit is properly designed. That is, a single electrode may have multiple uses, even simultaneous uses. Thus, the A electrode for the normals may be the B electrode for the laterals. It is common practice to use the M_{16} electrode for the SP measurement. Since measured borehole resistivity values usually need correction, the uncorrected value is called the apparent resistivity, R_a. For these discussions, the current required by a voltage measuring system is assumed to be zero. That is, the input impedance of the voltmeter is assumed to be infinite. This, of course, is not true. With modern circuitry, however, the input impedance is usually very high (10^8 to 10^{15} ohms) and is effectively infinite. This is something which should be checked, especially when working with older equipment and older data.

All unfocused resistivity arrays have at least four electrodes. Some may be at an apparent infinite distance away or may be combined with other electrodes. But, when considering downhole methods, only the downhole electrodes involved with a type of device are usually discussed. The other electrodes are considered to be an infinite electrical distance away. At this distance, they would not affect the performance characteristics of the downhole system. Thus, a normal system may be called a two electrode system, because it must have an *A* electrode and an *M* electrode downhole. Of course, it also has an *N* and a *B* electrodes. But, they may be so far away that they are at an apparent infinite distance. In this case, the apparent infinitely distant electrodes are so far away they may be moved without noticeably affecting the measurement.

The media in which we will be working, the formation materials, will contain electrically charged bodies. Metallic conductors have free electrons with single negative charges. Semi-metallic conductors have electrons and holes (which have apparent positive charges). Solutions have positive ions (cations) and negative ions (anions) which will have one or more charges. The charged bodies can be made to flow in response to either (or both) an electrical field or a magnetic field. When these charged particles flow under the influence of an electrical or a magnetic field, they constitute an electrical current.

Resistivity methods constitute the most important single group of logging methods in petroleum formation evaluation. These methods are among the few which allow us to directly detect the presence of hydrocarbon. In addition, they tell us something about the clay/shale content, formation waters, stratigraphy, era and period of the zones, and many other things. They are important in all types of geophysical investigation. They are most important in downhole geophysical investigations, especially in the search for and the evaluation of hydrocarbon deposits. Electrical resistivity methods (these will simply be called "resistivity methods") are extensively used in laboratory investigations and frequently used in surface measurements. The latter is often and routinely used with induced polarization techniques.

1.2   Nomenclature

To eliminate confusion between the resistance, r, and the radius, r, in this text the resistance and the radius will always have identifying subscripts. Thus, the radius of the electrode is r_e. That for the water resistance will be r_w.

When electrode methods are discussed in this text, conventional electrode designations will be used. The current electrode is called the "*A*" electrode.

Since a return path is required for the current, the return electrode, labelled "*B*", will be used. *B* may be close to the *A* electrode. This constitutes a dipole. Or it may be an apparent infinite distance away. Then the *A* electrode is treated as a monoelectrode (a monopole). The "apparent infinite distance away" is a distance so great that changing the position of the return electrode at that distance will negligibly affect the measurement signal. The electrodes may be conventional electrodes or they may be the cable armor, electrodes in the mud pit, metal stakes in the ground, or any other kind of terminal, junction point, or contact.

Similarly, the measure electrode will be designated *M*. The measure return is the *N* electrode. Because they work at lower signal levels than the current electrodes, more care is usually taken with the design of the measure electrodes. Again, if the *M* and *N* electrodes are close, they constitute a dipole. If the *N* electrode is an apparent infinite distance away, the *M* electrode is considered a monoelectrode. The *M* electrode is often made as small as possible, to be as near a point measurement as possible.

If there are several electrodes, they will have subscripts, such as N_1 and N_2 or A_1, A_2, and A_3. If the electrode has a special purpose, then the subscript will show that. Thus, a normal device electrode may be M_N or, if it is a 16-in normal measure electrode, it may be M_{16}. The distance between any two electrodes (their spacing) will be indicated by boxing the two electrode symbols; thus: [*AB*] or [*BN*].

Resistivity measurements require that an electrical current, *I*, be caused to flow through the material to be measured. We can put an electrical field, *E*, or a magnetic field, *H*, across or through the sample to accomplish this. The voltage drop or magnetic field strength required for a given current will be proportional to the resistivity of the formation material. The ratio of the voltage drop, *E*, to the amount of current, *I*, is the electrical resistance, *r*, of the material.

When the potential of an electrode is greater than zero, a current will flow from the electrode to a lower potential. Electrochemical reactions will take place at the surface of the electrode, which is the interface between the electrode and the medium the electrode is in. The reaction type and rate will be a function of the current density (among other things) on the electrode surface. The reactions are usually nonlinear. Usual resistivity system practice is to periodically reverse the polarity of the current to keep the average of the effects of the reaction at zero; i.e., to balance them out. Currents are kept as low as possible to minimize the reactions.

Most unfocused systems (all older systems) have passive control systems. That is, they do not use electronic control or monitoring systems. They are often characterized by their variable measurement geometry. Most modern focused systems (and many of the unfocused systems) have some degree of active control and/or monitoring. Focused systems will frequently have more than one measurement and/or one current electrode (although some of these may be combined). The additional electrodes are used to control the guard or focusing current and to monitor the fields around the tool. All modern focused resistivity systems use electronic (dynamic) control. They can be designed to do a number of special measurement tasks. And, they are characterized by their stable geometry and wide dynamic range.

Although this discussion will mostly be about downhole systems, the principles apply equally well to surface systems. The major differences between surface and downhole systems are that different arrays are popular in the two types of measurements because of the distances, positions, and volumes involved. Also, the surface methods usually measure a half of a spheroid of material and thus have the factor 2π in their mathematical definitions. On the other hand, the downhole methods usually have the factor 4π. They measure a full spheroid of material.

Also, the petroleum or hydrocarbon methods will primarily be examined in this book. These methods usually can be directly used or easily modified to use in a multiplicity of non-hydrocarbon applications. For example, oceanographic projects usually retain one of the major oilfield logging contractors. These methods are the basic ones from which other applications are derived.

1.3 Unfocused Resistivity Methods

Unfocused resistivity systems in present use are the single point, normals, laterals, inverses, and the micro-resistivities.

In an infinite, homogeneous, isotropic medium (a large tank of water, for example), a current of strength I will flow radially from a point electrode, A, to a return electrode an apparent infinite distance away. See Figure 1.1. Therefore, the current density, j, depends upon the distance (radius), (r_e), from the electrode:

$$j = \frac{I}{4\pi r_e^2} \tag{1.1}$$

This is essentially the inverse square law.

The incremental potential drop, $d(E)$, through the incremental distance $d(r_e)$, from an application of Ohm's Law, is

$$-d(E) = jR dr_e \tag{1.2}$$

where R is the resistivity of the medium surrounding the electrode. Substituting Equation 1.1 for j,

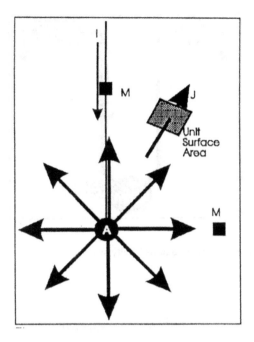

FIGURE 1.1
The electrical current flow around a spherical electrode in a homogeneous medium.

$$-d(E) = \frac{IR}{4\pi r_e^2}\, dr_e \qquad (1.3)$$

The potential gradient or electric field is,

$$\frac{-d(E)}{dr_e} = \frac{IR}{4\pi r_e^2} \qquad (1.4)$$

S_o, you can see, the response decreases exponentially with distance from the electrode. This is illustrated in Figure 1.1 and Equation 1.1.

The potential, E_M, at any point, M, in the field, at a distance $[AM]$, is

$$E_M = \int_{[AM]}^{\infty} -d(E) \qquad (1.5)$$

Substituting Equation 1.4 into 1.5,

$$E_M = \frac{RI}{4\pi} \int_{[AM]}^{\infty} \frac{1}{r_e^2}\, dr_e = \frac{RI}{4\pi[AM]} \qquad (1.6)$$

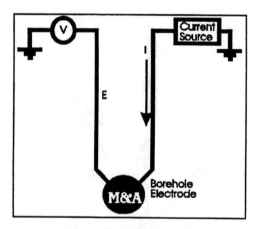

FIGURE 1.2
A schematic diagram of a single-point resistivity system.

or

$$R = 4\pi[AM]\frac{E_M}{I} = 4\pi[AM](r) \tag{1.6a}$$

where r = the resistance of the medium. This is the same form as the general expression for all resistivity systems:

$$R = G_R(r) \tag{1.7}$$

where G_R is $4\pi[AM]$, the geometrical conversion constant. This relationship is valid in a homogeneous medium.

1.3.1 Single-Electrode Systems

The simplest resistivity system is the single-point system (SGL), also called the monoelectrode or, erroneously, the point electrode system. You will seldom see this used in petroleum work, because of their large oil well hole diameters. It is, however, used extensively in other fields, in small-diameter boreholes and usually anywhere a quantitative resistivity value is not needed. It is the starting point for the analyses of all of the surface and downhole systems.

The single-point resistivity system consists of a single electrode downhole which is both the A and the M electrode. The circuit is shown in Figure 1.2. And,

$$R = K_G \frac{E_\ell}{I} \tag{1.8}$$

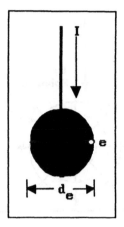

FIGURE 1.3
A representation of a single spherical electrode.

where

$$K_G = 4\pi r_e$$

r_e = radius of the spherical electrode

In practice, the electrical logging current, I, is kept constant. Thus, the resistivity of an infinite, homogeneous medium in which the electrode is immersed, is directly proportional to the electrode potential.

The potential of a spherical electrode, E_e, of radius $(d_e/2)$, will be that of the equipotential sphere which coincides with the surface of the electrode (point e is anywhere on the surface of the electrode) (Figure 1.3):

$$E_e = \frac{RI}{4\pi r_e} \tag{1.9}$$

E_e is the potential of the electrode, A. The ratio, E_e/I, is the resistance, r_g, in ohms, which would be calculated from the voltage and current measurements:

$$r_g = \frac{E_e}{I} = \frac{R}{4\pi r_e} \tag{1.10}$$

r_g is the effective resistance of the electrode to infinity. This is called the "grounding resistance". The resistivity of the medium, R, is

$$R = 4\pi r_e r_g \tag{1.11}$$

This is the same as Equation 1.7, where the geometrical factor, G_R, is

$$G_R = 4\pi r_e \tag{1.12}$$

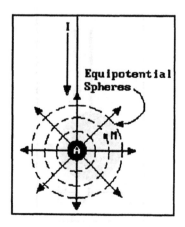

FIGURE 1.4
Equipotential spheres.

It describes the geometry of the measurement.

The grounding resistance to infinity of any equipotential sphere of radius, r_M (the distance [AM]) centered on the center of the electrode, Figure 1.4, is

$$r_{eM} = \frac{R}{4\pi} \frac{1}{r_M} \tag{1.13}$$

Spherical electrodes are difficult to build into a cylindrically shaped logging tool. Therefore, we usually work with cylindrical electrodes. The grounding resistance of a cylindrical electrode is

$$r_{e,cyl} = \frac{R}{4\pi L_e} \ln \frac{2L_e}{d_e} \tag{1.14}$$

or

$$R = 4\pi \frac{L_e}{\ln\left(\dfrac{2L_e}{d_e}\right)} \frac{E_e}{I} \tag{1.15}$$

where
    L_e = the length of the electrode
    d_e = the electrode diameter, in meters
    ln = the natural logarithm (to the base e)

The geometrical constant, G_R, in this case, is

$$G_R = \frac{4\pi L_e}{\ln\left(\dfrac{2L_e}{d_e}\right)} \tag{1.7a}$$

If the length of the cylindrical electrode is about the same as the diameter, the response will be very close to that of a spherical electrode. The size you will most often see in actual practice is about 1 in diameter by 1.5 in long (2.5 cm diameter × 3.75 cm long).

Note, in Equations 1.13 and 1.14, that the grounding resistance, for any given medium resistivity, is a function of the electrode length and diameter. The electrode diameter has not as great an influence on the resistance as the length, because the grounding resistance is a function of L_e (which can be any practical value) and of the $\ln(1/d_e)$ (which remains close to 2 or less). The length is a much more significant factor than the logarithm. Thus, we will find that the amplitude of r_e, for any given value of R, can be easily decreased or increased by changing the length of the electrode. The amount of detail "seen" by the system will decrease with an increase of length, L_e. The length of the electrode *must always* be on the heading of a single-point resistance log, because manipulating L_e to obtain a larger or smaller amount of detail is common practice.

The relationships for the single-point resistivity systems are the starting point for evolving and describing those for the multi-electrode systems. Example: the grounding resistance of a conventional petroleum type electrode can be calculated. The typical dimensions of the electrode are:

Diameter, d_e = 3 in (0.076 m)

Length, l_e = 4 in (0.102 m)

Surface Area, $A_e = d_e l_e$ = 0.0243 m

$R_e = [R/4\,(0.102)]\,\ln[2(0.102)/0.24)] = 1.7R$

Therefore, a 3 × 4 in electrode will have a grounding resistance, in ohms, numerically 1.7 times the resistivity. And, in practice, it will vary from about 0.17 to 1700 ohms. Normally, it will be 17 to 170 ohms.

1.3.2   Equivalent Circuit

With any unfocused resistivity system, a simple model is analogous to a parallel resistor circuit. If the borehole and the formation are considered as resistances in parallel, the grounding resistance, r_e, of an electrode is similar to the circuits shown in Figure 1.5.

The mud resistance, r_m, depends upon the resistivity of the mud, R_m, the borehole diameter, d_h, and the borehole length, L_m:

$$r_m = \frac{L_m}{4\pi \left(d_h\right)^2 R_m} \qquad (1.16)$$

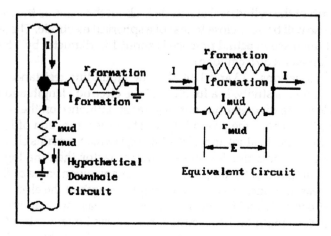

FIGURE 1.5
The approximate equivalent current circuit of the resistivity measurement.

If L_m is 25 in (63.5 cm.), this will include 98% of all of the mud contribution. And, if the hole diameter, d_h is 4 in (10 cm.), then $r_m = 0.2\ R_m$ (the conductance, $c_m = 1/r_m$ and is, therefore, $5(1/R_m)$, the conductance, C_m, of the mud).

The total grounding conductance, c_e, of these parallel resistors is

$$c_e = c_m + c_f \tag{1.17}$$

where c_m is the conductance of the borehole and c_f is the conductance of the formation material. Or, as resistance,

$$\frac{1}{r_e} = \frac{1}{r_m} + \frac{1}{r_f} \tag{1.18}$$

If the borehole is 4 in (10 cm.) diameter,

$$r_e = 0.25 \frac{R_m R_f}{R_m + 1.25} \tag{1.19}$$

The logging current, I, which is held constant, is

$$I = I_m + I_f \tag{1.20}$$

The voltage drop, E, is the same across each. Therefore, the current through the formation is

$$I_f = I - I_m = \frac{E}{r_f} \tag{1.21}$$

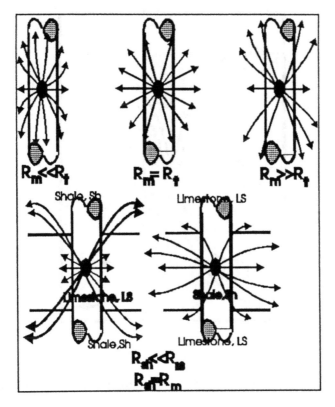

FIGURE 1.6
Current distributions in zones of various resistivities.

1.3.2.1 The Effect of Formation Resistivity

The current through the formation is inversely proportional to the resistance of the formation. The higher the resistance (the resistivity) is, the less current flows through the formation and the more through the mud column. Thus, the higher the formation resistivity is, the greater the contribution to the signal is by the mud column. See Figure 1.6. If the formation resistivity is high compared to the mud resistivity, the logging current will tend to flow in the mud column, and avoid the formation. Thus, the resistivity reading, R_a, will be lower than the true resistivity, R_t (or R_{xo}). See Figure 1.7. In an extremely high-resistivity formation, the reading will be primarily from the mud column. This can be especially misleading. If the hole is bottomed in a very-low porosity rock, for example, in a massive, tight carbonate or a granite, the reading is often entirely from the borehole.

The borehole effect can be especially aggravated if a high-resistivity bed is fairly thin and surrounded by low-resistivity beds. In this case even the focused systems can be affected. The current will tend to flow up and down the borehole into the adjacent beds. The apparent resistivity, R_a, value must be corrected for adjacent bed resistivities.

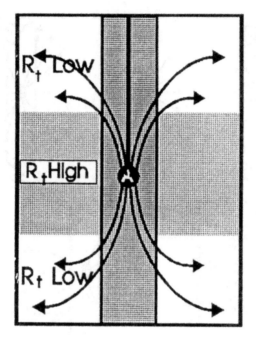

FIGURE 1.7
A very-high formation resistivity situation.

A single-point resistivity log from Wyoming is shown in Figure 1.8.

1.3.2.2   Volume of Investigation

The proportional resistance which is contributed by any sphere of formation (the resistance from the electrode surface to a sphere, M, of radius [AM]), is equal to the difference of the electrode grounding resistance, r_e (Equation 1.10), and the grounding resistance of the external sphere, M, of formation, r_{eM} (Equation 1.13) (the resistance from sphere M to infinity). For example, the size of the sphere which will contribute any portion of the measurement may be found:

$$F = \frac{R}{4\pi}\left(\frac{1}{\left(\frac{d_e}{2}\right)} - \frac{1}{\frac{d_M}{2}}\right) \tag{1.22}$$

where F is the factor indicating the proportion contributed by a volume of formation material (i.e., 0.5 or 0.7 or 0.9, etc.).

If we solve for the diameter of the sphere whose volume contributes 50% ($F = 0.50$) of the response, ($d_M/2$), we find

$$d_M = 2d_e \tag{1.23}$$

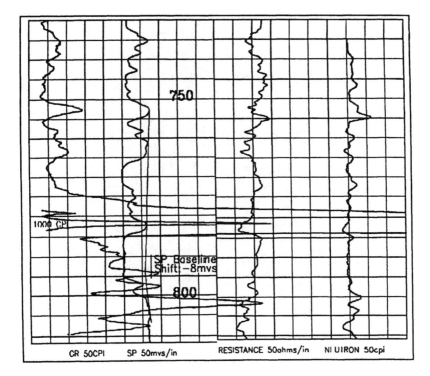

750

1000 CPI

SP Baseline
Shift -8mvs

800

CR 50CPI   SP 50mvs/in   RESISTANCE 50ohms/in   NI UIRON 50cpi

FIGURE 1.8
A Wyoming uranium log, showing the single-point resistance curve.

This indicates that the diameter of the 50% contribution sphere is two times the diameter of the electrode. Therefore, if the diameter of the electrode is 4 in (10 cm.), 50% of the signal will originate in a sphere whose diameter is 8 in (20 cm.), surrounding the electrode. Since a typical oil well has an 8 in diameter, this explains why monoelectrode resistivity systems are seldom used in hydrocarbon work.

Ninety percent of the signal would come from a sphere 10 times the electrode diameter. Thus, if the electrode diameter is 1 in (2.54 cm) (a common size in hydrocarbon work). Almost all of the signal originates from a sphere whose radius is 5 in (12.5 cm). The larger the electrode is, the deeper is its horizontal investigation. Also, the larger the electrode is, the smaller is its resolution of detail. This is because the larger sphere averages the response from a larger sphere of formation material.

If the investigating volume of a resistivity system is examined, we see that any system (non-resistivity systems included) samples a finite volume of space, either in reality or in effect. Thus, a lateral device samples a finite volume. A normal device samples an infinite volume in theory, but the vast majority of the contribution comes from the near neighborhood of the device; the rest, to infinity, is negligible.

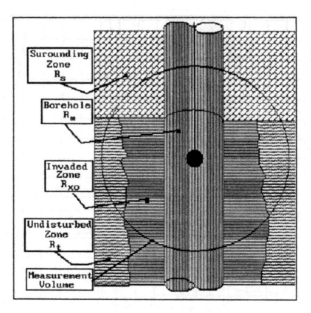

FIGURE 1.9
Contributions of the formation zones.

The volume of measurement of a logging tool may be pictured schematically as in Figure 1.9. The measurement volume may (but does not have to) include

1. The borehole, which is cylindrical and whose diameter is d_h. Its resistivity is R_m
2. The surrounding zone, which has an irregular geometry and whose resistivity is R_s
3. The invaded zone, which is an irregular cylindrical toroid in shape and is in electrical series with the deep zone. Its resistivity is R_i
4. The deep undisturbed zone, which is a cylindrical toroid and whose resistivity is R_t

Thus, every resistivity measurement is the sum of four contributions. These contributions come from the borehole, the surrounding zones, the invaded zone, and the deep, undisturbed zone. The proportional amount of each contribution will depend upon the geometry of the system being used. Therefore:

$$R_a = G_m R_m + G_s R_s + G_i R_i + G_t R_t \qquad (1.24)$$

where
- R = the resistivity
- G = the geometrical factor
- Subscripts a, m, s, i, and t represent apparent, mud/borehole, surrounding (adjacent) zone, invaded zone, and true (undisturbed) zone, respectively.

If we wish to measure the resistivity, R_t, of a particular zone, the influence of the other three zones must be eliminated or determined and subtracted from the total or apparent resistivity. The effect of the borehole ($G_m R_m$), the surrounding zone ($G_s R_s$) and invaded zone ($G_i R_i$) can be corrected for and R_t determined if we evaluate the effect of the borehole and invaded zone and subtract these from the bulk resistivity value.

The influence of a zone may be eliminated or reduced by tool design. For example, the lateral is a deep normal measurement from which a shallow normal measurement has been subtracted. Also, the 6FF40 induction log almost eliminates the borehole effect and the invaded zone effect by focusing the measurement beyond the borehole and invaded zone. The microresistivity uses an extremely short spacing to restrict the depth of measurement and an insulated shield to cut off the borehole effect. The single-point resistivity measurement normally does not extend beyond the borehole and the invaded zone. In some systems, the focused resistivities reduce the borehole and invaded zone contributions to proportionally small values.

The response curves for the unfocused resistivity systems will show this. These curves are called departure curves. They show the departure of the resistivity reading from the ideal case of an infinite, homogeneous, isotropic medium, where the formation and the borehole have the same resistivity. They attempt to correct for the presence of the borehole and the distorting effect of the resistivity unbalance (see Figure 1.10).

The influence of the unwanted zones may be eliminated by careful positioning of the measurement. Thus, the mud resistivity may be determined by taking and measuring a sample on the surface. The influence of the surrounding beds may be eliminated by positioning the tool measurement in the center of a thick bed.

1.4   Averaging within the Volume of Investigation

From the previous discussions, it can be seen that the farther away from the measure electrode an artifact is, the less influence it will have upon the measurement. Further, this decrease of response is not linear. In the case of

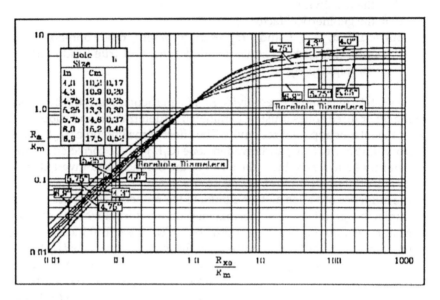

FIGURE 1.10
Departure curves for the single-point resistivity.

a spherical volume of investigation, the response is an inverse square function. The contribution of each infinitesimal volume within the volume of investigation forms the response of the system. The response is the weighted average of all of these contributions. Therefore, it is evident, that the larger the volume of investigation is, the less influence each infinitesimal volume has. In general, the larger the volume of investigation is, the less detail will be evident. We must always strike a compromise between volume and detail.

This is particularly evident when logged values are compared with core values. Because of the nature of the measurements upon core samples, the measurement at one depth has the same weight as that at any other depth. Wireline logs, on the other hand, will average values within some depth interval. In both cases, the value is an average value which is some function of the sample size. Thus, a single wireline log measurement should never be compared with a single core value. Rather, averaged intervals of each must be compared. This will be discussed further in a later chapter.

Also, because of the averaging effects of the wireline systems, in unfocused instruments one will find that the deep (horizontal) investigation systems tend to have more distortion when it comes to thin-bed response. Conversely, the thin-bed response systems tend to have shallow depths of investigation. Modern systems overcome some of these drawbacks through instrument design.

1.5 Position of the Return Electrode

The discussion, so far, has assumed that the current return and measure return electrodes, B and N, are infinite distances from the A and M electrodes. Thus, the return electrodes would have no effect upon the current distribution. Obviously, in actual practice, a real infinite distance cannot be obtained. The distance can, however, be made great enough that the influence upon the geometry of the current distribution is negligible. This is commonly done.

If the volume of investigation is small, as it is with a 1 in (2.5 cm) long electrode on the single-point resistivity, a return electrode 10 ft or 3 m away will have a negligible effect (~3%) upon a sphere of investigation which encompasses 90% of the response within a 10 in (25 cm) sphere, when compared to the infinite distance response.

An apparent infinite electrical distance away, in this context, then, can be described as that distance at which the position of the return electrode has a negligible effect upon the response of the system. Of course, this will vary, depending upon the system. In general, the greater the depth of investigation or the volume of investigation is, the greater the apparent electrical infinite distance must be. The designer of the system must decide if the response volume will encompass 50%, 90%, 99%, or some such value. Then, he can determine where the return electrode will be. Conversely, the return electrode can be put at a convenient place and then the geometrical constant can be modified to compensate. This latter was done with an early resistivity system.

1.6 Return Electrode Grounding Resistance

The measure return electrode, N, also has a grounding resistance. This grounding resistance is in series with the resistance of the measure electrode, M. Therefore, it is added to that of the M electrode. That is, the electrode resistance that is read in an infinite, homogeneous medium, is the total grounding resistance, r_e, and is

$$r_e = r_{eM} + r_{eN} \qquad (1.25)$$

Normally we are only interested in the value of r_{eM}. Thus, r_{eN} must be accounted for. This problem may be met in one of three ways:

1. On some of the arrays, the N electrode is downhole. It is part of the measurement system, and its grounding resistance varies as part of the measurement,

2. The N electrode is made small in diameter and long. It is placed in the constant environment of the mud pit. In this case, its grounding is small and constant,

3. The cable armor is used as the N electrode. In this case, the N electrode is relatively long and small in diameter. Therefore, its grounding resistance is not constant, but is negligible, compared to the measure or current electrode grounding resistance and changes.

In all cases, the grounding resistance, r_e, is the same as in Equation 1.14

$$r_e = \frac{R}{4\pi L_e} \ln\left(\frac{2L_e}{d_e}\right) \tag{1.26}$$

where

L_e = the length of the electrode
d_e = the diameter of the electrode
R = the resistivity of the medium

If the dimensions are in meters, the value of r_e is in ohmmeters.

In the second case, the grounding resistance is constant and is usually eliminated in the zeroing process. A typical return electrode is about 6 in (15 cm) long and about 3/4 in (2 cm) diameter. In a 1 ohmeter mud, in the mudpit, the grounding resistance of such an electrode is 1.438 ohms.

In the third case, the cable diameter is 0.1 to 0.6 in (0.25 to 1.5 cm.) and 100 to 3000 or more ft (30 to 1000 or more m) long, below the surface. The minimum length is set by the depth of the surface casing. The grounding resistance of the return electrode can be expected to be about 0.01 to 0.001 ohms or less.

Use of the cable armor as a return can cause trouble in two formation environments. The problem is known as the Delaware effect. The first situation is encountered when approaching a massive, very-low-porosity rock, such as a very tight carbonate or granite. The second is approaching the top of the mud or the surface in an uncased (without surface casing) hole. In both of these cases, the high-resistivity zone or the surface causes the effective length of the N electrode to become shorter as the high resistivity is approached. The grounding resistance becomes higher. Finally, it is no longer negligible and begins to affect the reading intolerably. Figure 1.11 illustrates this Delaware Effect in a log.

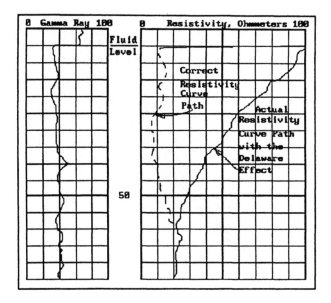

FIGURE 1.11
The Delaware effect upon the resistivity curve.

1.7 Multi-Electrode Systems

1.7.1 Normal Resistivity Devices

The definition of a normal system is that the current electrode, A, is close to the measure electrode, M (the distance [AM]), compared to the distances to the return electrodes, B and N (the distances are [AB], [AN], [MB], and [MN]). Most often, you will find B and N placed an apparent infinite electrical distance away. If [AM] is 16 in (40 cm.), the device is a 16-in normal (N_{16}) or a short Normal. On old logs you will commonly find, 1-, 2-, 4-, 10-, 16-, 20-, 32-, 40-, and 64-in normals. On modern logs, they will most often be 16-in normal and occasionally 64-in normal. Figure 1.12 shows a set of 16 in normal departure curves. Figure 1.13 is a set of 64-in normal departure curves.

If the measure electrode, M, is separated from the current electrode and placed some distance, [AM], away, the potential of that electrode will be that of the equipotential surface which coincides with the electrode surface. The radius of the surface, a spheroid, is equal to the distance from A to M, ([AM]). This array is called a normal resistivity system. Normal resistivity measurements, like those of the single-point resistivity system, are best suited to situations where the formation resistivity, R_{form}, is equal to or

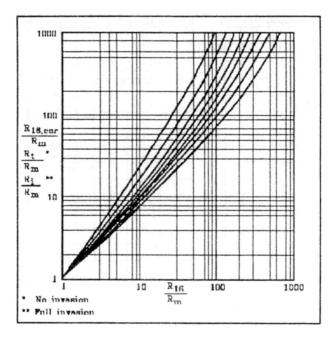

FIGURE 1.12
Short-normal resistivity departure curves.

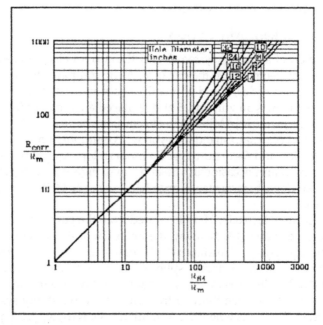

FIGURE 1.13
Long-normal resistivity departure curves.

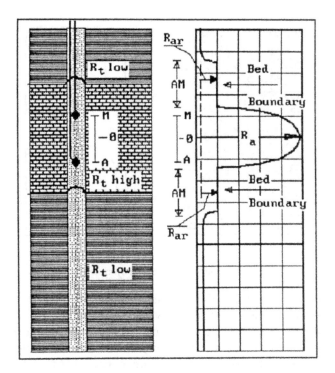

FIGURE 1.14
Thick-bed response of the normal system.

less than the borehole mud resistivity, R_m, and where the invasion diameter, d_i, is shallow to medium.

Until the advent of the induction log, both long and short normals were usually run. The short normal resistivity, R_{SN}, was used to estimate the resistivity within the invaded zone, R_i. The long normal, LN, was run to determine the extent of the invasion and/or to determine R_t. The LN has a longer spacing and may reach beyond the invasion. The only place you will now find the LN curve is on old electric logs. This is probably a mistake because, if the normal suite of logs will adequately suit your purpose, there is no good reason to use the more expensive and more complex modern methods. The normal curves, however, are not easy to use quantitatively.

During the passage of the array across the bed boundary, the deflection of the normal resistivity curve has a peculiar shape. See Figure 1.14. This can often be seen in small-diameter holes, but is difficult to see in larger holes. The normal curve has a ramp and platform effect. The length of the platform is equal to [AM], and is across the boundary. The level of apparent resistivity of the ramp, $R_{a,ramp}$, is

$$R_{ar} = \frac{R_1 R_2}{R_1 + R_2}$$ (1.27)

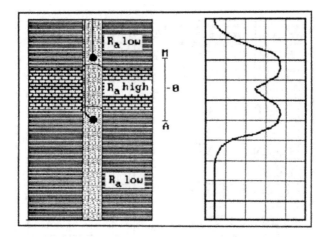

FIGURE 1.15
The thin-bed response of a 2-electrode normal resistivity system.

It is important to evaluate the ramp effect if you are trying to determine bed thicknesses using the normal curves. In a thick bed, a normal curve will indicate that a resistive bed is thinner than actual by a thickness equal to [AM]. If the bed has a lower resistivity than the surrounding beds, the determination will be too thick by an amount equal to [AM].

For example, if you attempt to determine the bed thicknesses from the N_{16} curve, use the conventional method of assuming that the bed boundary is at the midpoint between maximum and minimum deflection. Add 16 in (41 cm) to the apparent thickness of a resistive bed. Or, subtract 16 in from the apparent thickness of a low resistivity bed. If you are working with N_{10}, add or subtract 10 in (25 cm).

If the bed is thinner than [AM], the normal curves will show a "crater" or reversal in the center of the bed. Do not mistake this for two beds of lower contrast. This is shown in Figure 1.15.

The short normals are used to determine R_{xo} when the invasion is deep. The depth of investigation (50%) of a normal system is approximately 2[AM]. This is about 32 in (0.8 m) for N_{16}. The average effective depth of invasion is generally assumed to be 20 in (51 cm) radius. The SN is a shallow to medium investigation device.

1.7.2 Empirical Method to Estimate R_t

On old logs, the N_{64} value can very often be used to estimate R_t, if bed is thick, the invasion is shallow, and $R_s \simeq R_t$. This is an approximation and should be used with care. Read the value of the N_{64} resistivity, $R_{64,a}$, at least 3 ft (1 m) away from the bed boundary. If the bed is relatively thin and the N_{64} curve has a parabolic shape, do not rely on this method.

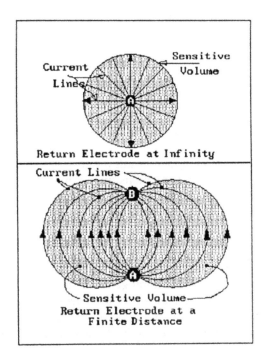

FIGURE 1.16
Current distribution in conductive media.

Schlumberger recommends that the following steps be considered for determining the value of R_t from the older electric logs. This procedure can easily be modified for suites of other contractor's logs:

1. Low resistivity (when $R_{16}/R_t < 10$ (invasion up to $2d_h$)): The shorter spacings, such as N_{16} and N_{64}, are most useful in finding R_t. Often, R_m is nearly equal to R_s. In this case, the apparent R_{64} can easily be corrected to R_t, depending upon the ratio of R_{64}/R_s and the bed thickness.

2. Medium resistivity (when $10 < R_{16}/R_m < 50$): In this case, the N_{64} is very useful in the lower range of resistivities.

We have talked about two electrode normal resistivity systems. These are systems with the measure electrode, M, and current electrode, A, downhole. See Figure 1.16. The return electrodes on some equipment may also be downhole. One solution is to put the return electrodes on an insulated piece of cable (a bridle cable) just above the tool. This is the method that Schlumberger uses. In fact, their return electrodes are not at an infinite distance from the A and the M electrodes. The resulting error is taken care of in the determination of the system constants.

The resistivity which a system measures is given by the relation

$$R = G_r(r) \qquad (1.28)$$

The value of G_R, for a 2 electrode normal, is

$$G_R = 4\pi[AM] \qquad (1.29)$$

If either the N is not at an apparent infinite distance from the M or the B from the A, use the 4-electrode relationship in Equation 1.32b to determine the value of G_R.

1.7.3　Departure Curves

The departure curves show the departure of the resistivity system response from that in infinite, homogeneous, isotropic conditions. The departures are due to the distortion of the current distribution by high resistivity in the formation forcing most of the current through the mud column. They represent the change in the geometry of the system. See Figures 1.12 and 1.13.

The departure curves correct only for borehole conditions. All of these charts are most accurate when the mud resistivity, R_m, equals or is greater than the formation resistivity, R_t. Readings must be taken well away (at least $5[AM]$) from the bed boundary. The departure curves are accurate enough when the formation resistivity is lower than the mud resistivity that this portion of the response is not usually shown. They begin to lose accuracy as the ratio of the formation resistivity to the mud resistivity becomes high. The actual point or range where the departure curves are not safe to use for quantitative purposes will depend upon the hole size and the array spacing. The single-point system, which can be considered a normal system with a zero $[AM]$ spacing, should not be used when the ratio of R_{sgl}/R_m is greater than about 10, especially at hole diameters. With N_{16}, the accuracy limit is at a ratio of R_{16}/R_m of about 10 to 100, again depending upon the hole diameter. With the N_{64}, the limit of accuracy of the curves is at a ratio of $R_{64}/R_m = 100$ to 1000.

1.7.4　Lateral Resistivity Devices

The lateral resistivity devices in general are deep depth of investigation devices. The use three or four electrode arrays and can be thought of as resulting in reading the difference between two normal devices. See Figure 1.17.

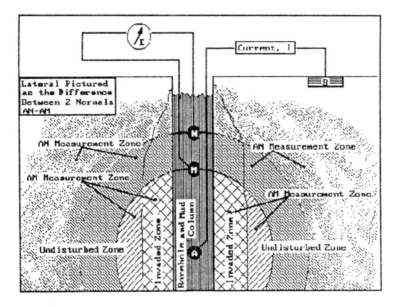

FIGURE 1.17
The lateral resistivity, pictured as the difference between two normals.

A lateral is defined as an array in which the current electrode, B, is closer to A than M is. The Lateral arrays are pictured on the accompanying figure. Occasionally, you will see the name "Inverse" used. An inverse array, Figure 1.17, is the inverse of a lateral array, and will give the same results. The "Principle of Reciprocity" states that the current and the measure electrodes of any array may be interchanged without changing the results. Inverses are used in place of laterals to allow a more convenient combination with the normal circuits. Figure 1.18 shows a set of departure curves for the 18 ft 8 in inverse.

The N electrode of the lateral or the B electrode of the inverse may be placed at an infinite distance, making it, effectively, a 3-electrode system.

The 3-electrode inverse can illustrate the idea that the system is the difference between two normals. There is one normal from A to M, a distance of [AM]. A longer one exists from a to N, a distance of [AN]. The voltages, E_M and E_N, measured at M and N, respectively are:

$$E_M = \frac{RI}{4\pi} \frac{1}{[AM]} \qquad (1.29a)$$

and

$$E_N = \frac{RI}{4\pi} \frac{1}{[AN]} \qquad (1.29b)$$

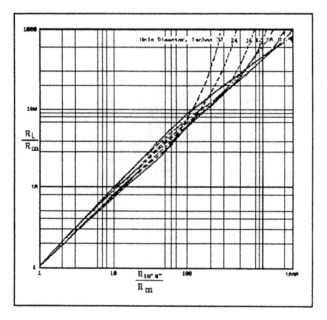

FIGURE 1.18
Departure curves for the 18 ft 8 in lateral resistivity.

If the difference of voltage between the two is measured, then

$$E_L = E_N - E_M = \frac{RI}{4\pi}\left(\frac{1}{[AN]} - \frac{1}{[AM]}\right) \tag{1.30a}$$

and

$$R = 4\pi \frac{\Delta V}{I}\left(\frac{[AM] \times [AN]}{[AN] - [AM]}\right) \tag{1.30b}$$

The laterals have very unsymmetrical response curves. This makes them quite difficult to use quantitatively and even qualitatively. You will find these curves on old logs, but they are not used much any more.

The response of the lateral in a thick resistive bed in a borehole of tool diameter is shown in Figure 1.19.

1.8 General Expression for Resistivity Devices

The general expression for any type of unfocused resistivity system, which will always have four electrodes, is

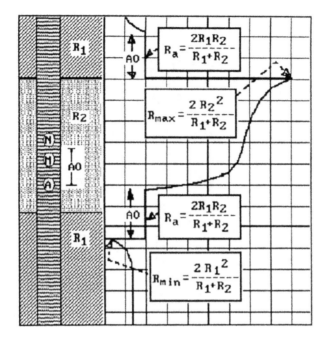

FIGURE 1.19
The response shape of the 3-electrode lateral resistivity curve.

$$E = \frac{RI}{4\pi}\left(\frac{1}{[AM]} - \frac{1}{[AN]} - \frac{1}{[BM]} + \frac{1}{BN}\right)$$ (1.31a)

and

$$R = 4\pi\left(\frac{1}{1/[AM]} - \frac{1}{1/[AN]} - \frac{1}{1/[BM]} + \frac{1}{1/[BN]}\right)r$$ (1.31b)

where the geometrical constant, G_R is

$$4\pi\left(\frac{1}{1/[AM]} - \frac{1}{1/[AN]} - \frac{1}{1/[BM]} + \frac{1}{1/BN}\right)$$ (1.32)

If the resistivity system is a surface system or if the device is a sidewall pad device, the factor will be 2π, instead of 4π.

Surface-resistivity arrays are usually 4-electrode systems. They follow the same principles that borehole resistivity systems do. They work, however, in a half medium, since the measurements are usually made from the surface, down. This is a big subject by itself and will not be discussed further at this point in the text.

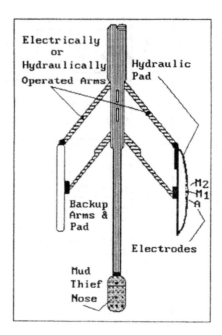

FIGURE 1.20
One type of microresistivity device.

1.9 Microresistivity Systems

The SWC MicroLog (MicroLog is the trade name of the Schlumberger Well Services microresistivity system) was the first microresistivity system and is still in use. It was an early attempt to accurately measure R_{xo} in a shallow invasion environment. A short spacing is used to obtain a shallow horizontal investigation. Since a shallow reading can be greatly affected by the mud in the borehole, a large insulating pad is used to isolate the measurement from the influence of the borehole. A drawing of a modern microresistivity tool is shown in Figure 1.20.

The usual unfocused *ML* device is a 3 electrode array mounted on a flexible, insulating pad which is applied against the wall of the hole. The electrodes are 0.5 in diameter and are spaced 1 in, center to center. The bottom electrode is the current electrode, *A*. The second electrode is the measure electrode, *M*, for the 1×1 in inverse. The 3rd electrode is the measure electrode, *M*, for the 2 in normal, N_2, and the measure return electrode, *N*, for the $I_{1 \times 1}$. The current return is on the back plate of the pad. See Figure 1.21. Figure 1.22 shows the principle of a sidewall measurement, compared to that of an omnidirectional method.

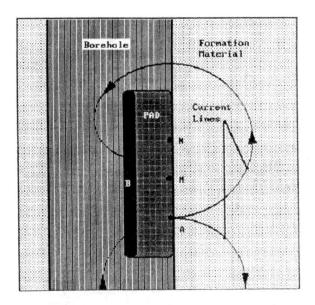

FIGURE 1.21
The current distribution for the microresistivity device.

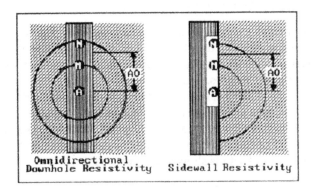

FIGURE 1.22
Measurement principles.

Unfortunately, the usual *ML* device is extremely sensitive to the thickness and resistivity of the mudcake between the array and the formation. It works well when the mudcakes are thin, as in saline and low waterloss muds. The departure curves for the Schlumberger MicroLog may be used to determine a probable value of the resistivity of the totally flushed zone, R_{xo}. Note that in this case, the normalizing factor is the mudcake resistivity, R_{mc}. The tool is mainly used now, because it is a very sensitive detector of the *presence* of permeability. It *will not/cannot*, however, quantitatively determine permeability. The value of R_{mc} is best obtained from the mud

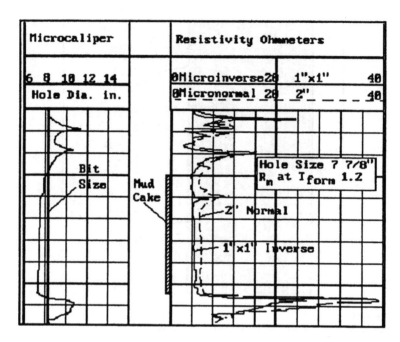

FIGURE 1.23
A microresistivity log. Source unknown.

sample and a mud press. R_{mc} can be estimated with the several charts available, but the result is usually little better than an educated guess. A set of departure curves for the *ML* can be found in any chart book of Schlumberger Well Services, Inc.

A caliper curve is always run with the microresistivity to determine the condition of the hole wall and the thickness of the mudcake. A line showing the bit diameter is usually traced on the log. The mudcake thickness is taken as one half of the reduction of hole diameter in the permeable zone. A sample microresistivity is shown as Figure 1.23.

1.10 Miscellaneous Resistivity Items

The horizontal investigation of electrical-resistivity devices may be very shallow to very deep, depending upon the device used and upon the formation resistivity. Numerous devices of different horizontal investigation depths and geometries are available. All resistivity systems are sensitive to the presence of hydrocarbon and gas. Resistivity systems (except induction logs) are very difficult to use in cased holes or in oil base muds. Induction logs cannot be used in metallic casing. A wide range of resistivity measurement capability is available, depending upon the device used.

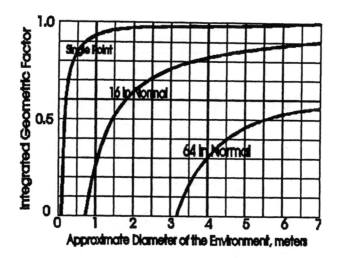

FIGURE 1.24
Approximate pseudogeometric curves for several systems.

None of the devices should be considered for high resistivity (>1000 ohm) use.

An unfocused resistivity method must be chosen on the basis of its depth of investigation. The depth of investigation must also be weighed against the resulting loss of resolution and the presence of thin beds. To aid in this, pseudo-geometric curves have been determined. A set for unfocused devices is shown as Figure 1.24. These curves show the apparent cumulative response of several devices. The value of each curve shows the fractional amount of the signal contributed by the environment between that depth of investigation and the logging sonde. The shape of the curve tends to indicate the geometrical shape of the sensitive volume.

Resistivity measurements are good methods under most open-hole conditions. They are probably the most important measurement for petroleum because they can be used to determine the water saturation of the porous zones. They have many uses in non-petroleum work, such as stratigraphy, pattern analysis, fracture detection and many other things.

One must make allowances for conductive solids (primarily shales) when using resistivity methods. The resistivity value depends upon the amount, salinity, and the geometry of the contained water, plus the considerable effects of the shales. Therefore, one must know the resistivities of the waters in the formation materials. Usual resistivity measurements are almost independent of lithology. This is because the conductivity of the rock matrix is negligible compared to that of the contained water. Since the fluid content is of prime importance in the petroleum industry, virtually all resistivity systems are designed to emphasize the response to the formation water.

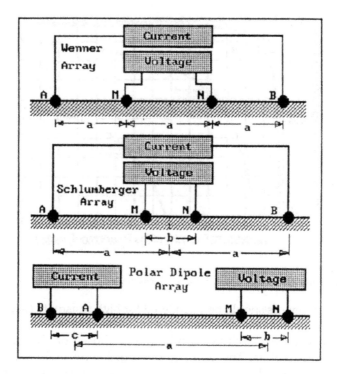

FIGURE 1.25
Common surface arrays.

1.11 Surface-Resistivity Methods

Methods of measuring the earth's electrical resistivity have been in use since before the 20th century. They were offered commercially by the Schlumberger brothers before they attempted downhole measurements. These methods usually involve a four electrode array on the surface. The arrays have been standardized and are known by various names. Figure 1.25 shows the array formats for the most commonly used arrays.

The principles involved and the determination of the geometrical constant of the surface arrays are identical to those of the downhole methods. The major difference is that the surface methods involve the factor 2π in the geometrical constant, G_r, rather than 4π found in most of the downhole arrays.

Figure 1.26 compares the investigated volume of a 3-electrode device downhole and on the surface. The other difference is the size of the volume of investigation. Downhole methods examine volumes of cubic centimeters to cubic meters. Surface methods examine volumes of hundreds of cubic meters and more.

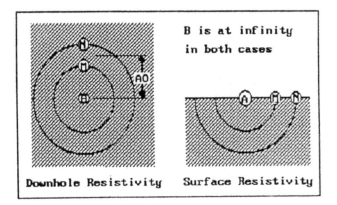

FIGURE 1.26
Investigation volumes for downhole and surface devices.

Because most of the surface arrays use four electrode, unfocused methods, the interpretation of them can be difficult. They are usually symmetrical, however. Thus, they may be somewhat easier to use than the downhole laterals are.

Surface measurements are conducted in discrete steps, rather than the continuous trolling used downhole. After the survey is finished, the individual measurements are combined by horizontal position. The result is a resistivity cross section of the apparent resistivity in the earth, along the track of the measurement. A sample is shown as Figure 1.27.

A number of different arrays are used. The choice depends upon the preference of the surveyor and upon the physical conditions of the surface terrain and the subsurface formations. All of the arrays are comparable with a symmetrical downhole lateral. In fact, the Schlumberger 18 ft 8 in (5.6 m) lateral is really an unsymmetrical Schlumberger array.

The depth of investigation of a surface array will depend upon the [MN] spacing. As with the downhole devices, approximately 50% of the signal in a homogeneous medium, originates from a volume whose radius is 2[MN].

The guarded-electrode systems usually use point-guard electrodes placed in strategic positions to force the measure current deeper into the earth and to obtain a higher resolution. The explanation for the focused downhole systems in the following chapter applies equally well to the surface methods. Figure 1.28 is a crude diagram illustrating a possible electrode placement and probable current lines for a focused system. An old, but comprehensive, discussion of surface resistivity methods is given by Roy and Apparao (Roy and Apparao, 1971). Some experimentation has been conducted using guarded electrode arrays. These systems are not as elaborate as they are in downhole systems, but effectiveness has been reported. The apparently are not used much.

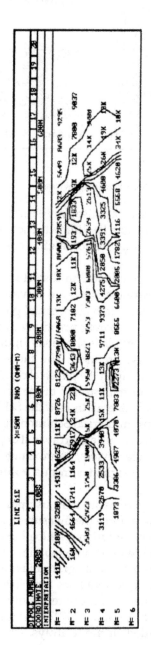

FIGURE 1.27

A surface resistivity cross section made with a dipole-dipole array.

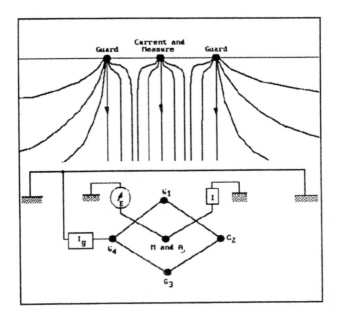

FIGURE 1.28
Probable current lines for a focused surface resistivity array.

2

Focused Resistivity Methods

2.1 Focusing Electrode Devices — Omnidirectional

Now we will take a look at some different and newer families of resistivity (and conductivity) systems. The first of these is the family of the omnidirectional focusing or guarded-electrode systems. They are offered by most logging contractors, including non-hydrocarbon logging contractors and have largely replaced the unfocused, passive methods, especially where ratios of R_t/R_m are high. The family of focused resistivity systems is large. They are known by various names; Laterolog,[SWC] Microlaterolog,[SWC] guard logs, focusing-electrode logs, guarded-electrode resistivity logs, spherically focused log, and focused resistivity. The generic name is guarded-electrode resistivity system and log. Guarded-electrode methods are common in laboratory measurements.

The guarded-electrode logging technique is an old method. It was a common laboratory method when it was suggested for borehole use and patented by Henri Doll (one of the original Schlumberger engineers, inventor of the induction log, and Chief Executive Officer of Schlumberger for many years). It was not acted upon at that time, however. The most common guarded-electrode devices are those of the Laterolog[SWC] family, abbreviated as LL. The devices should not be confused with the "lateral" log or lateral resistivity systems, which were discussed in the last chapter. The laterals are passive devices, while the Laterologs are active, focused systems and are entirely different. The trade names of both originated because the systems were designed to have deep lateral depths of investigation. The guarded-electrode systems use separate flows of electrical current through the formation material, in addition to the measure current, to shape (guard) the measure current. While all of the guarded resistivity system now in use have active control, the original suggestion by Doll used a passive control system. Schlumberger introduced all of these methods.

We saw in the last chapter that the unfocused-resistivity systems suffer from unstable geometry; that the geometry of measurement changes with the ratio, R_t/R_m, the resistivity of the formations to that of the drilling mud. Therefore, it is difficult to concentrate a measurement where it is desired with the unfocused systems. For the same reasons, it is also difficult to determine a correct resistivity value. In addition, bed boundary, bed response shape, and thin-bed effects are complex, with non-focused systems.

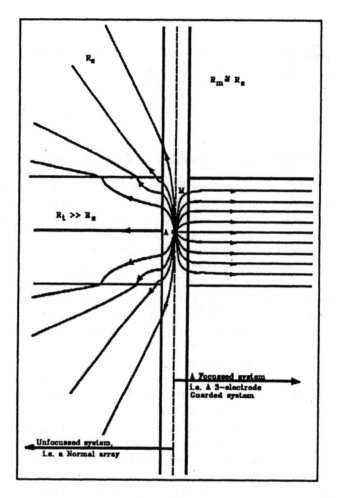

FIGURE 2.1
A comparison of the current distribution of unfocused and focused systems.(Courtesy of
S.P.E.- A.I.M.E.)

Focusing or guarded-electrode resistivity systems, on the other hand,
are designed to minimize or eliminate many of these stability problems.
The geometries of the systems are dynamically controlled by the tool and
circuit designs. Thus, a system can be designed for a specific purpose, such
as deep, medium, shallow, or very-shallow investigation, specific volume
investigation, thin-bed accuracy, high or low resolution, high-resistivity
use, and/or various combinations of these features. The investigative
geometries are determined and stabilized electronically. The limits of the
systems are set by the sensitivity of the monitoring system, the maximum
available guard current, and the electrode configuration. Figure 2.1 shows
a hypothetical comparison between a passive, unfocused system measure
current and that of a typical guard system.

TABLE 2.1

Omni-directional guarded-electrode systems

System	Zone of use	Conditions
Laterolog-3	R_t or R_{xo}	High R_t/R_m, no SP
3-electrode guard, mineral	R_t	High R_t and R_m, thin beds
Laterolog-8	R_{xo}	High R_t and R_m, Large d_i
Laterolog-7	R_t	High R_t and R_m
Laterolog-Deep	R_t	High R_t and R_m
Laterolog-Shallow	R_{xo}, d_i	High R_t and R_m, Large d_i
Spherically Focussed Log	R_{xo}, d_i	High R_t and R_m, Large d_i

All of the focused-resistivity systems use guard current electrodes in manners similar, in principle to those used in the laboratory, to force the logging current into the zone desired and to shape the current and voltage fields. The response of the tool is monitored electronically and the guard current is adjusted to shape the field properly and keep it stable. The result is that the measurement can be confined to a specific location. For example, the measurement can be concentrated deep into the formation and, at the same time, the resolution and range can be retained or improved. Table 2.1 lists several focusing electrode systems and their uses.

The focused resistivity systems will be referred to as "guard" systems. These systems were originally designed for use in very saline muds where none of the other types of resistivity systems work well. They are used now in a wide variety of environments. They are particularly useful when the ratio of R_t/R_m is high. Some of the systems may be used quantitatively in environments where the R_t/R_m ratios are much greater than 10,000. This makes them very interesting to the mineral sector, where formation water salinities are often very low (high R_w) and formation resistivities, R_t, are very high (very low or zero porosities) compared to the drilling mud, R_m, values. The stability of these systems in high R_t/R_m zones is one of the primary reasons they are attractive for non-hydrocarbon use. High resolution and wide range are usually desirable in these environments, also.

Figure 2.2 shows the approximate shapes of the measure volumes of three of the omnidirectional guarded-electrode systems. Figure 2.3 shows the high-resistivity response of a guarded-electrode system, compared to those of three of the unfocused systems.

The extent of the investigative volume is approximately shown by commonly published curves, known as "pseudo-geometric curves". These will be discussed later in this and later chapters.

2.1.1   3-Electrode Guard Logs

The simplest of the guard systems is the 3-electrode system (LL3), or Laterolog-3,* illustrated in Figure 2.4. The measurement component of the LL3 is essentially a single-point resistivity device (as are many of the guard

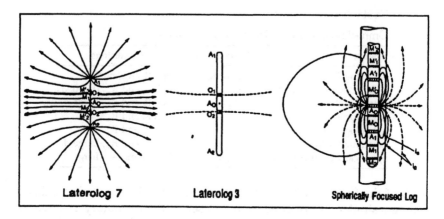

FIGURE 2.2
Some of the current patterns of guarded-electrode resistivity systems (after Schlumberger, 1987).

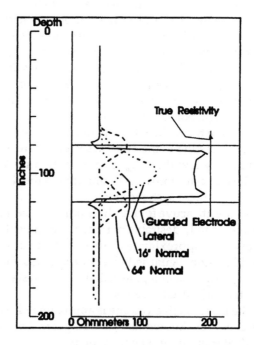

FIGURE 2.3
An approximate comparison of the response of unfocused and focused resistivity systems.

log systems). The logging current, I_0, from electrode A_0 of the LL3 is held constant. Some forms of this device measure conductivity, C_a, and are similar, except that the potential of the electrode A_0 is held constant and the current varied. The guard electrodes, A_1 and A_1' are electrically connected

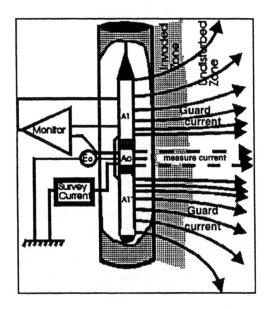

FIGURE 2.4
A schematic of the 3-electrode guarded resistivity system.

together. The current, I_a, from the guard electrodes is automatically adjusted by the monitoring circuit, so their potential is the same as that of A_0. Current from A_0, is thus forced to flow laterally, deep into the formation before it begins to diverge. No current will flow from A_0 toward A_1 or A_1'. The horizontal distance of I_0, before divergence, is generally determined by the length of A_1 and A_1'. The (vertical) resolution of the measurement (the height of the cylindrical volume of investigation) is equal to the length of A_0 plus one half the length of each insulator above and below A_0.

The apparent resistivity, R_a, is,

$$R_a = K_{LL} \frac{E_0}{I_0} \qquad (2.1)$$

where
- K_{LL} = the geometrical constant for the focusing electrode system (Laterolog)
- E_0 = the potential of the measure electrode
- I_0 = the current to the measure electrode

Notice that this is similar to the relationship discussed in Chapter 1 as applying to resistivity devices (Equation 1.8). Also, note that the ratio of E_0/I_0 is the grounding resistance of the electrode, A_0, and the measure system is a single-point resistivity system. The value of K_{LL} is constant for any given electrode arrangement and is stabilized electronically.

The apparent conductivity, C_a, of the similar conductivity device, is,

$$C_a = K_{LL}^l \frac{I_0}{E_0} \tag{2.2a}$$

where

$$K_{LL}^l = \frac{1}{K_{LL}} \tag{2.2b}$$

Any difference in potential between the measure electrode and the guard electrodes is monitored and the guard current is adjusted continuously to keep the difference at zero. Thus, the current pattern is as shown in the center diagram of Figure 2.2. This shape is maintained electronically through a wide range of resistivities. The range will depend upon the range and gain of the monitor amplifier. The potentials of A_1 and A_1' are kept equal to that of A_0. Thus, E_0 actually could be read from any one of the three electrodes. In general, mineral and scientific systems have much higher ranges than are needed in the hydrocarbon systems. The response across a bed boundary is symmetrical and mostly free of such irregular responses as the normals and laterals have. The measurement is a single-point resistivity measurement with deep investigation, high resolution, and stable geometry. The resolution typically is about 6 in (15 cm.). Thus, it is much better than the 16-in normal, which is about 32 in or 0.813 m. Mineral and other non-hydrocarbon guard systems have a finer resolution. Three to four in (7.6 to 10 cm) is typical. The hydrocarbon LL3 has a medium diameter of investigation. The lateral diameter of investigation depends upon the length of the guard electrodes. The commercial LL3 has a 40 in (1.0 m) diameter of investigation for 50% of the signal and 90% from 150 in (3.8 m). This tool is used to determine R_t in shallow to medium invasion environments. These approximate values can be determined from the pseudo-geometric factor charts. Since mineral systems do not need such deep investigations, they typically have 50% diameters of about 20 in (50 cm). Some of the mineral systems may be slightly asymmetrical across bed boundaries because some of their guard electrodes are not equal lengths.

An *SP* often cannot be recorded with the LL3 because of the large amount of metal (electrodes) in the measured zone. If an *SP* is needed, it must be taken from one of the bridle cable electrodes or with another tool on another run. A gamma ray correlation log is usually run with any guard log. In addition, the hydrocarbon guard systems are usually run in saline mud conditions where it is difficult to obtain an SP. Guard systems are effective and usually used in high R_t/R_m ratio environments where the *SP* is least effective.

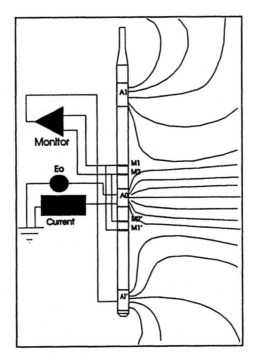

FIGURE 2.5
A schematic diagram of the 7-electrode guarded resistivity system.

2.1.2 7-Electrode Guard Logs

The most commonly used petroleum guard system is the Deep Laterolog (LLD) [the older device, the Laterolog-7 (LL7) is very similar]. This device is also sometimes erroneously called the point electrode guard log. These systems have seven electrodes downhole. The return electrodes are at a great distance. Their primary use is to determine R_t. A diagram of this system is shown in Figure 2.5.

The LL7 and the LLD are deep lateral investigation devices. About 50% of their signal comes from a cylindrical volume about 100 in (2.5 m) diameter. About 90% comes from about 350 in (8.9 m) diameter. The diameter of investigation is determined by the distance from the center of one monitor electrode pair to the guard electrode. The resolution of the LLD is equal to the distance from the center of one monitor electrode pair to the center of the other. The resolution, the height of the cylindrical volume of investigation, is about 30 in (0.9 m). See Figure 2.5.

In practice, the logging current, I_0, is usually kept constant. The control or guard current, I_a, flows from the guard electrodes, A_1 and A_1', which are connected together electrically. A pair of monitor electrodes M_1 and M_2 are

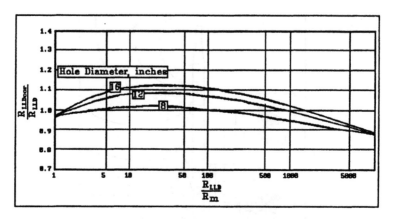

FIGURE 2.6
Borehole correction curve example for the LLD (after Schlumberger, 1988).

electrically connected to M'_1 and M'_2 respectively. These sense the potential field difference and the monitor adjusts I_a to keep the potential difference across the monitor at zero. The logging current, I_0, will then not flow past the monitor electrodes (since the hypothetical lines of current will not cross each other), but will be forced, laterally, out into the formation. Because of the parallel flow of measure current from A_0, the influence of the borehole and the invaded zone is minimized; it forms a small part of the total measured volume and, thus, is more easily corrected (refer to Figures 2.2 and 2.5).

Borehole and bed thickness correction curves are to be found in the Schlumberger Log Interpretation Charts book, the Atlas Wireline Log Interpretation Chart book, and the Halliburton Logging Services Open-Hole Log Analysis Reference Material book. These curves and their information are similar to those in Figures 2.6 and 2.7. In use, the ratio of the log reading, R_a (R_{LLD}, for example), to the value of the mud resistivity at the reading depth, R_m, is entered on the horizontal axis of the chart, such as Figure 2.6. Move upward to the line of the hole diameter. Interpolate, if necessary. Then, move horizontally to the left vertical axis and read the correction factor. Multiply R_a by this factor to correct for borehole effects. This, then, is R'_{LLD}, a new value of R_{LLD}. Note that the system is least influenced by the borehole when R_m and R_{LLD} are equal. The borehole influence increases as R_{LLD}/R_m increases, until R_{LLD} becomes large enough that R_m becomes inconsequential.

Determine the ratio of the borehole corrected reading (R'_{LLD}) to the log reading of the nearest different zone (R_s). Determine the thickness of the bed in which the log reading was made. As an alternate, when taking a reading in a thick bed, near the bed boundary, use the distance to the nearest different bed. Enter the horizontal axis of a chart, such as shown in Figure 2.7, at that value. Move upward to the ratio line of R'_{LLD}/R_s. Then

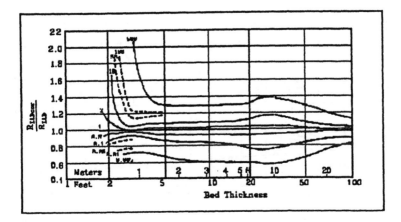

FIGURE 2.7
A bed-thickness chart for the LLD (after Schlumberger, 1988).

move horizontally to the left and read the bed thickness correction factor from the left axis scale. Multiply the borehole corrected R'_{LLD} by this factor to get the corrected value of R''_{LLD}. This is the corrected value which the chart books label R_{LLD} and which will be used in all subsequent processes. In actual use, be sure to refer to the proper chart book and the chart designed for that specific tool.

Note that the DLL is least affected when the bed thickness influence exceeds the center-to-center spacing of the monitor electrodes and that the influence of the adjacent bed is great until a bed is thicker than the resolution of the tool (about 3 to 4 ft or 0.9 to 1.2 m). After that, the response is quite stable with respect to bed thickness. There remains some small influence of the resistivity of the adjacent bed because of its small effect upon the guard current field.

2.1.3 Shallow Investigating Guarded Electrode Device

The shallow investigation guarded-electrode devices are the eight electrode guard logs. The newer systems are called the Shallow Laterolog (LLS) and the (older) Laterolog-8 (LL8). They are the medium investigation diameter devices of the Dual Laterolog service, DLL. These may also be used as shallow investigation devices with the dual induction log, DIL, when the invasion is very deep. The main purpose of this family of shallow or medium depth systems is to help define the horizontal depth of the invasion of the borehole fluid into a permeable zone. The operating principle of the LLS is the same as that of the LLD, except that the guard current return electrodes are placed close enough to the guard electrodes, A_1 and A_1', that the current flow is sharply bent around and the logging current, I_0, pattern diverges a short distance from the wellbore. About 50% of

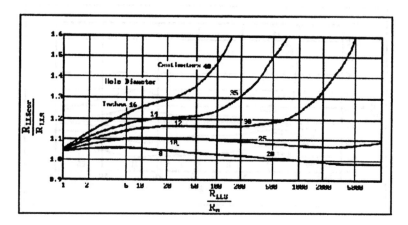

FIGURE 2.8
An example borehole correction chart for the shallow Laterolog (after Schlumberger, 1988).

the signal comes from a diameter of about 30 in (0.74 m) for the LL8 and about 35 in (0.89 m) for the LLS and 90% from 100 in (1.65 m) for the LL8 and about 120 in (3.0 m) for the LLS. Their resolutions are about the same as that of the LLD. See Figure 2.10. Note that these systems, because of their shallow lateral depth of investigation (shallower than the LLD), are much more sensitive to the borehole characteristics than is the LLD.

The borehole and bed thickness correction charts (which were adapted from information for the Schlumberger LLS) are shown as Figures 2.8 and 2.9. The corrections methods of other contractors are similar. This author wishes to emphasize that these illustrations are adapted from the contractor's charts and are simplified. *Be sure to use the correct actual chart from the proper chart book for any actual analysis program.* Note, in Table 2.1, the LL8 and its similarity to the LLS.

The LLS has about the same vertical resolution as the LLD (Figure 2.9) but, because of its smaller volume of investigation and more sharply curved guard current field, it is less influenced by the adjacent bed resistivity, R_s, after the whole array is within the bed, than is the LLD (about 9 ft or 2.7 m bed thickness).

2.1.4 Spherically Focused Systems

The spherically focused log (SFL) system was developed as an improvement over the N_{16} and the LL8. When invasion is medium to deep, it is sometimes also used as a short spacing companion to the dual induction log, DIL.

The SFL system uses two independent current systems one of which establishes constant potential, spheroid guard shells around the measure electrode. The guard current system prevents a flow of measure current up and down the borehole and establishes the equipotential spheroids. A first equipotential spheroid is established about 9 in (0.23 m) from the survey

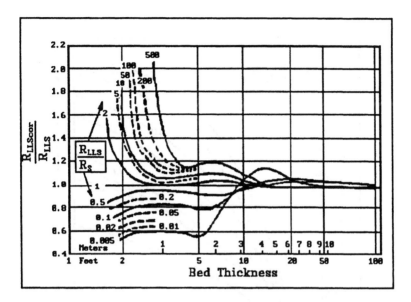

FIGURE 2.9
An example bed-thickness chart for the shallow Laterolog (after Schlumberger, 1988).

current electrode. A second is established about 50 in (1.27 m) away. A constant potential of 2.5 millivolts is maintained between the spheroid shell surfaces. The independent measure current flows through a fixed measurement volume and is proportional to the conductivity of its volume. See Figure 2.10. The borehole and bed thickness correction charts for the SFL will be found in the chart book of the contractor or manufacturer of the particular system in use.

2.1.5   Radial Pseudo-geometrical Factors

The volume and horizontal depth of investigation of each of the resistivity and conductivity devices is important. In the next several chapters we will see that this is a major consideration in designing a geophysical instrument and choosing one for a particular purpose. To describe this feature, pseudo-geometrical charts are published in the literature concerning these devices and their uses. Figure 2.10 is a chart of this type for the guarded-electrode devices. There are similar ones for the electromagnetic devices. A later figure combines the geometrical responses of many of the devices, allowing a direct comparison.

The pseudo-geometrical factor describes the contribution to the signal of the device, in a direction away from the tool axis, by the volume whose diameter is indicated on the horizontal axis on the chart in Figure 2.10. It also has a tendency to show some of the shape of this volume. It assumes that the signal has been corrected for borehole and bed thickness effects. These curves can be used for approximating the geometrical factors for the

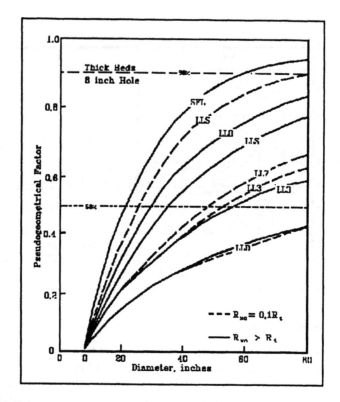

FIGURE 2.10
Pseudogeometrical factors for some of the omnidirectional guarded-electrode devices (after Schlumberger, 1989).

invaded zone (G_i) and the undisturbed zone (G_t). This is done by first determining the diameter of invasion by previous calculation, estimation, or some other reliable method. The factor at that diameter is G_i. The value of G_t is $(1 - G_i)$. It is better, however, to use the charts specifically designed to correct for these factors. But, sometimes we do not have enough information to do this. This will be covered later, in detail.

2.2   Sidewall Guarded-Electrode Systems

The same operating principles that we saw with the deep investigation, omnidirectional, guarded-electrode systems apply to the guarded-electrode sidewall tools. The sidewall tools are designed to make a geometrically controlled measurement of a function of R_{xo}. These systems do not measure R_{xo} directly, because of the complex environment of the measurement. The focusing principle, however, allows the measurement to be concentrated

TABLE 2.2

Available sidewall focusing electrode systems

System	Zone of use	Conditions
SFL/LL8	R_{xo}	d_i deep
SFL	R_{xo}	d_i medium
MLL	R_{xo}	d_i medium, t_{mc} <3/8 in
PL	R_{xo}	d_i medium, t_{mc} <3/4 in
MSFL	R_{xo}	d_i shallow, t_{mc} <3/4 in

in a given, zone at an optimum depth in the formation for the determination of R_{xo}. Also, it allows overcoming or minimizing adverse effects of high formation resistivity and thick mudcake. An insulating pad, on which the electrode array is mounted, protects the measurement from most of the adverse effects of the borehole. Mudcake corrections must be made, however, because even the least affected of the systems will be influenced to some degree by the mudcake thickness and resistivity. The operation capabilities of the focused sidewall resistivity tools cover a wide range of hole and invasion conditions. Table 2.2 shows some of the available sidewall devices.

Figure 2.11 shows a comparison of the current flow opposite a thick, low resistivity mudcake and high-resistivity formation zone for the unfocused microresistivity system and for the guarded-electrode sidewall devices. These sidewall guarded-electrode devices are seldom used in non-hydrocarbon applications for several reasons. First, the usual non-hydrocarbon borehole is not as safe, mechanically, as petroleum holes are. And, they are expensive. Second, and more important, invasion processes are not as important in non-hydrocarbon environments, even in permeable sediments, as they are in petroleum investigation. The significantly higher vertical resolution of the sidewall systems, however, means that they should be seriously considered for some applications. They are becoming standard for quantitative non-hydrocarbon resistivity measurements, especially in thin beds and coal exploration.

2.2.1   Microlaterolog (MLL)

The original sidewall focusing electrode system was the Microlaterolog, MLL. An unofficial name, at the time, was the "Trumpet" log, because the measure current pattern is shaped like a trumpet. The Atlas Wireline device is called the Micro Laterolog. Forxo is the older name for the Dresser Atlas device. The Halliburton equivalent is the Micro Focused system. They are all very similar to the MLL. These devices use a sidewall pad, similar to the Microlog pad, to isolate the measurement as much as possible from the borehole influence. The pad has a ring guard electrode, about 5 in (12 cm) diameter, a pair of monitor electrodes centered at about 2.5 in (6 cm), and a button measure electrode in the center. The depth of investigation is

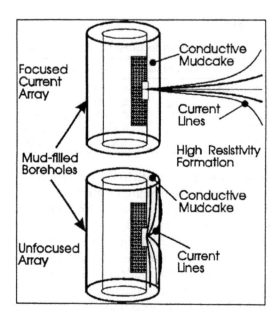

FIGURE 2.11
A comparison of the current distributions for focused and unfocused sidewall devices.

less than 8 in (0.254 m) for 50% of the measurement and approximately 13 in (0.28 m) for 90%. The MLL may be used for high ratios of R_{xo}/R_{mc} and mudcake up to 3/8 in (1 cm) thick.

These systems replace the older MicroLog device. Their advantage is that the focusing current confines the measurement to the formation material, rather than being severely influenced by the borehole and, especially, the mudcake, as with the MicroLog. Thus, a much more reliable determination of R_{xo} may be made, especially when use is combined with other resistivity methods. Figure 2.12 shows the MLL pad configuration and approximate current flow.

The mudcake correction chart is shown in Figure 2.13 for the Schlumberger Microlaterolog. This device is fairly sensitive to the influence of the mudcake when the thickness is greater than 3/8 in (1 cm). Note the great sensitivity to the value of R_{mc}. Notice however that the MLL is very much less sensitive than the unfocused MicroLog. A similar chart is available for the Atlas Wireline (Dresser Atlas) Microlaterolog in their chart book.

2.2.2   Proximity Log (PL)

The Proximity Log, PL, is used for thicker mud cakes (up to about 3/4 in or 2 cm). It uses a principle similar to the MLL, but the pad is wider and the electrodes are differently shaped than those of the MLL. Vertical resolution is about 6 in (0.15 m). Bed thickness corrections are not needed if the thickness is twice that. It has a depth of investigation of about 11 in

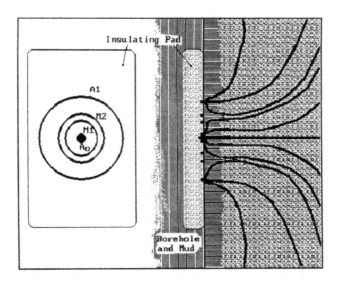

FIGURE 2.12
The pad and current flow of the Microlaterolog.

(0.28 m) for 50% measurement volume and about 37 in (0.94 m) for 90%. If invasion is shallow (<40 in or 1 m) its measurement may be affected by R_t. Its measurement is

$$R_{PL} = G_{XO}R_{XO} + (1 - G_{XO})R_t \qquad (2.3)$$

The Dresser Atlas chart for mudcake correction for their Proximity log is shown in Figure 2.14. You can see that the effect of the mudcake upon the PL is much less than upon the ML. The depth of investigation is somewhat greater, however.

2.2.3 Microspherically Focused Log (MSFL)

The Microspherically Focused Log measures a small, sharply bounded volume, several inches into the zone and is excellent for use in front of thick mud cakes, because of the use of the spherically focusing principle (see the discussion on the SFL system). It is usually combined on the same downhole tool as the DIL and the DLL, obviating the need for a second run. It may be used in shallow invasion situations, however.

Figure 2.15 shows a diagram of the pad and electrode design of the MSFL. The survey current, I_0, flows from electrode A_0. The guard currents flow between A_0 and A_1. Equipotential spheroids are maintained by the monitor electrodes, M_0, which monitors the measure voltage, and M_1 and M_2, which maintain the outer equipotential sphere. The survey current is measured, as a function of the formation conductivity.

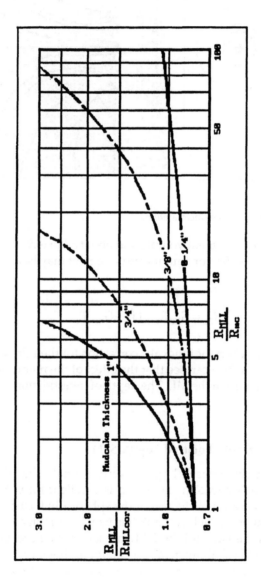

FIGURE 2.13
A drawing of a mudcake thickness correction chart for the Microlaterolog (after Schlumberger, 1988).

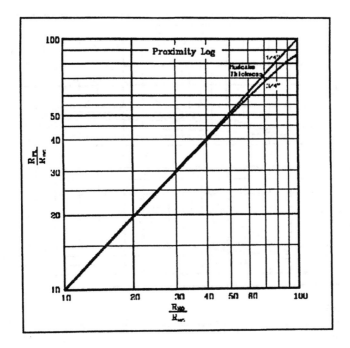

FIGURE 2.14
The Proximity Log mudcake correction chart. (Courtesy of Western Atlas Logging Services Division of Western Atlas International, Inc.)

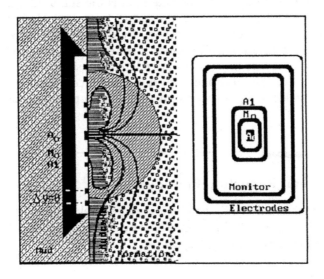

FIGURE 2.15
The pad and current distribution of the MSFL. (After Schlumberger.)

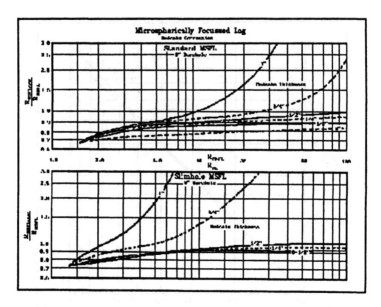

FIGURE 2.16
A drawing of the mudcake correction charts for the Schlumberger MSFL systems. (After Schlumberger.)

The mudcake correction charts for the Schlumberger MSFL are shown in Figure 2.16. This illustration shows the curves for both the standard diameter tool and the slimhole tool. Note that the slimhole system is much more sensitive to the presence of mudcake than is the standard system. This illustrates well the need to match the system, and particularly the downhole tool, to the environment in which it will be used. The logging tool (downhole) should always fill as much (diameter) of the borehole as is safe to run. Also note that the systems are relatively insensitive to changes in the thickness of the mudcake until thickness exceeds 1/2 in (1.27 cm) and/or the value of R_{mc} is very nearly that of the formation resistivity.

2.3   General Reduction Procedure

The apparent resistivity, R_a, of a typical guarded-electrode resistivity system, the LLD, (Figure 2.17) is influenced by and is a function of the resistivities and effective, relative volumes of

1. The borehole
2. The surrounding zones
3. The invaded zone
4. The deep, undisturbed zone

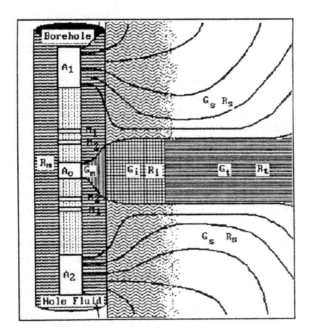

FIGURE 2.17
An approximation of the geometry of the LLD.

$$R_{a,LLD} = f(R_m, R_s, R_i, R_t) \tag{2.4}$$

Since this is a deep horizontal diameter investigation tool, the only contribution we are interested in, with this deep investigation system, is R_t. The system response equation is shown in Equation 2.5 and we must solve it for R_t.

$$R_{a,LLD} = G_m R_m + G_s R_s + G_i R_i + G_t R_t \tag{2.5}$$

where G_m describes the geometry of the borehole component of the measurement. G_s describes the component due to the surrounding or adjacent bed. G_i is the contribution of the invaded zone and G_t is the geometrical factor of the undisturbed zone measurement component. R_m, R_s, R_i, and R_t are the resistivities of those zones, respectively, as noted before. Notice in the Equation 2.5 general relationship, if we have chosen a suitable device and wish to know R_t, we must evaluate the first three terms and determine G_t. There will be three charts, a borehole correction chart (for G_m), a bed thickness correction chart (G_s), and a pseudo-geometrical factor diagram for (G_i) or some combination to evaluate this. The value of G_t can be assumed to be equal to $(1 - G_i)$, if the other terms are evaluated. Note that these charts are designed empirically or semi-empirically and can be used *only* with the system for which they have been designed. They cannot be

interchanged with the charts of another contractor or instrument manufacturer. Samples of these charts were shown earlier in this chapter as
Figures 2.6 and 2.7 (also Figures 2.8 and 2.9). A pseudo-geometric diagram
for the focusing electrode devices is Figure 2.10. *Be sure, however, to refer to
the original charts for actual use. The drawings in this text are for illustration
only.* Some of the charts by some contractors may be combined. Regardless
of the chart combination and notation, Equation 2.5 or its equivalent, must be
evaluated. Notation can be a problem and one should decide carefully what
a label indicates. Older, unfocused systems, such as the 16-in normal, do not
use most of these charts. Often, with the normals, only a borehole correction
is made. Bed thickness corrections generally are not made. And, the charts for
bed thickness corrections are not generally available. Therefore, thick beds
(>20 ft) must be used and values must be picked sufficiently far from the bed
boundaries that they are not affected by the bed thickness.

If a Laterolog (or any other system) is run with a standoff from the wall
of the hole, it will always result in a different response from that run
against the wall of the hole. This *must* be carefully checked and the appropriate chart used. It is important, even with the use of computer processing, to read the log heading and be aware of the tool configuration. It is
extremely important to use the correct chart. It is equally important for the
logging engineer to fill in the heading *completely,* even if some of the data
appear, to him, to be obvious, routine, and/or inconsequential. Most of the
logs will be in use for many years after the on-site details have been lost
from personal memory.

The Microlaterolog is more affected by the mudcake than is the Proximity
Log because the Proximity Log is designed for deeper lateral investigation.
Thus, the Microlaterolog would be chosen for conditions where the invasion
is shallower and the mudcake is thinner. The combination of thin mudcake
and shallow invasion frequently occur together because of the mud quality.
The Microspherically Focused Log devices have a more severely restricted
geometry than do the R_{xo} devices which have been discussed previously.
These devices are used in very shallow invasion situations.

The major difficulty encountered is that of determining the diameter of
the invasion. This problem will be addressed after the induction logs have
been discussed. If the diameter of invasion is known or can be estimated,
then this type of chart can be used to find the value of the geometrical factor of the third term, G_i.

The data reduction of the sidewall focusing electrode device information
proceeds along the same principle as outlined for the Laterologs. Since
these are shallow investigation tools, they are affected by mud cake thickness, t_{mc}, and resistivity, R_{mc}. Because they have a shallow investigation,
they are not affected by R_t. Also, since their measured volume is small, vertically, they are generally not affected by the surrounding bed resistivities.
Thus, their logged response, R_a, is

$$R_{a,sidewall} = G_m R_m + G_i R_{xo} \tag{2.6}$$

To use the charts for the sidewall devices, enter the ratio of the log reading to the temperature-corrected mudcake reading, R_a/R_{mc} (which is, in this case, R_{MLL}/R_{mc}) on the horizontal axis. Move vertically to the mud cake thickness and read the ratio of R_{corr}/R_a (which is, in this case, $R_{MLL,corr}/R_{MLL}$) on the vertical axis. Multiply this ratio by the log reading to get R_{corr}. Under many conditions, $R_{corr} \simeq R_{xo}$.

Some of the sidewall tools, such as the MLL, are usually run on a separate run. Not all of the focusing electrode tools, sidewall or omnidirectional, are designed to record an SP. This is because in the petroleum business they are usually used where the SP is poor, such as in saline muds and high resistivity formations. Some will measure R_m or take a mud sample.

If we are logging in a thick bed and/or the adjacent bed resistivities are near R_a, then G_s becomes zero and we need only solve for the first term and use this to correct R_a. This will only require one chart. Figures 2.7 and 2.9 show the bed-thickness correction charts for the deep and the shallow Laterolog devices. Notice, if the surrounding bed resistivity is near that of the target zone, there is virtually no correction to make for a bed thickness of 3 ft (or 1 m) and thicker.

If we wish to determine information about the flushed zone (i.e., R_{xo}), we would pick a device which has a shallow depth of investigation. This would probably be a sidewall tool system. These systems usually have a smaller volume of measurement and a more shallow horizontal investigation than that of the deeper, omnidirectional investigation systems. There are usually no bed thickness correction charts for the sidewall devices, such as the MLL, PL, and the MSFL because it is assumed that it is easily possible to get sufficiently far from the adjacent bed to get a valid thick bed reading because of the small volumes of investigation of these tools. Also, if the device is a sidewall device with a pad, the borehole is effectively shielded from the measurement, except for the mudcake influence (R_{mc} and t_{mc}). Note that even though these sidewall tools are isolated from the borehole by an insulating pad, they are still affected by borehole conditions. In this case, the correction is for the thickness of the mudcake, rather than for the borehole diameter. Thus, G_s and G_t would be zero or near zero, if the device was picked correctly. Therefore, only the first two terms must be evaluated. Only two charts are needed to do this, the mudcake (borehole) correction chart and the pseudo-geometrical factors chart. These charts will be found in the chart books from the contractor who offers the particular device.

Remember that a deep investigation device does not measure R_t, nor does a shallow device measure R_{xo}. These systems measure *functions* of R_t and R_{xo}. Because of the uncertainty of the diameter of invasion and the resulting unknown factors, combinations of measurements must be made. Then, they must be used to eliminate the undesired values and determine the useful ones (i.e., R_t, R_{xo}, and d_i).

3

Induction Methods

3.1 Introduction

The induction log (IL) was developed about 1948 to handle oil-base mud conditions of the Texas Permian Basin. It is now used in any type mud (except very saline muds) and may be used in any type hole except metal-cased holes. It was first used in the saline sediments of the Texas Gulf Coast.

Induction logs operate on a much different principle than the resistivity tools. They are electromagnetic devices using alternating magnetic fields to induce logging currents in the formation material. They measure the conductance (the reciprocal of resistance) of the volume (formation) around the downhole tool. Conductance is measured in Siemens (S). Conductance is usually converted to resistivity electronically by applying the proper geometrical factors and reciprocating the conductivity. Many logs, especially the older ones, present the conductivity curve along with the reciprocated resistivity curve. A few of the oldest logs may have only the conductivity curve. One of the resistivity devices is usually run with or simultaneously with the induction log.

Conductivity is the reciprocal of resistivity. It is labelled in milliSiemens per meter, mS/m, in geophysical work.

Induction logging systems give the best results when the ratio of R_t/R_m is low (high formation conductivity). Logs made with older tools, before the automatic "skin effect correction, are not useable when R_t is greater than 100 Ωm ($C_t < 10$ mS/m). Those with automatic correction can be used to about 200 Ωm ($C_t < 5$ mS/m). The Phasor systems are valid at higher resistivities. Note, however, virtually all of the commercial, presently available induction systems are intended for hydrocarbon use. This means that they are designed for the lower range of resistivity (<1000 Ωm or > 1 mS/m) sediments. All induction logging systems measure conductivity, rather than resistivity. The resistivity curve on the log is a computed curve. It is the reciprocal of the measured signal. Modern logs may have had various corrections applied before printout.

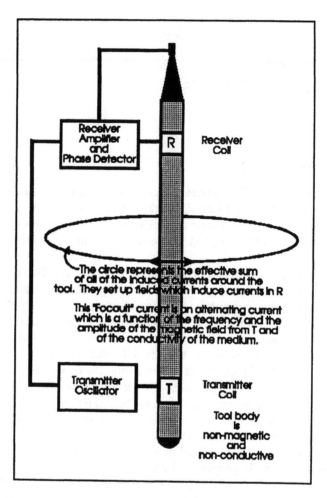

FIGURE 3.1
Illustrating the principle of the induction log.

3.2 Principle

The principle of IL systems is illustrated in Figure 3.1. IL systems use no
electrodes and do not require galvanic contact with the borehole fluid.
Instead, they set up high frequency (10 kHz or more) alternating magnetic
fields in the formation with one or more coils through which high-frequency electrical currents flow. The resulting alternating magnetic fields
induce electrical currents in the conductive formation when they move

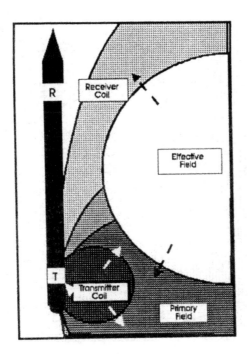

FIGURE 3.2
The operational principle of the induction log.

through it. The current is proportional to the magnetic field strength (which is held constant), the rate of propagation of the magnetic field, and the conductivity of the formation material. It is 90° out of phase with the exciting current. The currents in the formation flow in toroids, normal to and around the transmitter coil. The net induced currents are functions of the conductances of the formation material within the toroid.

The toroidal formation currents set up secondary magnetic fields which induce electrical signal currents in the tool receiver coils. The signal currents are functions of the formation conductance and are 90° out of phase with the induced currents (180° out of phase with the original current).

There exists an in-phase signal component which is a function of the magnetic susceptibility of the formation material. This susceptibility signal is not used in the older models of the induction log. Some of the newer devices, however, use it for control purposes. Some magnetic susceptibility tools use this component. Figure 3.2 is a schematic diagram of the operating principle of a two-coil induction log.

Figure 3.3 illustrates, crudely, the field shaping and cancellation principle. This method allows the effective field to be shaped to cancel unwanted response and to concentrate the resultant field in the desired volume.

The IL systems have several advantages:

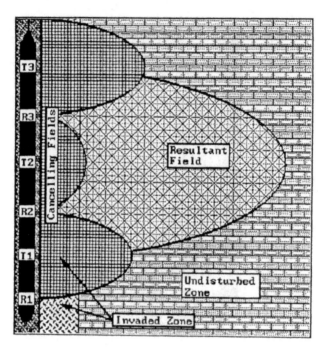

FIGURE 3.3
The field cancellation principle of the induction log.

1. They perform best when in low R_t/R_m environments. This is the usual sedimentary environment.

2. The IL may be used in holes cased with fiberglass or other electrically nonconductive casing.

3. The formation signal can be made almost independent of the borehole fluid character.

4. The IL response can be restricted to specific zones. A suitable combination of coil turns, positions, polarities, phase detection, field strengths and currents will allow the measurement to be concentrated deep within the formation. Thus, the 6FF40 induction log is almost insensitive to an average-diameter invaded zone.

5. The IL does not use electrodes nor require electrical contact with the drilling fluid. Thus, there are no adverse electrode effects.

Table 3.1 shows some of the factors needed for the induction log data reduction. Table 3.2 lists some of the induction log systems and their uses.

TABLE 3.1

Information needed for induction log data reduction

Information needed for data reduction:

1. Mud resistivity, R_m, and temperature from the log heading or another reliable source
2. Temperature at the formation level, T_{form}
3. Borehole diameter, d_h, from the log heading (bit size) or the caliper curve
4. Bed thickness, h, from the log
5. Resistivity of and distance to the closest adjacent or shoulder bed, R_s, from the log
6. Tool type and contractor
7. Tool position (i.e. centered, eccentered, amount of standoff, etc.), from the log heading
8. Log reading, R_a, from the log

TABLE 3.2

Induction logging tools and their uses

System	Use	Qualifications
6FF40	R_t	Low R_t/R_m
ILD	R_t	Low R_t/R_m
6FF28	R_t	Low R_t/R_m, Small diameter borehole
ILM	d_i	Low R_i/R_m
5FF40	d_i	Low R_i/R_m
IDPH	R_t	Low R_t/R_m, Low distortion
IMPH	d_i	Low R_i/R_m, Low distortion

3.3 Conventional Induction Logs

The common forms of the IL have various combinations of the coils and spacings downhole. The conventional (older) models are the 6FF40, deep induction (ID), 6FF28 (a smaller diameter, deep induction), medium induction (IM) and the 5FF40. The newer devices are the Phasor^{SWC*} systems (IDPH and IMPH for the deep and medium investigation devices, respectively). The coil currents and the resulting magnetic fields usually are fixed, by tool design, at about 20 kHz.

The 6FF40 and the ID are very similar. They make deep (laterally) measurements, with a cancellation of the invaded zone signal component. Their measurements are concentrated within the undisturbed zone of the formation and past the average invaded zone. The 6FF28 is similar to the 6FF40, but of a smaller tool diameter. The necessary corrections are small. Thus, their values often are very near R_t. Often, the log reading can be estimated to be R_t, as a first approximation.

---

* Phasor is a registered trade name of Schlumberger Well Services, Inc.

The 5FF40 and the IM are similar and are medium lateral depth of investigation devices. They are designed to allow an estimate of the diameter of invasion and their measurements are largely within the invaded zone. The IM is used in combination with the ID on the dual induction log (DIL).

The dual induction log (DIL) is usually run in combination (on one run) with one of the shallow focused resistivity tools, such as the LL8, MLL, PL, N_{16}, or the MFSL. All of the IL tools and combinations measure the *SP* also. Occasionally one will see tools with other numbers of coils, up to 7. The original IL system (still occasionally used) was the IES. It was a combination of the 6FF40, N_{16}, and the SP.

Oceanographic work sometimes employs a two coil system which is analogous to a normal resistivity system. Non-electrode mud testers also employ a two coil sensor. These are mostly specialty tools and must be handled separately.

The accuracy of the Schlumberger 6FF40 induction logging system is published as ± 2 mS/m. This will result in a true resistivity error as high as 20% at 100 Ωm (10 mS/m). The newer systems (i.e., the ILD), with automatic skin effect compensation, have appreciably lower uncertainties, thus may be safely used at much higher formation resistivities.

Data handling, transmission, and processing is done in analog form on all of the older systems. Some of the later revisions of the older analog systems incorporated digital data processing after an analog transmission. This consisted mostly of digitizing the analog signal on the surface. It was then fed into a small computer. The computer usually was a dedicated one. This, of course, was inflexible and not very satisfactory. Most of the newer systems are fully digital.

3.4   Phasor Induction Systems

The Phasor dual induction logging system is a modern version of the standard Dual Induction Log. It is a design which is the result of computer modelling. Its designations are IDPH for the deep investigation device and IMPH for the medium horizontal depth device. It is run in combination with the SFL (spherically focused log) and the *SP* (spontaneous potential) on one sonde. Improved response characteristics make this system a major improvement over the older systems.

Digital signal transmission and digital data processing are standard with the Phasor systems. This allows the various corrections to be applied to the finished log, at the logging site. It also allows continuous calibration verification. This effectively make cable frequency and resistance effects negligible. It also allows multiple use of a cable conductor. Finally, it allows direct communication with the data processing system.

The Phasor systems operate nominally at a frequency of 20 kHz, as do the other types of induction logs. It can operate at 10 kHz, however, to reduce skin effect in extremely low-resistivity (high-conductivity) environments. Also, it can operate at a frequency of 40 kHz for greater accuracy in high-resistivity (low-conductivity) situations.

In contrast to the conventional induction systems, the Phasor systems measure the out-of-phase quadrature signal (which Schlumberger calls the "X-Signal"). It uses the "X-signal" to derive a component to improve the accuracy of the skin effect correction and allows better deconvolution of the Phasor resistivity curve and correction for the shoulder effect. These last improve the thin bed response. The uncertainty of the Phasor system, according to Schlumberger, is ±0.75 mS/m at 20 kHz and ±0.40 mS/m at 40 kHz.

3.5 BPB Array Induction Logs

BPB Instruments designed and operates an induction logging system using a sonde which departs from the conventional multi-transmitter, multi-receiver coil system which was originally introduced by Doll in 1948. The BPB Array Induction sonde uses one transmitter and four receiver coils. The signals from all coils are transmitted digitally to the surface. They are then manipulated, combined and enhanced by the truck computer system. The four-coil design allows the possibility of generating vertical and radial response functions (Elkington 1993). Several options result. One option in the processing is to synthesize the standard medium and deep induction curves. The detail from the short spacing coil (18 in or 45 cm) is used to enhance the detail of the deep and medium induction curves.

Several presentations are available. The process used is the Vectar* process which is described further in the next chapter. A standard presentation is similar to the conventional induction log format. It features a deep induction, a medium induction, and a focused-resistivity (FE) curve. The 4 curve high-resolution presentation (VIP) displays the signals from all four of the receivers. This will give a good picture of the invasion of the formation. The invasion profile is a graphical presentation of the invaded zone and six resistivity curves (four raw curves and two synthesized curves). Another presentation (VIVID) gives a visual image of the resistivity profile at horizontal depths up to 2 m (6.5 ft) into the formation from the borehole. Figure 3.4 shows a comparison between the deep induction log (ILD) and the Phasor log.

---

* Vectar is a registered tradename of BPB Wireline Services.

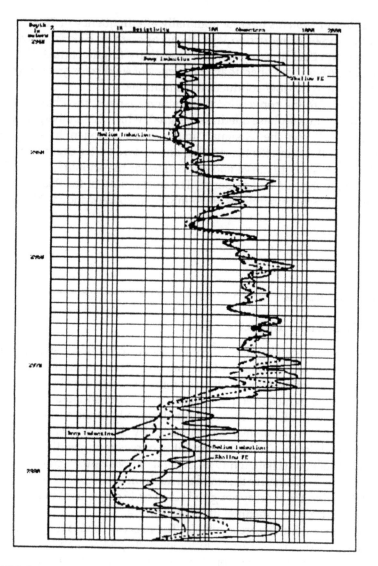

FIGURE 3.4
Part of a Vectar Induction Log. (After BPB Instruments.)

3.6   Factors Affecting Induction Logs

3.6.1 Skin Effect

An electromagnetic field moving into or through a conductive medium
will induce an emf (electromotive force or potential). This is the effect

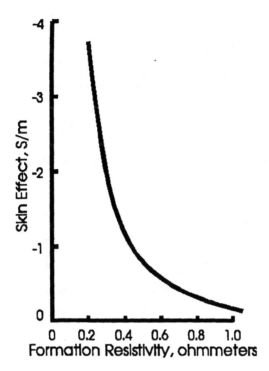

FIGURE 3.5
The result of skin effect with the 6FF40 (after Schlumberger, 1987).

which produces the induced currents of the induction log. The resulting magnetic fields (from the induced currents), if they are strong enough, can induce out of phase additional emf in the formation. Thus, they will oppose the primary emf and reduce the net field. This will reduce the induced current in the same manner that an increase of resistivity of the formation material would. Thus, a low-resistivity zone will appear to be more resistive than it actually is. This phenomenon is called the "skin effect." It is well known in electrical and in electronic work. It is quite predictable and formulae are given in many handbooks for calculating it. Most modern induction logs have been automatically corrected for skin effect. Figure 3.5 shows the result of the skin effect upon the log of a 6FF40. Figure 3.6 compares the standard and the Phasor induction logs.

3.6.2 Bed Boundary Effects

For the purposes of recording, analyzing, and reference, an apparent point of measurement (commonly called the "zero point") is assigned to a system. This zero may occasionally be arbitrary, but will usually have a sound basis for its position, such as a plane at the center of the sensitive volume, normal to the sonde axis.

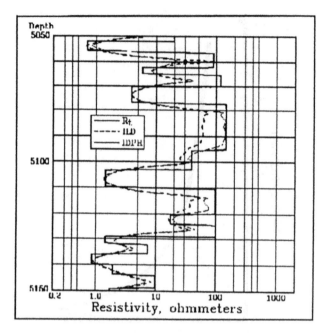

FIGURE 3.6
A comparison of the ID and the IDPH (after Schlumberger, 1989).

Since any measurement system (including a borehole system) detects a volume of material, the logged measurement will anticipate a bed boundary. That is, as the tool approaches the boundary the shape of the curve will begin to respond to the bed before the zero arrives at the bed boundary. The shape will also be affected after the measure point leaves the bed. This is illustrated for a simple spherical measurement in Figure 3.7. This volume effect is the reason why any measurement, particularly a large volume measurement, must be corrected for the influence of adjacent beds. It accounts for the "thin-bed" effects. Physical measurements are not point nor planar measurements. They *always* encompass a finite volume.

If the measurement is a simple sphere (measured by a point detector), the bed boundary response shape of the logged curve will be sinusoidal. This is very nearly the situation for the gamma ray and the single-point resistivity measurements. A complex geometry system, such as the induction log, will have a much more complex bed boundary response shape. In any case, when examining a tool response near the bed boundary, the response shape across the bed boundary must be considered. Remember, too, that the "nearness" of the bed boundary is relative. It depends upon the size and shape of the measured volume (the longitudinal and lateral extent and distribution of the measured volume).

Bed boundary responses are measured empirically either in a physical or a computer model, or both. Figure 3.8 shows the response of an induction log,

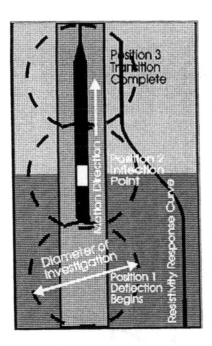

FIGURE 3.7
The effect of measurement geometry on curve shape.

ID, in a vertical borehole, through a horizontal, sharply bounded bed. This information comes from Schlumberger Well Services, Inc. Note that the curve shape across the bed boundary is not a simple sinusoid and the bed boundaries are not at the curve inflection point. This indicates that the measured volume is not a simple sphere. Note, also, that the curve within the bed is not simple. This indicates further, that the volume geometry is complex. Note, also, that the recorded curve of Figure 3.8 drops below the adjacent bed resistivity volume. This is symptomatic of the cancellation of the invaded zone response, with the ID and the 6FF40.

3.6.3   Thin-Bed Response

The thin-bed response of any system is a variation of the bed boundary response. Figure 3.9 shows the thin bed response of the Schlumberger ID in a thin, horizontal bed. The reduced amplitude and rounded or sharp top is a typical thin bed shape. It indicates that the measured volume has started to leave the bed before it has fully entered it.

If a bed is thinner than the height or diameter of the measured volume parallel to the sonde axis, the amplitude of the signal will be reduced because of the bed boundary effect (for *any* system). This must be taken into account. The need for this, with induction logs, is shown in Figure 3.9.

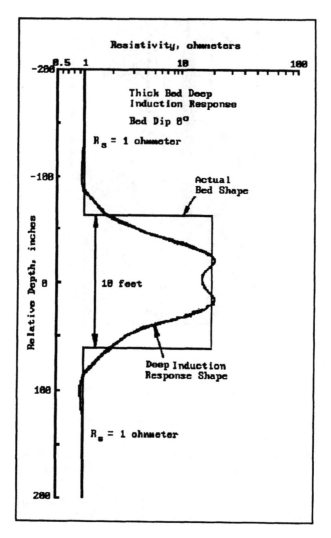

FIGURE 3.8
The bed-boundary effect upon the ID (after Schlumberger, 1989).

3.6.4 Dipping Beds

In our initial analyses of tool response, we usually make many simplifying assumptions. The most common of these are that the borehole is perfectly vertical (0° deviation), the bed is perfectly horizontal (0° dip), and that the tool is centered in the hole. These things are seldom true in actual practice. If the bed is dipping and/or the borehole is deviated (the bed dipping with respect to the borehole axis), the logged measurement will be affected.

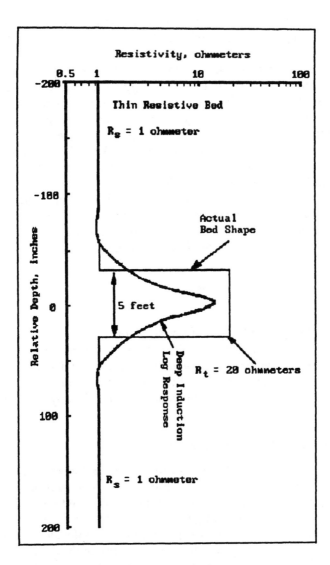

FIGURE 3.9
The thin-bed response of the ID (after Schlumberger, 1989).

Since the measuring volume of the induction log (or any other system) extends up and down the borehole from the "measure point", The measurement will not only anticipate the approaching of the bed (as described in the last section), but the volume will extend some lateral distance (normal to the sonde axis) into the formation. This is illustrated in Figure 3.7 for a hypothetical simple spherical system measurement from a vertical hole through a horizontal, sharply bounded bed. The measurement detects the bed boundary approximately one measurement radius for a spherical measurement (or thickness for a non-spherical volume), $d_v/2$, before the zero or "measurement point" arrives at the intersection of the bed boundary with the borehole. This results in the bed boundary response described in the last section.

Note that the "measurement point" with the simple spherical geometry is not arbitrary, but is the point on the logging tool which passes the bed boundary when the inflection point occurs on the logged curve. Thus, where it can be seen on the logged curve, the inflection point will always indicate the bed boundary. The inflection point is the second derivative of the response function of the curve. In Figures 3.8 and 3.9 you can see that the complex geometry of the induction systems results in the bed boundary and inflection point *not* coinciding.

If the bed is dipping, with respect to the borehole, then the measurement will detect the bed boundary and bed before $d_v/2$. The distance from the start of the bed deflection to the intersection of the dipping bed boundary with the center of the borehole (Figure 3.9), $d_v'/2$, is

$$R_{a,IDPH} \tag{3.1}$$

where

b = the apparent depth of investigation
d_v = the apparent vertical thickness of the volume of measurement or the diameter, the maximum anticipation distance
θ = the angle of bed dip with respect to the borehole axis
d_v' = the instantaneous anticipation distance for any bed boundary

3.6.5 Net Bed-Boundary Response

The net bed-boundary response for the induction log system (or for any logging system) will be a combination of the bed-boundary response (including the thin-bed response) and the dipping-bed effect. It will be a function of the measurement geometry (the size and shape of the measured volume). The effect of the dip of the bed will effectively be to extend the normal bed-boundary response by a distance of ($b \tan \theta$).

There will be effects, however, due to the geometry of measurement of the particular tool being used. The ID, for instance, will detect dipping-bed boundaries at a relatively great distance because of the great lateral extent of its investigation. As the measure point approaches the bed boundary, however, the response will be affected because of the substantial insensitivity of the system to the zone a short distance from the borehole (the cancellation of the measurement within the "invaded zone"). Figure 3.10 illustrates this change in response in a thick bed, as a function of the distance of the measure point from the bed boundary. Figure 3.11 shows the ID response to dipping thin beds. Note that the net effect is quite important in logging in 90° dip (horizontal) boreholes.

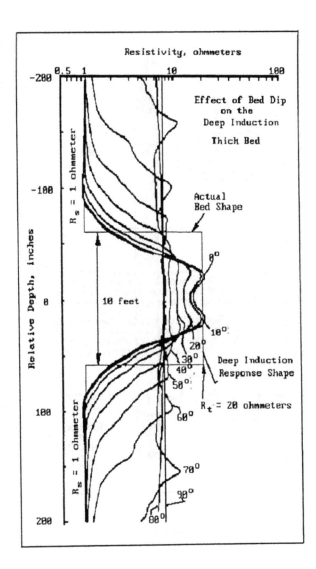

FIGURE 3.10
The response of the ID in thick dipping beds (after Schlumberger, 1989).

3.7 Data Reduction

Data reduction for the induction logging systems measurements proceeds along much the same lines as with the focusing electrode systems. That is, corrections must be applied sequentially starting with the borehole correction. Then, the bed thickness (adjacent bed) influence must be accounted

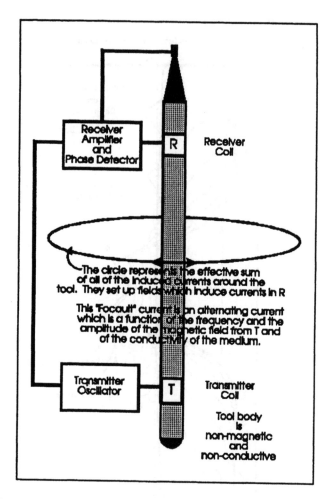

FIGURE 3.1
Illustrating the principle of the induction log.

3.2   Principle

The principle of IL systems is illustrated in Figure 3.1. IL systems use no electrodes and do not require galvanic contact with the borehole fluid. Instead, they set up high frequency (10 kHz or more) alternating magnetic fields in the formation with one or more coils through which high-frequency electrical currents flow. The resulting alternating magnetic fields induce electrical currents in the conductive formation when they move

Induction log response in high-resistivity (low-conductivity) zones is a problem with some of the older induction log (Rocky Mountains and similar areas), where the operators were unaware of the IL problems with high-resistivity formations. On these old logs, check the low-conductivity zones very carefully.

Note that R_a is called various things, depending upon the type tool, the contractor, and the chart being used. This designation means the "apparent resistivity" and can mean any resistivity value that may or may not need further correcting. It is most often used to designate the uncorrected log reading. The actual designation (for the deep Phasor induction reading, for example) should be of this order:

$$R_{a,IDPH}$$

Except that the IL measures conductivity, the data reduction process for the IL tools is very similar to that for the focused resistivity tools. For the ID,

$$C_{ID} = f\left(C_b, C_s, C_i, C_t\right) \tag{3.2a}$$

$$\frac{1}{R_a} = \frac{1}{R_{ID}} = \frac{G_b}{R_m} + \frac{G_s}{R_s} + \frac{G_i}{R_i} + \frac{G_t}{R_t} \tag{3.2b}$$

where C_b and G_b are the borehole conductivity and geometrical constant, C_s and G_s are the adjacent bed conductivity and geometrical constants, C_i and G_i are the invaded zone conductivity and geometrical constants, and C_t and G_t are the true conductivity and geometrical constants. Note the similarity to Equations 2.4 and 2.5.

The charts which you use have been designed by the contractor whose system was used and for that particular tool. The contractor (or the logging system) will usually correct for tool factors (design factors, such as skin effect). The borehole effects must first be eliminated to give a corrected value, C_a'. The reciprocal of C_a' must be determined to obtain R_a'. The shoulder zone or bed thickness (adjacent bed resistivity, R_s) effects will be handled next to give a new corrected value, R_a''. Then the invasion corrections will be made next to give a value for R_a'''. R_a''' can then be used to derive R_t plus some other values.

3.7.1 Borehole Corrections

Use the borehole correction chart, Figure 3.12, or similar, to obtain a value of G_B (analogous to G_m, which we used in the last chapter). Enter the value of the borehole diameter on the horizontal (x) axis. Move vertically to the

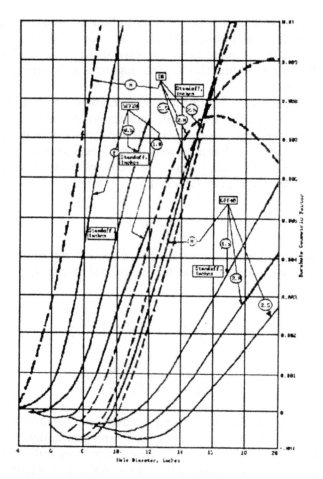

FIGURE 3.12
An example of a borehole correction chart for the IL (after Schlumberger, 1988).

line or point for the proper tool type and standoff. The value of G_b will be read on the vertical (y) axis. The borehole signal, BHS, is the product of G_B multiplied by the mud conductivity, $1/R_m$, and is

$$BHS = G_B * \frac{1}{R_m} \tag{3.3}$$

Then, subtract Equation 3.3 from Equation 3.2b:

$$\frac{1}{R'_a} = \frac{1}{R_{ID}} - \frac{G_b}{R_m} \tag{3.4a}$$

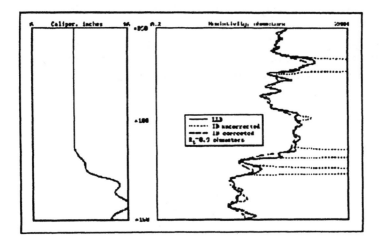

FIGURE 3.13
A comparison of the IL response with and without borehole correction (after Schlumberger, 1989).

$$\frac{1}{R'_a} = \frac{G_s}{R_s} + \frac{G_i}{R_i} + \frac{G_t}{R_t}$$ (3.4b)

Some charts have a nomograph on the right side, which may be used with a lesser accuracy, to determine BHS. This value will usually be in mS/m.

The charts, from the contractor's chart books, for this step will derive a value which frequently is a conductivity value. If you choose to continue using the contractor's charts for subsequent steps (which is the best way to proceed) you must take the reciprocal of the conductivity value in order to use the charts for the subsequent steps. If the conductivity value is in mS/m, it must be multiplied by 1000 before taking the reciprocal. The reciprocal will then be in ohmmeters.

Figure 3.13 shows the result, to the log, of removing the borehole effect from the deep induction curve.

3.7.2 Adjacent Bed Correction

The next step is to correct for the effect of the adjacent beds. This is necessary because the measured volume of the ID is quite large and will probably be influenced by the adjacent bed well after leaving it. This step is a bed-thickness correction.

The value, $1/R'_a$, of Equation 3.4 has been corrected for the influence of the borehole and contains only 3 terms. We would proceed, from this point, by obtaining a value of G_s from the appropriate chart, divide it by the resistivity (multiply it by the conductivity) and subtract that value from $1/R_a'$ to obtain a value for $1/R_a''$.

This process is equivalent to determining the geometrical factor of the adjacent bed contribution, multiplying it by the conductivity of the adjacent zone, and subtracting the product from $1/R_a'$:

$$\frac{1}{R_a''} = \frac{1}{R_a'} - \frac{G_s}{R_s} \qquad (3.5a)$$

or

$$\frac{1}{R_a''} = \frac{G_i}{R_i} + \frac{G_t}{R_t} \qquad (3.5b)$$

In actual practice, the charts supplied by the contractors will do this for you. One must be careful, however, because of the existing nomenclature. The borehole corrected value of Equation 3.4 must be reciprocated to obtain a resistivity value, R_a'. The charts from the contractor call this R_a. There will be several charts to choose from. The choice will depend upon the resistivity of the adjacent bed. In the Schlumberger chart book, the choice is between adjacent bed resistivities, R_s, of 1, 2, 4, or 10 ohms. This correction is most sensitive at high values of R_a' and/or thin beds. It is probably best to consider interpolating between charts, when in thin beds and high resistivities.

Figure 3.14 shows one of the bed thickness correction charts for the ID (Rcor-5) from the Schlumberger *Log Interpretation Charts* book, 1986 edition. Enter the bed thickness in the horizontal axis, proceed upward to the value of R_a'. Move to the left axis and read the corrected value, R_a'''(R_t in the Schlumberger *Log Interpretation Charts*).

3.7.3 Invaded Zone Correction

Note that, in the chart book, the value of R_a'' is labelled R_t. *This value is not R_t.* This value has not been corrected for the effect of the invaded zone. The invaded zone effect is probably small. But, a final step is needed before R_t can be determined. Even though the deep induction log (ID or 6FF40) is relatively insensitive to the invaded zone influence, there are many cases where the influence cannot be ignored. These cases are when the measurement is made with a system other than the ID or 6FF40 and/or when the invasion is very deep.

Figure 3.15 show the geometrical factors for the invaded zone for the Schlumberger conventional induction logs.

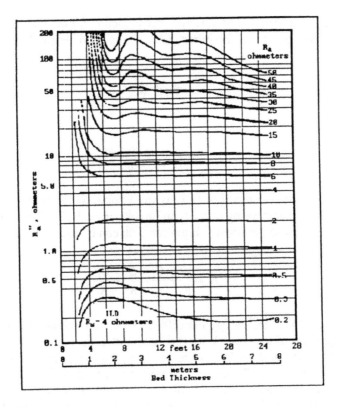

FIGURE 3.14
A sample bed thickness correction chart for the ID (after Schlumberger, 1988).

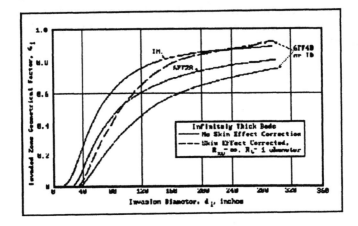

FIGURE 3.15
Integrated radial geometric factors for the ID, IM, 6FF40, and 6FF28 (after Schlumberger, 1988).

Since R_a'' contains only the geometrical factors for the invaded zone and the undisturbed zone, the other factors have been eliminated and we can say that

$$1 = G_i + G_t \qquad\qquad (3.6a)$$

Therefore,

$$G_t = 1 - G_i \qquad\qquad (3.6b)$$

Because the resistivity of the invaded zone, R_i can be measured and determined from the log and G_i can be determined, if the effective diameter of the invasion is known, Equation 3.5b can be solved for R_t:

$$R_t = \frac{1}{R_a''} - \frac{G_i}{R_i} - \frac{G_i}{R_a''} + \frac{G_i^2}{R_i} \qquad\qquad (3.7)$$

In actual practice, it is usually difficult to accurately determine the effective diameter of invasion. With the information provided by three curves, the ID, the IM, and an R_{xo} device, however, the invasion diameter and the corrected values for R_t, and for R_{xo} can usually be determined. This is also performed by the contractor's charts. One popular version of this step is called the "tornado" chart, because of the appearance of the data lines. Figure 3.16 shows a sample of such a chart (after Schlumberger) for the Dual Induction — Laterolog 8. There are similar charts for the various combinations of induction logs and Phasor induction logs with shallow investigation resistivity or conductivity devices. These charts are also similar in utility and information to those described in the guarded-electrode resistivity chapter.

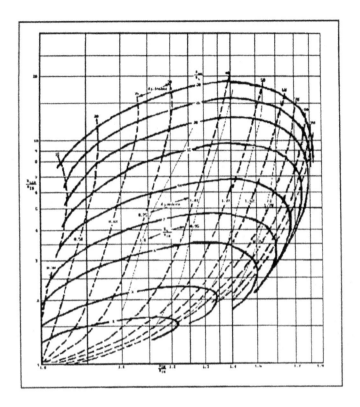

FIGURE 3.16
LL8 "Tornado" chart for the DIL (after Schlumberger, 1988).

4

Spontaneous Potentials

4.1 Introduction

The Spontaneous Potential (SP) (also called the Self Potential) log is a measurement of the naturally occurring potentials in the earth. It is usually simply denoted "the SP."

The SP measurement appears deceptively simple. It requires only a pair of stable electrodes and a high-quality voltmeter circuit. In reality, it is probably one of the most complex of the commonly used methods. The measured signal is made up of a large number of components. Fortunately, the dominant four or five components are normally the most useful ones. The primary uses of the SP are to determine the distributions, motions, states, and compositions of the aqueous fluids in the borehole/formation complex. In borehole surveys, the borehole fluids are also involved.

All of the present hydrocarbon applications of the SP measure potentials due to electrical currents in the borehole, as well as currents and potentials of both the invaded and the undisturbed permeable zones of the formations. Some of the scientific, engineering, and mineral applications do not involve the borehole. Also, some of these latter focus on the impermeable zones, in addition to the permeable. Some of the petroleum uses of the spontaneous potential curve are shown in Table 4.1. Table 4.2 shows non-hydrocarbon uses of the same measurement.

SP measurements are subject to many disturbing factors. It is important to recognize these for correction and/or elimination. The presence of metallic bodies, electrical storms, corrosion protection electrodes, fence lines, pipe lines, power lines, and electrical generators all can disturb the SP measurement. It is common, for example, to ask the driller to shut down his rig generator during logging, especially if it is a dc system. Earth currents due to sun spot and borealis activity contribute to and are disturbing components of the total SP signal, especially in the north and south near-polar regions.

Mineral exploration, in the 19th century, used surface SP measurements to locate metallic mineral bodies which lay imbedded in the ground water/arid zone interface. This is illustrated in Figure 4.1. A similar method was successfully used in Wyoming in 1950 to 1970 to locate near-surface uranium geochemical cells. This uses the redox component of the SP.

TABLE 4.1

Petroleum uses of the SP curve

1. R_w determinations
2. Differentiate permeable from impermeable zones
3. Bed boundary locations
4. Shaliness determination, qualitative
5. Aid in correlation
6. Oil/water contact locations
7. Water entry location
8. Determining depositional history
9. Location of "fish"
10. Location of lost circulation

TABLE 4.2

Non-hydrocarbon uses of the SP

1. Determination of water salinity
2. Show stratification in sediments
3. Determine depositional history of sediments
4. Determine degree of shaliness
5. Correlation
6. Determine relative redox state
7. Locate water entry or loss
8. Locate "fish"
9. Determine permeable and impermeable zones
10. Locate some mineral bodies

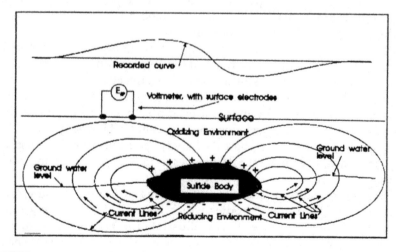

FIGURE 4.1
The *SP* around an ore body at the alteration interface.

Modern mineral exploration makes downhole use of the redox component of the *SP* curve to locate the portion of a geochemical cell intersected by the borehole in reduction-oxidation type mineral deposits (i.e., some uranium, silver, and copper deposits).

The primary use of the *SP* curve in the petroleum business involves the electrochemical components. The change of potentials of this component are caused by the separation of ion types because of their different mobilities and their different electrochemical potentials. The curve is used to determine the formation water resistivity, R_w, and salinity. This depends, to a large extent, upon the invasion of the permeable formation by the drilling mud filtrate. It is an excellent curve to use to determine bed boundaries, as it is almost a point measurement. It often correlates well with the drilling time log. The response of the *SP* curve (and resistivity) to changes in grain size in sands and silts is used with increasing regularity in geological studies for petroleum, and engineering purposes.

The measurement techniques of the *SP* in a borehole are somewhat different from those of the surface methods. Except for the influence of the borehole, however, both types of *SP* measurements share many of the same causes. All borehole *SP* measurements measure the potential drop in the borehole due to the flow of electrical current caused by the various *SP* phenomena. All borehole *SP* values are compared to a fixed reference potential at the surface. Surface *SP* measurements usually use a fixed spacing, differential array which is moved across the prospect in discrete steps.

The *SP* contrast between permeable zones (particularly sands) and impermeable shales is widely used in all modes of geophysical logging. The changes in the redox component from one zone to another (most noticeable in the shales), are used in mineral exploration to determine the presence and status of some mineral deposits (primarily sedimentary uranium, silver, and copper). The effects of the electrofiltration component are frequently used in trouble-shooting borehole problems.

The *SP* curve is used to determine bed boundaries because the response is a symmetrical, predictable point response. The bed boundary is always at the inflection point of the traced curve. The *SP* is regularly used to detect permeability (and especially, excessive permeability). Note however, that this use is not *quantitative* because the controlling permeability is that of the mudcake, not the formation. In geothermal work, the *SP* is used to locate the influx of hot waters. The *SP* method and systems have been used to locate lost outboard motors, a lost airplane, and a sunken submarine (the redox component). A very common use is to locate lost drillpipe, tools, casing, and tubing ("fish") in boreholes. This application also uses the redox component of the SP. In petroleum work, the *SP* is frequently used to determine or verify the "oil/water contact." The *SP* measurement is also used for those petroleum applications shown in Table 4.1. Table 4.2 is a similar list for non-hydrocarbon applications. Note the similarity of the two lists.

4.2 Principles

The borehole *SP* system reads the voltage in the borehole, due to the flow of current through the drilling mud. These currents are due to chemical and physical actions in the borehole and in the formation zones. In surface work, the differential *SP* due to earth currents, is read on the surface. Four primary components to the *SP* are normally used in borehole logging: the potentials due to diffusion, E_{SP_d}, absorption, E_{SP_a}, electrofiltration, E_H, and to oxidation/reduction phenomena (redox), E_{ox}. E_{ox} is not important to petroleum logging because the deeper environments are usually reduced. But E_{ox} is used frequently in mineral exploration. Other causes include those resulting from telluric currents, power line noise, lightning, and anti-corrosion devices.

In the borehole, the *SP* measurement is made with a single stable electrode. The electrode is normally on the sonde and may be shared with the resistivity systems. Sometimes however, it is located up on the bridle cable. For surface surveys, a fixed spacing, differential array is usually employed.

The measure circuit of the *SP* is a voltmeter which has a very high input impedance ($\geq 10^9$ ohms). Any current drawn by the measurement will result in a resistivity component being added to the real *SP* signal. The impedance is purposely made high so that a negligible current will be drawn for the measurement. Thus, the signal will not be noticeably distorted. Old logs made use of a manually balanced potentiometer to achieve (usually poorly) an "infinite impedance" measurement (see Problems with the SP, later in this chapter).

The *SP* reference or return electrode is usually on the surface, in the mud pit, in borehole logging. It is virtually an infinite electrical distance from the measure electrode. This electrode should be in an environment which is as nearly like the downhole environment as possible. The *SP* measurement is a differential electrode potential measurement. If the environments of the two electrodes are identical, then the fixed difference of voltage between them will be zero.

In practice, the *SP* system reference electrode is in a stable, constant environment. Its potential will remain constant. The downhole electrode is in a similar environment. Its potential can change, however, due to factors within the formation. Thus, any change of potential registered by the voltmeter will be due to a downhole change. Measurements are referred to the "shale baseline", which will tend to eliminate constant differences of the two electrode potentials. Only the changes in the downhole environment will be regarded.

4.2.1 Physical Principles

When an electrode is put into an electrolyte solution — that is, a solution in which the solute exists wholly or partially in the form of ions — the electrode

Spontaneous Potentials

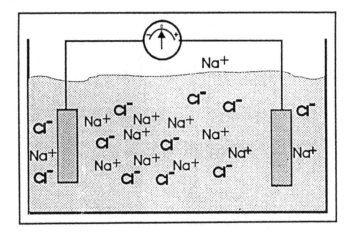

FIGURE 4.2
A pair of identical electrodes in an aqueous solution.

will react with the electrolyte. Metallic ions (cations, positive ions, +) will go into solution. This will leave an excess negative charge on the electrode. This charge is called the "electrode potential". If the measure electrode and (identical) reference electrode are in the same or identical solutions and are at the same temperature, they will be at the same potential. See Figure 4.2. Any *SP* "signal" will be due to a local change in the neighborhood of the measure electrode.

4.2.1.1   Absorption or Shale Potential

In shales, the silica components of the clays exhibit an apparent negative charge which will bind cations, such as Mg^{++}, Ca^{++}, etc. When solutions of NaCl are in proximity to the clay, the resident cations can easily be exchanged for Na^+ ions of the solution. This will result in the solution in the shale becoming more positive. This is the Nernst or absorption component. See Figure 4.3.

Also, because of the layered structure of the clay and the associated charges on the layers, shales are much more permeable to the cations than to the anions. Only the positive ions are able to move through the shale from a more concentrated to a less concentrated solution, similar to the action involving a membrane. This too results in a potential across the shale. Because of this action, this potential is called a membrane potential.

4.2.1.2   The Electrochemical or Diffusion Potential

Within the permeable formation, the invading fluid will usually have a different concentration or salinity than the native fluid. There will be a difference in potential across the interface. See Figure 4.4.

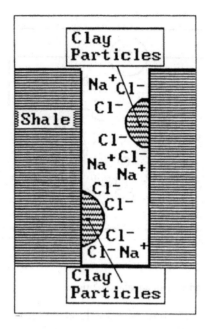

FIGURE 4.3
A representation of the probable ion distribution in a shale.

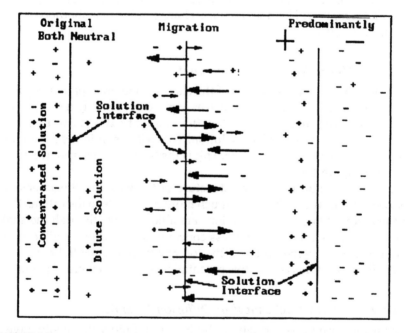

FIGURE 4.4
The diffusion principle of the SP.

In a solution of NaCl, the chloride ions, Cl⁻, have a higher mobility than the cations, Na⁺. Therefore, when dilute and concentrated solutions are adjacent, more of the chloride ions will cross the interface between the concentrated to the dilute solution and the lower salinity will acquire a more negative charge. The remaining sodium ions will give the more concentrated solution a more positive charge. This is the diffusion component of the SP. It occurs at the interface between the formation water and the invading hole fluid. This is the component used to determine the resistivity of the formation water.

If a_w is the chemical activity of the formation water and a_{mf} that of the mud filtrate, we can also say,

$$E_D = -K_D \log \frac{a_w}{a_{mf}} \tag{4.1}$$

4.2.1.3   Effect of Ion Types

The diffusion potential, E_D is, in terms of resistivity,

$$E_D = -K_D \log \frac{R_d}{R_c} \tag{4.2}$$

where R_d is the resistivity of the dilute solution, and R_c is the resistivity of the concentrated solution.

And,

$$K_D = 2.3 \frac{R_g \left(I_c N_c - I_a N_a \right)}{F_a \left(I_c Z_c N_c + I_a Z_a N_a \right)} \tag{4.3}$$

where
  R_g = the universal gas constant
  K_D = the diffusion coefficient of a liquid junction cell
  F_a = Faraday's Constant
  I = the ion mobility
  N = the number of ions
  Z = the valence of the ion
  $c, a,$ = subscripts, denote cation and anion, respectively

If the solution is other than NaCl, Equation 4.3 can be modified to account for the addition of Ca⁺⁺ and Mg⁺⁺ ions (the most common ones encountered) at 75°F (24°C) (Gondouin et al., 1957):

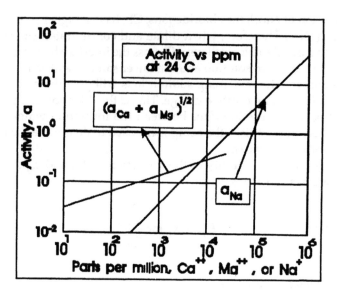

FIGURE 4.5
Activity vs. concentration (after Schlumberger, 1989).

$$E_D = -K_D \log \frac{\left(a_{Na^+} + \left(a_{Ca^{++}} + a_{Mg^{++}} \right)^{1/2} \right) w}{\left(a_{Na^+} + \left(a_{Ca^{++}} + a_{Mg^{++}} \right)^{1/2} \right) mf} \tag{4.4}$$

assuming that the anion is Cl⁻. If the concentrations are known or can be reasonably estimated, the solution activities, $a_{Na^+} + (a_{Ca^{++}} + a_{Mg^{++}})^{1/2}$ can be determined from Figure 4.5.

Since conductivity, and thus the resistivity, is a function of activity, equation (4.4) can be written

$$E_D = -K_D \log \frac{R_{mfe}}{R_{we}} \tag{4.5}$$

where R_{mfe} and R_{we} are the resistivities these solutions would have, if the solution is dilute NaCl. Figure 4.6 shows the actual effects of the various common ion types on the solution resistivity, by laboratory measurement.

At the dilute concentrations, Equation 4.4 describes the departure of the potential, E_D, when other cations than Na⁺ are involved. When the solution becomes very concentrated, there is a departure, also. This is shown, for NaCl solutions, in Figure 4.7. This departure at high concentrations is due to the restriction of the thermal movement of the ions and molecules, due to crowding.

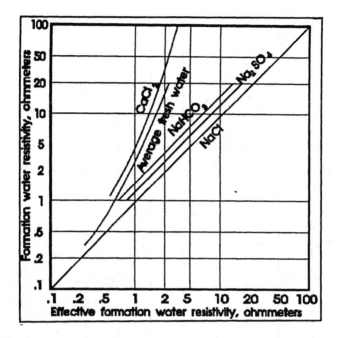

FIGURE 4.6
R_w vs. R_{we} for salt solutions (after Schlumberger, 1989).

The total picture, with a permeable sand or carbonate, and a borehole is shown in Figure 4.8. The electrode "sees" the potential due to the flow of current in the mud column. Therefore, the electrode detects a potential opposite a shale and a different one opposite a sand or a carbonate. This combination of the diffusion and the absorption components is described in the equation for the static *SP* potential:

$$E_{SSP} = -K_{SP} \log \frac{R_{mfe}}{R_{we}} \tag{4.6}$$

$$K_{SP} = 0.133 \frac{mvs}{°F} T_{f,F} + 60.77 \, mvs \tag{4.7}$$

where T is the temperature in degrees Fahrenheit (see Figure 4.8). If the temperature is in degrees Celsius, the relationship is

$$K_{SP} = 0.24 T_{f,C} + 65.03 \tag{4.7a}$$

The value of E_{SP} is read from the log for a *unit* permeable zone. If the temperature is in degrees Kelvin, the relationship is

$$K_{SP} = 6.43 \times 10^{-2} T_{f,K} + 64.97 \tag{4.7b}$$

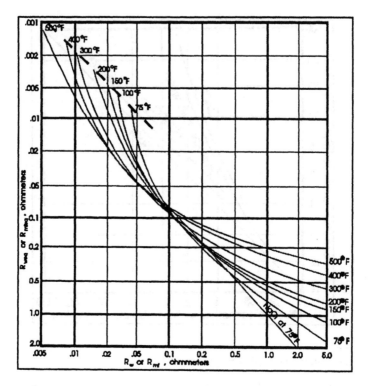

FIGURE 4.7
The probable R_w value, as a function of R_{we} and T_{form} (after Schlumberger, 1988).

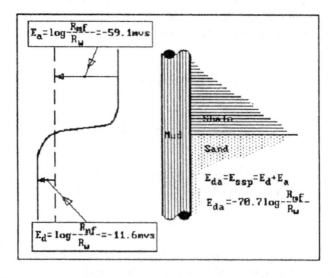

FIGURE 4.8
The combined Nernst and diffusion potentials.

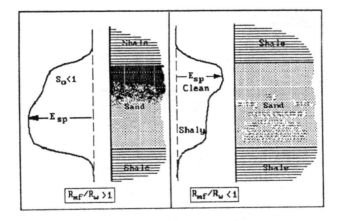

FIGURE 4.9
Illustrating the measurement of E_{SP}.

$$K_{SP} = \frac{460°R + °F}{537°R} \qquad (4.8)$$

where R indicates Rankine degrees.

Many texts use a value of 60 mvs, instead of 60.77, but rounding off to 61 mvs is more nearly correct. Some contractors use the form which is the same as Equation 4.7a. The result is shown in Figure 4.9.

Notice that these equations only deal with the temperature and the diffusion potential. The SP Equation 4.7 describes the departure of the SP from the shale potential. By convention, the SP is measured from the shale potential (shale line or shale baseline) and would be measured as shown in Figure 4.10. Note, however, if the shale baseline shifts, as due to redox potentials, the line of SP measurement would shift with it.

The value of E_{SSP} is the maximum value the SP would read in an infinitely thick, clean, low R_t/R_m bed. This, also, is illustrated in Figure 4.9. This is particularly critical if the upper part of the zone has a high hydrocarbon saturation, S_o. Measure the SP deflection in a zone that has a high water saturation, S_w.

4.3   Thin Beds

If the bed is thin or if R_t/R_m is high, then a bed-thickness correction must be made to E_{SP} to get E_{SSP}. See Figure 4.10. There are several types of charts for correcting the SP curve for bed thickness effects. These are available in the contractor's chart books. One of the best methods is that suggested by Segesman (1962). A copy of a part of the Segesman charts is shown as

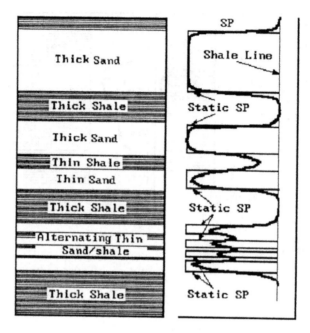

FIGURE 4.10
The effect of the bed thickness upon the *SP* curve.

Figure 4.11. The zone must not have any shale breaks or other permeability barriers within the unit which would prevent the free mixing of the formation waters.

4.3.1 Formation Water Resistivity, R_w

After reading E_{SP}, correct for bed thickness. The greatest deflection value within the unit permeable zone should be read in plus or minus millivolts, with the sign. *The sign is important.* This is shown in Figure 4.9. After the value of the *SP* is read from the log, it must be corrected for bed thickness. There are several correction charts available from the petroleum logging contractors. Experience will tell which suits your needs best. One of the better methods is that of Segesman (Dec. 1962)(Figure 4.11) which is to be found in the Schlumberger chart book and others. Then, from the *SP* equation, you can solve for R_{we} with Equation 4.5. If the permeable zone is broken by a shale stringer or any impermeable barrier, it may not be one unit. Each unit must be treated separately.

Notice that the formation water resistivity found this way is called R_{we} (or, sometimes, R_{weq}). It is the equivalent formation water resistivity, for a solution of NaCl. If it contains any other electrolyte, a correction must be made for the other ion types before the calculated resistivity is a true electrical resistivity value of R_w or R_{mf}. R_{mfe} (or R_{mfeq}) is similarly the NaCl equivalent value of R_{mf}, at formation level temperature, T_{form}. Since the

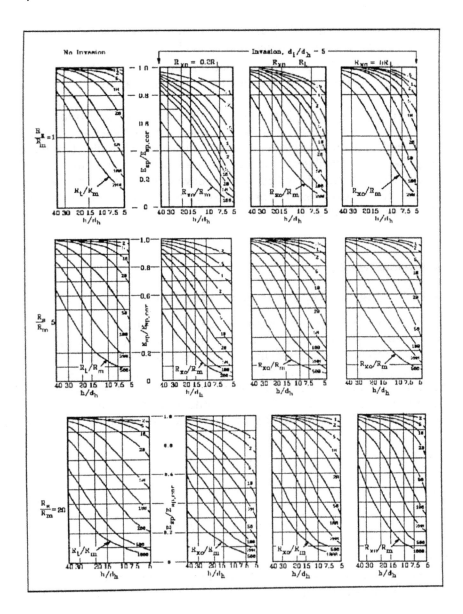

FIGURE 4.11
Bed thickness correction curves for the *SP* (after Schlumberger, 1988).

value of R_{mf} is usually measured or calculated from a measured value of R_m, this value is the actual real value of R_{mf}, not R_{mfe}. It must be converted to the equivalent NaCl, R_{mfe}, using the proper conversion chart, such as Figure 4.7 or similar, if R_{we} is to be correctly calculated. Also, these calculations **must be made with all solutions corrected to a temperature of 75°F (24°C). *If that last correction is not made, the calculations will not be correct.*** For predominantly NaCl muds, Schlumberger recommends,

a. If R_{mf}, at 75°F (24°C), is greater than 0.1 Ωm, correct R_{mf} for formation level temperature and use $R_{mfe} = 0.85 \, R_{mf}$.

b. If R_{mf}, at 75°F (24°C), is less than 0.1 Ωm, use chart SP-2 to find R_{mfe}, after correcting to formation level temperature.

When the value of R_{we} is found, the value of R_w may be estimated by using the correction chart, Figure 4.7. This is a partially empirical chart which will apply to average conditions. Only experience will determine if it applies in a particular situation. In general, it will be satisfactory for the middle to higher salinity waters. It does not work well in very low-salinity waters. Refer to Chapter 5 for information to estimate the value of R_{mf} from the measured value of R_m.

4.4   Taking Ion Types into Consideration

Since many of the logging systems cannot distinguish between ion types, it is customary to denote the salt in solution as eNaCl. Since NaCl is, by far, the most common salt present in formation waters, eNaCl can often represent the ion content satisfactorily. At other times, it is valuable to know the actual ion content. Since this problem is intimately involved with the *SP* measurement, some relationships are worth mentioning here.

The salinity, *S*, in ppm, of the formation water (for example) is

$$NaCl_e = 10^x \qquad (4.9)$$

where

$$x = \frac{3.562 - \log(R_w - 0.0123)}{0.955} \qquad (4.9a)$$

The chart of Figure 4.12 shows the relationship between the water salinity, resistivity, and temperature.

4.5   Dunlap Multipliers

Dunlap Multipliers form a convenient method for making ion content corrections or converting to equivalent NaCl solution values.

If the ion content of each part is known in parts per million, add the total ppm of each type in the solution (i.e., HCO_3^-, $SO_4^=$, $CO_3^=$, Cl^-, Na^+, Ca^{++},

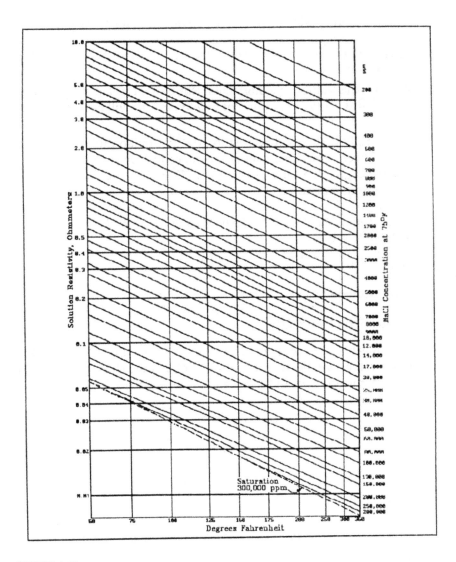

FIGURE 4.12
A chart showing the relationship between solution resistivity, salinity, and temperature (after Schlumberger, 1986).

Mg^{++}, and K^+. Enter the value of the total dissolved solids (TDS) in the horizontal scale of the Dunlap chart, is shown as Figure 4.13. Extend a line upward, vertically. As the line intersects the line of each of the contained ion types, read the multiplier from the vertical scale. Multiply each ion content value by its respective Dunlap Multiplier. Add them to get an effective or equivalent ppm. Enter the value in Figure 4.13 to find the effective NaCl resistivity at any temperature. *Note that this conversion, as with most of the solution conversion problems must be made at the resistivity when the solution is at 75°F (24°C).*

To use the Dunlap Method to find the resistivity of a water sample,

1. Add all of the concentrations of the individual ion types in parts per million (ppm).
2. Enter this value in the horizontal axis of the Dunlap chart, then find and list the multiplier for each type. This will be on the left vertical axis.
3. Multiply each ion concentration by its Dunlap multiplier and find their sum. This sum is the concentration of NaCl which would give the same resistivity at 75°F (24°C) as the original solution of item 1.

Of course, the relationship can be worked in the other direction. That is, from eNaCl to NaCl.

4.6   The Effect of Shale

If the permeable zone contains shale, the *SP* value will be lowered. The thick bed values will be the pseudo *SP* (PSP), not the static *SP* (SSP) that we need. A correction must be made to obtain the SSP. The Doll method (Doll, 1949) is widely used:

$$\alpha = \frac{PSP}{SSP} \qquad (4.10)$$

$$\alpha = 1 - \frac{\log \frac{u+g}{1+q}}{\log u} \qquad (4.10a)$$

where

$$u = \frac{R_t}{R_i} \qquad (4.10b)$$

and

$$q = \frac{R_t}{R_{sh}} \qquad (4.10c)$$

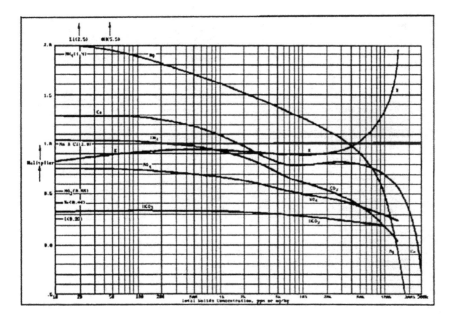

FIGURE 4.13
The Dunlap Multiplier correction chart (after Schlumberger, 1988).

4.6.1 Direction of Deflection

When the SP equation is examined, notice that the conversion constant, K_{SP}, has a negative sign. The negative sign is part of the constant. It is usually separated out for emphasis.

If R_{mfe} is more resistive (less saline) than R_{we} (the usual petroleum situation), then the ratio, R_{mfe}/R_{we} is larger than 1.0. Therefore, its logarithm is positive. The positive logarithm times the negative constant will indicate a negative deflection of the SP. See Figure 4.14. The more saline the formation water is (lower resistivity), the larger the logarithm will be and the greater the SP deflection in a negative direction will be.

If R_{mf} is less resistive (more saline) than R_w, then the ratio is less than 1.0 and the logarithm is negative. This times the constant indicates a positive deflection of the SP curve. This happens occasionally. In shallow holes in the Rocky Mountains in the Spring during runoff, the shallow formation waters are quite low salinity (<< 100 ppm). The positive deflection then becomes very common. In this case, the salinity is so low, that changes in the relative salinities occur easily. Thus, the ratio of R_{mf}/R_w is likely to change. If $R_{mf} = R_w$, then the ratio is equal to 1.0 and the logarithm is zero. Therefore, the SP deflection is zero. This can, and frequently happens during the Spring runoff in the snow country. Notice that the direction and amplitude of the SP deflection are each a function of the ratio of the salinity of R_w to R_{mf}. The shape of the transition of the SP curve across a bed

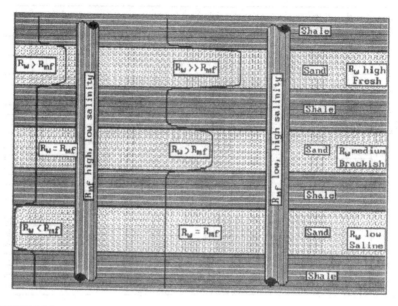

FIGURE 4.14
The effect of the ratio, R_{mf}/R_w on the *SP* Curve.

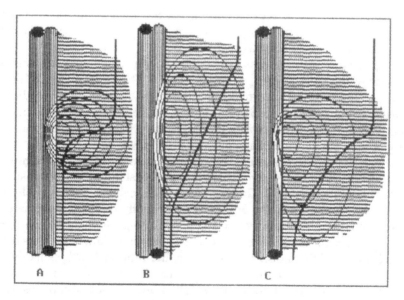

FIGURE 4.15
The effect of formation resistivity upon the *SP* curve.

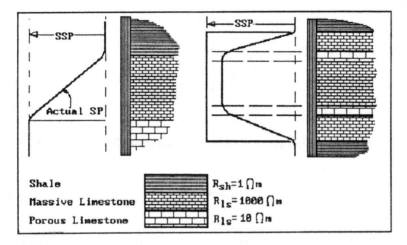

FIGURE 4.16
SP curve shape and response in massive, resistive zones.

boundary is a function of the relative and absolute resistivities of the formation and the drilling mud. See Figure 4.15.

4.6.2 Formation Resistivity Effects

An impermeable zone (i.e., a massive, zero porosity limestone) will contribute nothing to the value of the SP curve. Across such a zone, the SP value will be only the linear transition from one permeable zone or shale to the next. This is illustrated in Figure 4.16. Notice that the SP curve is responding only to the shales and the permeable zones.

4.7 The Static SP

The sum of the potentials constituting the value of Essp can be considered to be,

$$E_{SSP} = I(r_s + r_{sh} + r_m)$$ (4.11)

where

r_s = the resistance of the sand, $(r_s = R_s/G)$
r_{sh} = the resistance of the shale
r_m = the resistance of the mud column
I = the current in the mud column

As the electrode moves from on zone to the other,

$$E_{SP} = I r_m \qquad (4.12)$$

Combining Equations 4.13 and 4.14,

$$E_{SSP} = \frac{E_{SP}\left(r_s + r_{sh} + r_m\right)}{r_m} \qquad (4.13)$$

If the borehole is plugged with an insulating plug, so that no current can flow,

$$\lim_{r_m \to \infty}\left(E_{SP}\right) = E_{SSP} \qquad (4.14)$$

The value of the SP, at the bed boundary between a shale and a permeable zone is

$$E_{SP} = h\left(E_{SSP}\right)(4h+1)^{1/2} \qquad (4.15)$$

where h = the thickness of the permeable bed in units of the hole diameter. Also,

$$\frac{E_{SP}}{E_{SSP}} = \frac{(h+1)^{1/2}}{(4h+1)^{1/2}} \qquad (4.16)$$

4.8   The Calculation of Salinity

The calculation of R_w and the subsequent determination of the probable salinity of the formation water from the *SP* curve can be an effective tool in mineral exploration, engineering, and environmental work. Mineral deposits and, especially engineering projects, are often sensitive to salinity changes of the formation waters. Leakage of salt water, salinity plumes, and unexpectedly low or high salinities can all have great significance in engineering and environmental investigation. Note the discussion of salinity earlier in this chapter.

4.9   Reduction of Data

The steps to use to reduce the data from the *SP* curve to obtain R_w are,

1. Identify the shale zones above and below the zone of interest. Draw a line on the log from the top shale value to the bottom shale value. This is the shale line.

2. Read the number of millivolts and direction (±) from the shale line to the greatest deflection in a single formation zone unit. This is the value of E_{SP}.

3. Correct the value for bed thickness. Use the chart from the contractor which corresponds to Figure 4.11. These Segesman curves are actually quite suitable for use on the *SP* curves of any contractor's system. This, then, is the value of the shaly sand SP, E_{SP}.

4. Correct for the presence of shale. Use the Doll relationship (Equation 4.10). The result is now E_{SSP}.

5. Determine the value of R_{mf}, from the log heading, if it is available:

   a. If R_{mf} is not available, calculate the probable value from the value of R_m (found on the log heading). Correct this to the value at 75°F (24°C) and determine the probable value of R_{mf} at 75°F. Information to do this can be found in Chapter 5 of this text and/or in any contractor's chart book. Then, correct R_{mf} to the value at T_{form}.

   b. R_{mf} may also be estimated from one of a number of specially designed charts. These are frequently the result of studies in a particular field by one drilling company.

6. Correct R_{mf} to the proper value at T_{form}, if you have not already done so.

7. Determine the value of R_{mfe} from the SWC recommendation or from the ion type correction chart, (Figure 4.7).

8. Calculate the value of K_{SP} from T_{form} (Equation 4.8).

9. Calculate R_{we}, using the *SP* Equation 4.6.

10. Determine the value of R_w from the ion type correction chart (Figure 4.6).

The value, E_{SSP}, is the static SP, obtained by correcting the log reading for thin bed effects. The static SP, E_{SSP}, is the value that the *SP* would read if all borehole current were blocked by insulating plugs or if the formation zone were infinitely thick. It is the potential to be found at the center of a very thick, clean, permeable zone.

To use the thickness correction charts, Figure 4.11,

1. Divide the invasion diameter by the hole diameter and pick the diagram column which comes closest to that number. You can interpolate, if you wish, but it is probably not necessary.

2. Divide the resistivity of the closest shale, R_{sh} (R_s), by the temperature-corrected mud resistivity, R_m. Pick the diagram row closest to that value. Again, you can interpolate, if you wish, but it probably is not necessary.

3. Bed thickness is given in hole diameters. Determine the bed thickness from the log and divide it by the hole diameter (or bit size). *Do not mix units.* Enter this value on the horizontal axis.

4. Read R_t and R_m from the log, calculate the ratio, R_t/R_m. Pick the curve of this value.

5. Extend a line from the R_t/R_m ratio to the vertical axis. Read the ratio of the log reading to the corrected reading. Read the correction factor on the vertical axis.

6. Divide the value of E_{SP} (from the log) by the correction factor to get E_{SSP}. If the zone is shaly, the bed thickness corrected SP value (PSP) must still be corrected for the presence of the shale. This will be covered again, in the chapter containing the shaly sand interpretation. This corrected value is the E_{SSP} (SSP).

   (The Segesman charts are very good. Figure 4.11 is a copy of the one from the Schlumberger Well Services, Inc. Log Interpretation Charts. Be sure to refer to the charts in the chart book. This copy may have some inaccuracies. There are many other charts which are designed to correct the SP for bed thickness effects. Use the one which best suits your purposes.)

7. The value of E_{SSP} will be used in the SP Equation 4.6. Determine the value of K_{SP} with Equation 4.7.

8. Correct R_{mf} (from the log heading, chart, or calculation) for temperature effect (correct to formation level temperature, T_{form}).

9. The value of R_{we} must be corrected for the ion type at high resistivities and for concentration effects (restriction of activity) at low resistivities. If the value of R_{we} is approximately 0.08 ohms, then no correction is needed. If the value is lower than 0.08 ohms, then the value must be corrected for concentration effects. The ion type correction chart may be used. This chart is based on the average of 10,000 mud samples from the U.S. mid-continent, Texas Gulf Coast, and Rocky Mountain regions. You can see the spread of the curves, due to the temperature of the solution and the curve due to the concentration.

At higher resistivities, (above 0.08 ohms) R_{we} must be corrected for the dominance of the Ca^{++} and Mg^{++} ions. This correction is less reliable than the correction at lower resistivities. The author's experience in the Rocky Mountain area was that most of the time, above R_{we} = 1.0 ohmmeter, the assessment was better without correction. These were shallow, very low-salinity sands, however. Experience is necessary.

4.10 Methods of Determining R_w and Use of R_w Calculations

4.10.1 SP vs. R_{xo}/R_t — The Ratio Method

The *SP* values can be used to verify saturation. The equation is:

$$E_{SSP} = -K_{SP} \log\left(\frac{R_{mf}}{R_w}\right) + E_H \tag{4.17}$$

where E_H = the electrical filtration potential (see page 111).

It can be said that the value of R_{mf} is

$$R_{mf} = \frac{R_{xo,s}}{F_R} \tag{4.18}$$

Also, the value of R_w is

$$R_w = \frac{R_o}{F_R} \tag{4.19}$$

Therefore,

$$E_{SSP} = E_H - K_{SP} \log\left(\frac{R_{xo,s}}{R_o}\right) \tag{4.20}$$

$$R_o = \frac{R_t}{I_R} \tag{4.20a}$$

and

$$R_{xo} = \frac{R_{xo,s}}{I_{xo}} \tag{4.20b}$$

Thus,

$$E_{SSP} = E_H - K_{SP}\left(\log\frac{R_{xo}}{R_t} + \log\frac{I_R}{I_{xo}}\right) \tag{4.20c}$$

and

$$E_{SSP} = E_H - K_{SP}\left(\log\frac{R_{xo}}{R_t} + n\log s_W - n\log s_{XO}\right) \tag{4.20d}$$

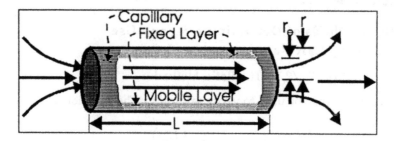

FIGURE 4.17
The electrofiltration potential, E_H.

At 100% water saturation, $\log S_w = \log S_{xo} = 0$. Then,

$$E_{SSP} = -K_{SP} \log \frac{R_{xo}}{R_t} + E_H \qquad (4.21)$$

Since the equation is a semi-logarithmic one of the y-intercept type, the plot of this relationship will result in straight lines when plotted on a semi-logarithmic paper. This is illustrated in Chart SW-2 in the Schlumberger Log Interpretation Chart Book. Note that,

1. The y value is E_{SSP}
2. The x value is $\log(R_{xo}/R_t)$
3. The y intercept is E_H
4. The slope is $-K_{SP}$

In the handbooks of the logging contractors, this chart is (i.e., SW-2) usually rotated 90 degrees from the mathematical convention of "x" as the horizontal axis and "y" as the vertical axis. Thus, E_{SSP}, the y-axis is the horizontal axis and R_{xo}/R_t [$\log(R_{xo}/R_t)$] is the vertical axis.

If S_w is some other value than 100%, then $\log(R_{xo}/R_t)$ is reduced by the amount of $\log(I_r/I_{xo})$. If the line of $S_w = 100\%$ goes through $E_{SSP} = 0$ and $\log(R_{xo}/R_t) = 0$ or $R_{xo}/R_t = 1$, then the y-intercept is zero and the value of E_h, the streaming potential, is 0. If it does not, then the value of the streaming potential is equal to the y-axis displacement (the y-intercept), in millivolts. A modification of them, a cloud of cations will form in the solution, near the surface of the metal, finally saturating the solution and impeding the process.

1. Most of our work will be in eNaCl. That is, the characteristic that the electrolyte and/or solution would have if it were NaCl in water. A conversion will be made, if it is needed. We do not normally have effective means of determining the ion types

within the formation in a down hole tool. Core and fluid sample analysis are usually used for this information. Specific ion electrodes are available and quite good. They, however, measure in the borehole fluid, which is usually not very representative of the formation fluid. Measured solution resistivities are always eNaCl resistivities, unless it is stated otherwise.

2. Electric fields will cause a migration of the cations to the negative direction and of the anions to the positive direction. This constitutes an electrical current flow in the solution. The rate at which these ions migrate will depend upon their respective valences and masses. This aspect is termed the mobility of the ion.

3. The electric field may be set up by any one or more of several natural, induced, or interfering mechanisms. It may be due to an unbalance of ion types in adjoining or the same zones. It may be due to an electrical circuit and electrodes we put into the formation.

The electrical resistivity of a given solution will depend upon the concentration of the ions of the salts in solution, upon the temperature of the solution, and upon the types of ions in solution. All three of these factors can affect the mobilities and activities of the ions.

The resistivity of any solution of salt may be determined if the ion types and their concentrations are known. The process uses "Dunlap Multipliers". See the Figure 4.13.

Enter the sum of the concentrations of the ion types on the horizontal axis. Move upward, along the constant salinity line to the line denoting that ion type. Move horizontally to the left axis and determine the multiplier for that ion concentration. Multiply the concentration of each ion type by its appropriate multiplier. The sum of these new concentrations will denote the equivalent sodium chloride concentration.

4.10.2 R_w from Resistivity Values

The value of R_w can be calculated from the resistivity values by using Archie's relationships. If the porosity is known from the core analysis, the neutron porosity log, or the acoustic log and the value of R_0 is known from the resistivity or density log, then the value of R_w can be calculated:

$$R_w = R_0 \phi^m \qquad (4.22)$$

Quite frequently, also, reliable values of R_w can be obtained from records in established areas. In the U.S., the State Geologist's office may have such information on file for explored areas.

4.10.3   The Use of R_{wa}

If the value of R_w is calculated from the resistivity value, R_t, then R_t must be equal to R_o to obtain a correct value. If, however, the value of R_t is influenced by the presence of hydrocarbon or gas, the value of R_{wa} will be erroneous. If R_t is used instead of R_o, Equation 4.22 becomes

$$R_{wa} = R_0 S_w^n \phi^m \tag{4.23}$$

and R_{wa} is erroneous by a factor of S_w^n. A comparison, however, of the values of R_{wa} and the real value of R_w (i.e., from the SP) will often reveal zones of hydrocarbon saturation.

4.10.4   The Dual Water Model

The dual water model can be used to obtain a better value for R_w in shaly sands. The value of R_w is only correct when calculated from the *SP* value of a clean sand. If the value is calculated from the *SP* value of a shaly sand, the value is incorrect (too low) and is designated R_{wa}:

$$\log\left(\frac{R_{wa}}{R_w}\right) = (1-\alpha)\log\left(\frac{R_{mf}}{R_w}\right) \tag{4.24}$$

The value of alpha, a, (Doll, 1949) is

$$\alpha = 1 - \left(\frac{\left(\frac{\log\left(R_t + R_{sh}\right)}{\left(R_i + R_{sh}\right)}\right)}{\log\left(\frac{R_t}{R_i}\right)}\right) \tag{4.24a}$$

where

R_t = the corrected, measured, deep resistivity
R_{sh} = the resistivity of the contained or adjacent shale
R_i = the resistivity of the invaded zone

Determine the value of α from log values and then solve Equation 4.24 for R_w.

4.10.5 R_w from R_{xo} and R_t

If the value of R_{mf} is accurately determined for the formation level, it can be used, with the resistivity values to determine R_w:

$$\frac{R_o}{R_{xo,s}} = \frac{R_w}{R_{mf}} \qquad (4.25)$$

If there is hydrocarbon or gas present,

$$\frac{R_{mf}}{R_w} = \frac{R_{xo} S_{xo}}{R_t S_w} \qquad (4.26)$$

A "Quick Look" version of this uses the R_{xo} log values and the Induction Log values.

Example

Now, we can determine the value of R_w from the *SP* curve:

SP deflection	= –70 mvs,
T_{form}	= 105°F,
R_{mf}	= 0.8 Ωm at T_{form},
R_t	= 10 Ωm,
R_m	= 1.0 Ωm at T_{form},
Hole diameter, d_h	= 12 in = 1.0 ft,
Bed thickness, h	= 6.0 ft.

Then, the bed thickness = $6.0/1 = 6d_h$.

$R_t/R_m = 10/1.0 = 10$.

Therefore, the correction factor, from Figure 4.11, $F_{ac} = 1.25$.

$$E_{SSP} = E_{SP} \times F_{ac} = -70 \times 1.25 = -87.5 \ mvs$$

$$K_{SP} = 105°F \times 0.133 \ mvs/°F + 61 \ mvs = 75 \ mvs$$

$$E_{SSP} = -87.5 \ mvs = -75 \log \frac{0.8}{R_{we}}$$

$$\frac{-87.5}{-75} = \log \frac{0.8}{R_{we}} = 1.17$$

Therefore,

$$10^{1.17} = \frac{0.8}{R_{we}}$$

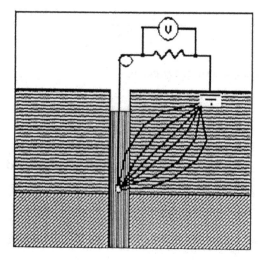

FIGURE 4.18
A schematic diagram of the resistivity component of the SP measurement.

or

$$R_{we} = \frac{0.8}{10^{1.17}} = 0.054\Omega \; mat \; 105°F$$

From Figure 4.12,

$$R_w = 0.06\Omega mat \; 105°F(40.5°C)$$

Correct the resistivity to 75°F (24°C) *first*. Then, the salinity in ppm at 75°F, S_{75}, is,

$$S_{75} = 10^x$$

$$x = \frac{3.562 - \log(R_w - 0.0123)}{0.955}$$

$$Salinity \approx 4500\,ppm$$

4.11   The Electrokinetic Component

The electrokinetic potential, E_H, is an unwanted component of the SP. It is also known as the electrofiltration potential, the Helmholz potential, and the streaming potential. It interferes with the quantitative use of the SP. It is set up primarily by the passage of small amounts of mud filtrate through the mud cake. It can, however, be caused by any flow of fluid into or out of the permeable formation zone. Thus, we will see it also at lost circulation zones and water entry zones.

The mudcake acts as a semipermeable membrane. The potential developed across it can be represented by Helmholz's equation:

$$E_H = \frac{\xi R \zeta}{4\pi\mu} \Delta P \qquad (4.27)$$

where

ξ = the dielectric constant of the moving fluid
R = the resistivity of the moving water
ζ = the zeta potential, which is the potential due to the existence of fixed and moving layers in the moving water
μ = the viscosity of the moving water
Δ = the pressure drop across the membrane

The polarity of the zeta potential can be either positive or negative, depending upon the direction of flow of the fluid.

Notice that the value of E_H is directly proportional to the resistivity, R, of the moving fluid. Therefore, the streaming potential is very sensitive to the entry of low salinity water into the borehole. Notice also that E_H is proportional to the dielectric constant, ξ, of the moving fluid. The dielectric constant of water is very high compared to those of other fluids and solids. E_H is inversely proportional to the viscosity of the moving fluid. The viscosity of water is rather low.

Pirson (1935) evaluated the "streaming potential" as

$$E_H = 0.0391\left(R_{mc} t_{mc} WL\right) \qquad (4.28)$$

where

R_{mc} = the resistivity of the mudcake at formation temperature
t_{mc} = the mudcake thickness
WL = the water loss in cubic centimeters per 30 minutes at 100 psi differential pressure

The potential, E_H, is added to the SP potential which we have already examined. Therefore, the resulting SP equation is

$$E_{SP} = E_{DA} + E_H \qquad (4.29)$$

where

E_{da} = the diffusion-absorption component
E_H = the electrofiltration component

The assumption is usually made that $E_H = 0$, unless we know better. This, of course, may not always be a safe assumption, especially in low salinity sands. The electrofiltration potential or *streaming potential* is a good indicator of water entry or loss. The movement of drilling mud, in one case, and of the formation water, in the other, will result in a large and formless anomaly where the SP curve should have a predictable form. The presence and magnitude of E_H can be found with the R_{xo}/R_t crossplot. It will appear as an offset of the saturation family of curves. The amount of the offset, in =/− millivolts is the amount of E_H.

4.12 Redox Component

When an oxidation-reduction (redox) reaction occurs, it involves an interchange of electrical charges (electrons); an electrical current flow. In fact, oxidation is defined as the gaining of an electron. Oxidation is a common geological process and is one form of alteration. oxidation is not common in the deep petroleum deposits, however.

The redox component of the SP is

$$E_{OX} = \frac{R_g T_K}{ZF} \ln \frac{kC_{OX}}{C_r} \tag{4.30}$$

where

R_g　= the universal gas constant
T_K　= the temperature in Kelvins
Z　= the valence
F　= Faraday's Number
k　= the reaction constant
C_{ox}　= the concentration of the more oxidized component
C_r　= the concentration of the less oxidized component

Note that because of the concentration values, E_{ox} can be either positive or negative.

The total SP equation, then, is

$$E_{SP} = E_{DA} + E_H + E_{OX} \tag{4.31}$$

The redox process is especially important in the deposition, solution, and solution transport of many types of metallic deposits, notably sedimentary uranium and copper. The process occurs with other metals, also. This process is associated with uranium, silver, copper, iron, and many other deposits in the world.

Uranium, for example, occurs in low concentrations in many types of source rocks. A good example is granite. In the weathering process, uranium compounds are oxidized, becoming very soluble (on the order of grams per liter, for some compounds). They are then leached out of the granitic body by the oxygen-charged surface waters. They may be carried, in their oxidized state, into the neighboring sands by the meteoric waters. As a general rule, uranium compounds are very soluble in the oxidized state. When the uranium-rich waters encounter a reducing environment, such as reduced uranium, humates, sulfides, organic trash, reduced gases, or hydrocarbons, the solution compounds are reduced. In the reduced state they may have a low solubility in cold water (often on the order of 10^{-5} g/l or less). Thus, they drop out of solution and enrich the zone in uranium.

A uranium geochemical cell or *roll-front* is a special form of this redox reaction, where the oxidizing solutions pick up additional uranium at the initial cell interface, carry them forward a short distance until they become reduced, at which point the uranium is dropped out of solution. As a result, we see a "rolling", concentrating action of the cell, as long as it is active. In a roll front, one can also see the adsorption of the uranium ions by the clays of the adjoining shales. As long as the more concentrated mineral solutions and mineral body are in contact with the shale, the uranium will diffuse into and be adsorbed by the clay. After the main body has "rolled" forward, the uranium is retained for a time in the oxidized zone because of the low permeability of the shale, until it can migrate out. This results in the typical "C" shape of a roll front. The same type of reaction can cause a metallic and radioactive halo near petroleum deposits. A seepage of hydrocarbon will have a tendency to reduce the surrounding solutions. If these are, or have been in the oxidized, weathered zone, the waters flowing past will deposit a halo of metals, including uranium around the seepage.

4.13 Problems with *SP* Measurements

4.13.1 Resistance Component

Ideally, the *SP* measure electrode and the surface reference electrode (or the two electrodes of a surface array) are identical in composition, state, temperature, and are in identical solutions. Thus, their electrode potentials are identical (excluding any external signal) and the *SP* signal is zero, except for any real *SP* signal. Refer to Figure 4.2. This ideal condition is seldom achieved in actual practice. The electrodes are never identical, the temperatures are seldom the same, the measure electrode is usually in motion, and the solutions are almost never exactly the same. Since these phenomena are functions of the chemical activities of the solution components, the ambient temperature is important. Differences are usually small and constant, however.

The *SP* measurement is a direct current (dc) measurement. It is sensitive to any dc flow in the vicinity. It is, however, easily separated from the resistivity signal, because the resistivity current is usually alternating (ac). Any spurious dc signal, however, will be read by the *SP* circuit.

When a current flows through an electrode, to or from an electrolyte, a reaction due to the current flow will occur at the electrode surface. A (polarization) potential becomes evident. The use of ac in the resistivity system alternates the reactions rapidly so they cancel each other out and will not noticeably affect either the *SP* nor the resistivity measurement. The current of the *SP* measurement is kept so small as to be negligible. Thus, there is no significant polarization potential influencing either one.

Voltage measurements require some current to operate the measure circuit. Any current drawn by the measure circuit will distort the measurement. In the case of the *SP* measurement, the current will result in a single-point resistance component being added to the *SP* signal. This resistive component, E_R, is

$$E_R = RG_R I_m \tag{4.32}$$

where

 R = the resistivity of the medium surrounding the electrode,
 G_R = the geometrical constant of the resistive component, and
 I_m = the current drawn by the measuring device.

The geometrical constant, G_R, is

$$G_R = 4\pi \left(\frac{L_e}{\ln \dfrac{2L_e}{d_e}} \right) \tag{4.33}$$

The *SP* voltage, E_{SP}, must be measured with a voltmeter whose impedance is as high as possible. Modern voltmeters have high input impedances, on the order of 10^9 ohms and higher. Thus, they drawn little current and the resistive component is negligible. Assuming that the minimum *SP* signal is about 1 millivolt, a typical resistive component will be about 2×10^{-9} millivolt. This range is not true of some older equipment, where the voltmeters had input impedances on the order of 100 to 200 ohms.

4.13.2 Sensitivity to Motion

Another problem which can occur with the *SP* measurement is caused because the electrode in motion is in a pseudo-stable state.

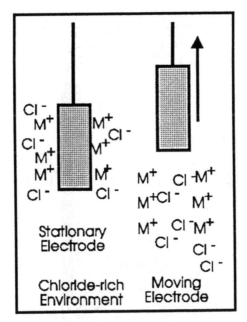

FIGURE 4.19
The electrochemical effect of moving an electrode.

As soon as a metallic electrode is immersed in the mud, it begins to react with the electrolyte, resulting in a changing electrode potential. A reaction product coating will form and cling to the surface of a well-designed electrode. Also, a cloud of metallic ions will form around the electrode. The reaction will proceed until the surrounding environment becomes saturated with the metal ions. At this time, the potential of the electrode will stabilize. See Figure 4.19.

As the electrode moves upward during logging, some of the ions are left behind in the mud. The environment of the electrode then is no longer saturated. Reaction begins again and the electrode potential changes. As long as the rate of logging motion is constant, however, the reaction rate (and the electrode potential) will remain relatively constant: a pseudo-stable state is achieved (Figure 4.20).

4.13.3 Electrode Touching the Sidewall

A more serious situation occurs when the electrode is allowed to touch the sidewall or any other solid object during logging. In that case, some of the reaction products on the surface of the electrode can be rubbed off and the electrode potential will change sharply. As the coating builds up again, the electrode potential will slowly return to normal. The result is a series of sharp and large sawtooth forms on the log. This problem can be prevented by recessing the electrode below the surface of the sonde.

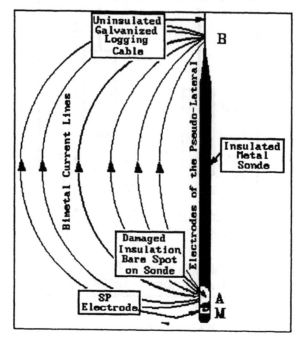

FIGURE 4.20
The effect of bimetallism.

4.13.4 Bimetallism

The core of most logging sondes is steel. It is covered with an insulating material, such as neoprene or polyurethane. During the course of logging, the coating can be damaged and the steel core exposed. The steel core is electrically connected, internally, to the logging cable head and to the zinc coated armor of the cable. Exposed steel on the sonde will constitute an electrochemical (bimetal) couple with the zinc of the armor. Differences in potential between iron and zinc is almost two volts. A large exposed area can result in electrolytic currents of amperes flowing in the borehole between the spot and the armor. This couple and the *SP* electrode form a lateral resistivity device and many millivolts of resistivity signal can be superimposed upon the *SP* signal. See Figure 4.20.

The sonde *must* be inspected carefully before each run. If bare spots are found, they must be taped. Upon returning to the shop, the bare spots must be covered permanently. It helps to insulate the cable for 15 ft (4.5 m) above the tool. This can be done with tubing, as the cable head is installed.

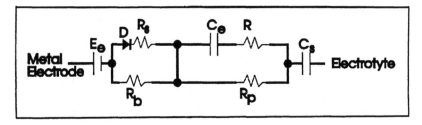

FIGURE 4.21
A probable equivalent circuit of an electrode in a solution of an electrolyte.

4.13.5 Improper Electrode Material

If the proper electrode material is not used, the *SP* signal can be badly distorted or destroyed entirely. Many factors — mechanical, electrical, thermodynamic, economic — are involved. All must be considered.

Figure 4.21 shows a probable electrical model of the electrode-solution interface. This is the surface of the electrode metal and reaction products in contact with the electrolyte solution and the saturated or partially saturated solution around the electrode. Note that the electrode potential, E_e and the diode, D, may have either polarity, depending upon the materials used.

The usual electrode material is lead. It is satisfactory mechanically. It has only two thermodynamic states and one is not likely to occur under logging conditions. Also, the resistances of the model, R_b and R_s are normally small. The diode, D, does not exist. The capacitance, C_s large and C_e does not exist. Therefore, the current, ac or dc, can easily flow in either direction.

If stainless steel, aluminum, or tantalum are used, they form a highly insulating coating on the surface. This can have resistances, R_s and R_b, of 10^1 to 10^{15} ohms. This can be of the same order as that of the voltmeter input impedance and can drastically reduce the *SP* signal.

If cuprous metal is used (copper, brass, bronze, etc.), the diode, D, will exist in one direction or the other. The coating formed will be low resistivity to one polarity of current and high to the other. These coatings are rectifiers. The *SP* signal will be badly distorted.

If iron is used, it has the advantage that the surface area is very large because of corrosion. Thus, the current density, due to any reaction, is small. On the other hand, iron is very reactive and will not likely achieve a stable state. It also has several thermodynamic states which can occur under logging conditions. Any change of state will have a change of potential accompanying it.

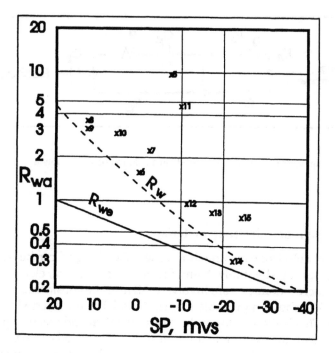

FIGURE 4.22
The relationship between the *SP* value and the resistivity of the formation water.

4.14   The *SP* Measurement in Fresh Water Sands

Alger and Harrison (1939) state that, in fresh water, the divalent cations are usually a significant part of the TDS (total dissolved solids), and these cations have a very strong effect upon the SP (Figure 4.22). As seen from Equation 4.1, it was pointed out that the E_{SP} is a function of the activity of the hole fluid and the formation water components; that the static SP, E_{SSP}, is

$$E_D = -K_D \log \frac{a_w}{a_{mf}} \qquad (4.1)$$

As presented earlier in Equation 4.7, K_{SP}, in fresh waters in Tertiary sediments, could be smaller than indicated by the relationship,

$$K_{SP} = 0.133 \frac{mvs}{°F} T_{f,F} + 60.77\,mvs \qquad (4.7)$$

because the shales do not behave exactly like semipermeable membranes. In offshore Pleistocene Tertiary sediments, Alger and Harrison found values of K_{SP} nearer 60 than the anticipated 80. Also, the presence of bicarbonates tends to lower the value of K_{SP}.

If these fresh water sands contain clay, Alger and Harrison suggest the relationship,

$$E_{SSP} = \frac{E_{PSP}}{\left(1-V_{sh}\right)} \tag{4.34}$$

Gondouin, et al. (1957) explain that well-fluid filtrates behave like NaCl solutions because base exchange tends to displace divalent cations in the mud filtrate. This is more likely if Aquagel additives are used and/or long shale sections have been drilled. Verification of the relationship between R_{mf} and a_{mf} should be made by chemical analysis of the formation water. This will allow calculating a_w and back-calculating a_{mf} from the E_{SP}.

Figure 4.23 shows a typical spontaneous potential (SP) curve on a field log.

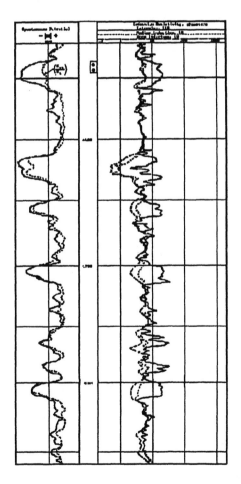

FIGURE 4.23
A typical electric log with the SP curve.

5

Resistivity Log Interpretation

5.1 Introduction

The use of the resistivity systems, suites and techniques and their interpretation are most important in hydrocarbon usage. These are not quite as important in non-hydrocarbon usage (but they are widely used). We will examine non-hydrocarbon usage, but most of this chapter will be confined to petroleum interpretation.

The goal of hydrocarbon interpretation procedures is to determine the location of and the probable amounts of hydrocarbons in suitable sedimentary formation zones. Secondary are zone thickness, bedding, rock type, and rock mechanics.

We have briefly examined, in Chapters 1, 2, and 3, each of the conventional resistivity/conductivity measuring systems and in Chapter 4, the spontaneous potential measurement. In those chapters, we also examined the data reduction procedures. We have seen that, in the petroleum industry, a deep investigation system (Figure 5.1) is usually used to determine the value of the true resistivity of the undisturbed reservoir zone, R_t. From this, S_w, the water saturation and the hydrocarbon saturation, S_o are determined. Also, a shallow investigation system is used to discover the value of the resistivity of the flushed zone, R_{xo}, from which the flushed zone water saturation, S_{xo}, and the residual hydrocarbon saturation, S_{or}, are calculated. Modern hydrocarbon analysis also requires a medium depth of investigation device to define the "effective" depth of invasion of the borehole fluid into the formation. This, then, is the "conventional" hydrocarbon resistivity suite of logs.

The resistivity systems and techniques are important to petroleum evaluation, of course, because of the usual extremely high resistivity contrast between hydrocarbons (10^{10} to 10^{15} ohms) and the accompanying formation waters (10^{-2} to 10^3 ohms). In the formation, this contrast is typically on the order of 2:1 to 1,000:1. It will vary because of the varying resistivity (salinity) of the formation water (10 to 250k ppm), the different proportions of the water/hydrocarbon mixture (0 to 100%), and the resistivity of the formation solid material (10^{-2} to 10^{15} ohms). Mostly, we can expect a useable resistivity contrast between the hydrocarbon and the rest of the formation components, but not always.

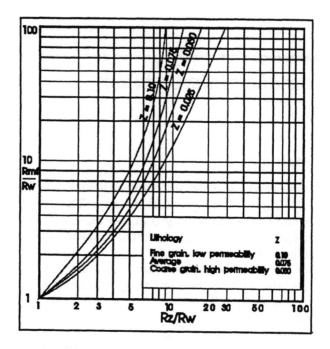

FIGURE 5.1
Rocky Mountain Method, value of R_Z (Courtesy of Atlas Wireline Services Division of Western Atlas International, Inc.)

Conventional resistivity interpretation, in hydrocarbon detection, is primarily aimed at determining the hydrocarbon saturation, S_h (oil saturation, S_o or gas saturation, S_g) (the fraction of the pore space occupied by the hydrocarbon oil or gas), the amount of hydrocarbon present (total reservoir amount or total reserves), the producible hydrocarbon amount (reserves), and the ease with which the hydrocarbon can be produced (permeability, k [only an estimate with resistivity methods], water cut, and shaliness).

These absolute parameters must be related to the economic factors, before a decision to produce or abandon is made. This will be discussed in a later chapter.

Of course, other information, than just the resistivities, is necessary. This information can sometimes come from the resistivity measurements, but is often much more reliably obtained from other, independent sources. Thus, porosity, permeability, and reservoir parameters will probably come from other logs, core analyses, cuttings analyses, experience, surface geology and geophysics. These will be assumed to be available. They will be discussed in later chapters. It is good practice to examine information from *all* available sources before making any decision.

In this chapter, we will cover *mostly* the conventional resistivity methods of determining, the water saturation, S_w, the hydrocarbon saturation, S_h, the flushed zone water saturation, S_{xo}, the residual hydrocarbon saturation,

TABLE 5.1

Systems choice for several environments

System	Use	Qualifications
IES	R_t&R_{xo}	Low R_t/R_m, N_{16} geometry unstable
DIL	R_t&d_i	Low R_t/R_m, Some shape distortion, corrected in Phosor
DLL	R_t&d_i	High R_t/R_m Needs water-base mud

S_{or}, the porosity, ϕ, and the permeability, k, in clean, shale-free, sedimentary zones. When necessary, we will assume that reliable supporting information is available from other sources. Shaly, fractured, and carbonate environments will be examined in later chapters. Table 5.1 shows the systems choices for several environments.

Upon examining a model of an invaded, permeable, potential hydrocarbon-bearing zone, it quickly becomes apparent that we need resistivity values at, at least, two different depths of investigation to satisfactorily determine the reserves in a zone. We need, of course, the deep resistivity value, R_t, to determine the water saturation, S_w (and thus, the hydrocarbon saturation, S_h). The shallow R_{xo} value provides a means for checking the formation resistivity factor. The value of S_{or} also allows us to estimate how much hydrocarbon we can remove from the reservoir. We will also need the shallow measurement plus some idea of the diameter of invasion to determine how the deep resistivity measurement has been affected by the invasion of the reservoir by drilling fluids. It turns out too, as we will see later, that there are other things this shallow measurement will show us.

The accuracy of the deep and the shallow resistivity determinations is controlled by the effective depth of invasion of the borehole fluid. By examining resistivity data from devices of several depths of investigation, we can begin to see a profile of the invasion situation and correct for the effects upon the flushed zone and the true resistivity. A series of tests were conducted using thirteen different depths of investigation (Campbell and Martin, 1955) (Winn, 1958). A minimum number of different depths of investigation turns out to be three. Fortunately also, most of the time three curves will serve.

5.2   The Archie Method

The first successful quantitative use of resistivity measurements was proposed by Gerald Archie (Archie, 1942). It addressed the deep resistivity determination separately from the shallow determination. The two sets were compared, but were not combined, as has been done with later methods. The Archie Method is used to determine the several formation reservoir parameters. They are assumed to be ideal or must be corrected for disturbing factors.

For any given porosity system, the ratio of R_o to R_w is constant. That is, if R_W changes, R_o will change proportionally. The Formation Resistivity Factor (or simply, Formation Factor), F_R, is

$$F_R = \frac{R_o}{R_w}$$

(5.1)

where

R_o = the deep, undisturbed zone resistivity in the special case when the formation water saturation is 100%

R_w = the formation water resistivity filling the pore space

If R_t is measured in a zone which is 100% formation water saturated, then $R_t = R_o$. Further, if $R_o = R_w$, then $F_R = 1$. Under any other circumstances, $F_R > 1$. Under *all* normal circumstances $R_0 \geq R_w$.

Archie further determined that the Porosity, ϕ, is related to F_R:

$$F_R = \phi^{-m}$$

(5.2)

where

ϕ = the fraction of the total zone which is pore space. This is often expressed in percent. It is the "porosity."

m = the *cementation exponent* and is a function of the shape of the pore space; the *tortuosity, T*.

Since, in a real situation, the value of "m" changes with porosity (because the shapes of the pore and the interconnections change), Winsauer (Winsauer et al., 1952) suggested that Archie's equation, in actual practice, needs a factor "a" which is frequently different from 1. This helped the problems of getting a better relationship between porosity and the formation factor. The modified Archie relationship is

$$F_R = a\phi^{-m}$$

(5.3)

This is an estimate, however.

One usually sees the final equation as the combination of Equations 5.1 and 5.3:

$$\phi = \frac{V_p}{V_b}$$

(5.4)

There are two schools of thought among log analysts. One school believes that the value of "a" should be altered to allow a constant value of "m" for any given zone. The other believes that the value of "*a*" always equals 1 and the value of "*m*" should vary with the tortuosity, *T*, of the pore space. While each school has its merit, the varying "*m*" school seems to be gaining in favor as methods of estimating or determining *T* are developed. In this text, we will assume that a = 1.00, unless it is otherwise stated.

A rigorous value of ϕ can be found from density, acoustic, neutron logs, and from the core samples; the value of R_w can be found from the SP. Then, a good value for R_o can be calculated. This is often done by the computer continuously during logging and the values plotted on the log.

The value of R_t is the "true" resistivity of the deep, undisturbed formation zone. If the zone is 100% saturated with formation water (contains no hydrocarbon) then $R_t = R_o$. If the zone contains hydrocarbon, R_t will be greater than R_o. The ratio of R_t to R_o is called the *resistivity index, I_R*:

$$I_R = \frac{R_t}{R_o} \qquad (5.5)$$

Since the difference between R_t and R_o is due to the hydrocarbon content of the zone, I_R is a function of the *water saturation, S_w,* (or *oil saturation, S_o*):

$$I_R = S_w^{-n} \qquad (5.6)$$

where *n* = the *saturation exponent*. The value of "*n*" is usually very near 2.0. S_w is related to S_o:

$$S_o = 1 - S_w = S_h \qquad (5.7a)$$

Note that the porosity value is a fraction of the total volume under investigation. The saturation values, on the other hand, are fractions of the pore volume. Both of these are commonly expressed as percent. Remember that percent means "parts per hundred" and that the fraction must be used in calculations.

Gas has much the same electrical effect as oil. Both are extremely resistive. These resistivity methods normally cannot tell the difference between types of hydrocarbons nor types of rock. Non-hydrocarbon gases give the same effect as hydrocarbon gases. And, there are many non-hydrocarbon uses for the saturation calculation. *Any* gas saturation is S_g or S_h:

$$S_g = S_h = 1 - S_w \qquad (5.7b)$$

Actually, the hydrocarbon saturation value, instead of representing only oil or gas, is usually a mixture of oil, condensate, hydrocarbon gas, and inorganic gas.

The totally flushed (by mud filtrate) zone has a set of relationships similar to those of the deep zone:

$$F_R = \frac{R_{xo,s}}{R_{mf}}$$ (5.8)

when the flushed zone contains no hydrocarbons nor gases. ("$R_{xo,s}$" is not an officially recognized symbol. It is used in this text, for convenience, to designate the 100% mud filtrate-saturated value.) R_{mf} is the resistivity of the mud filtrate (the liquid portion of the mud), at the same temperature as $R_{xo,s}$.

This value of F_R, in a uniform zone, normally should be nearly the same as that determined for the deep, undisturbed zone (Equation 5.1). This, then gives an alternate method of determining and checking some parameters.

The *flushed zone* water saturation value corresponding to Equations 5.5 and 5.6 is

$$S_{xo} = \left(\frac{R_{xo}}{R_{xo,s}} \right)^{-n}$$ (5.9)

where "n" usually should be the same value as in Equation 5.6.

The corresponding *hydrocarbon* saturation value is the residual oil (or gas) saturation:

$$S_{or} = 1 - S_{xo}$$ (5.10)

This represents the fraction of the hydrocarbon which has not been flushed out by the invading fluid.

In a uniform, 100% water saturated zone, the formation factor and the porosity value in the flushed zone should be approximately the same as those determined in the deep zone. Therefore,

$$F_{R,deep} = F_{R,xo} = \frac{R_o}{R_w} = \frac{R_{xo,s}}{R_{mf}}$$ (5.11)

Since R_o, $R_{xo,s}$, and R_{mf} can all be measured, Equation 5.11 is a convenient way to find the value of the formation water resistivity, R_w. This value should be the same as that determined from the SP curve in Chapter 4. The value of ϕ can be determined from the "porosity" log. The value of R_{mf} can be measured on the surface. Thus, both R_o (Equation 5.5) and $R_{xo,s}$ can be determined. Since the logs in a petroliferous zone will read R_t and R_{xo}, the values of S_w, S_o, S_{xo}, and S_{or} can be determined.

TABLE 5.2

Approximate values of the Cementation
Exponent, "m"

m	Material
1.3	Unconsolidated sandstone
1.4–1.5	Very slightly cemented sandstone
1.5–1.7	Slightly cemented sandstone
1.7–1.9	Moderately cemented sandstone
1.9–2.2	Highly cemented sandstone or limestone
2.1–	Carbonates

There are many other values which can be determined with these Archie relationships. Many of these require other curves and will be covered in later chapters.

5.2.1 Values of the Cementation Exponent

As a first approximation, the value of $m = 2.0$ is generally accepted. In fact, some texts and most contractor's literature mention no other value. We already have seen, as the pore geometry becomes simpler, the value of "m" approaches 1.0. Archie (1942) found that the value of "m" for an unconsolidated sand is about 1.3. Chombart (1960) noted that "m" generally had values, for crystalline and granular media, of 1.8 to 2.0, 1.7 to 1.9 for chalky limestones, and 2.1 to 2.6 for vugular carbonates. Pirson (1935) and Lynch (1962) suggested that the variations in the values of "m" were due to pore geometry (tortuosity). The value of "m" is actually determined almost entirely by the degree of tortuosity of the pore space. This is the complement of the rock matrix geometry, which is controlled by the rock texture. The values shown in Table 5.2.

$$F_r = \phi^{-m_k} \tag{5.12}$$

If the value of "m" is plotted against the value of "F_r", using Archie's porosity equation, the plot is shown in Figure 5.2.

Raiga-Clemanceau (1976) suggests a porosity relationship which takes permeability (a function of both porosity and tortuosity) into account:

$$m_k = 1.28 + \frac{2}{\left(\log_{10} k\right) + 2} \tag{5.13}$$

where (k is the permeability in millidarcies). Table 5.2 shows some approximate values of "m".

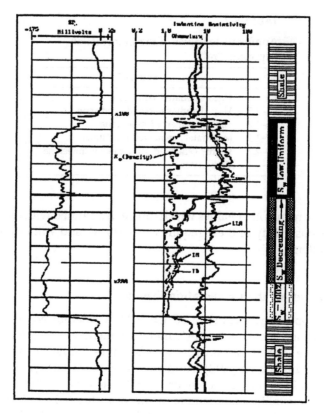

FIGURE 5.2
A sample log showing the migrated hydrocarbon effect.

5.3 The Ratio Method

When the zone being examined is uniform and 100% water saturated, the ratio of R_{xo}/R_t examines a function of the salinity difference between the invading mud filtrate and the formation water. This is the same relationship that the SP examines. Obviously, these two functions are related. If, however, hydrocarbon or gas occupy part of the pore space, then, of course, the ratio of R_{xo}/R_t will not be directly related to the SP value. The difference can be equated to the saturation situation. This method also tells about some SP effects. This will be discussed in detail in the SP chapter. The Ratio Method is another, excellent way to use the Archie Method to obtain good saturation values.

5.4 The Rocky Mountain Method

The model we have used so far pictures a sharp interface between the invading fluid and the native fluid. This is very seldom the situation, even without the presence of hydrocarbon. If the boundary is a sharp interface, it never lasts for more than a few minutes. The interface is nearly always diffuse. When we speak of a sharp interface or a diameter of invasion, we use the "effective" diameter of invasion. That is the diameter of a hypothetical sharp boundary which would produce the same gross results as the real diffuse zone. This is the zone whose resistivity is called R_Z. The Rocky Mountain Method is the only interpretation procedure which was used in this zone with unfocused devices.

The Rocky Mountain method uses the shallow and deep investigation systems of the Electric Log (N_{16}, N_{64}, SP) or the IES (6FF40 induction log, N_{16}, SP). The SP is used to estimate the resistivity of the formation water, R_w. The chart shown in Figure 5.3 is then used to estimate value of the ratio R_z/R_w from the ratio R_{mf}/R_w and an estimate of the type of lithology.

The ratio of R_i/R_t (R_i from the N_{16} or the LL8 and R_t from the deep induction log) is used with the ratio of R_z/R_w to find the probable water saturation, S_w:

$$S_w = \left(\frac{R_i/R_t}{R_z/R_w} \right) \tag{5.14}$$

Dresser Atlas (now Atlas Wireline) used a nomograph which shows the solution to these problems. It will be found in their chart book.

The Rocky Mountain Method (proposed by Tixier), served a useful purpose. It is not a very precise method, but can be used in a very difficult situation especially with old logs.

5.5 The Migrated Hydrocarbon Method

The Migrated Hydrocarbon Method makes use of the separation of hydrocarbon fluids from water because of the differences in their densities. Over a long period of time the oils and gases in a permeable formation may migrate upward within the zone and separate from the water in the pore space. This is shown in the log of Figure 5.2.

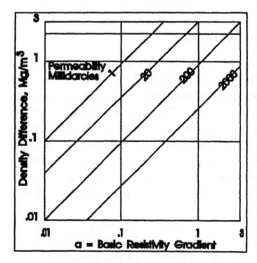

NaCl	Solutions
1 Atmosphere	60°F (16°C)
ppmx10^3	g/cm^3
0	1.000
50	1.034
100	1.071
150	1.109
200	1.148
250	1.189

The value of the permeability, k, in millidarcies, is,
The basic resistivity gradient, a, is,
This chart provides an estimate of rock permeability from the resistivity gradient over the transition zone.

To estimate the permeability, the difference in the densities of the saturating hydrocarbon and water phases is plotted against the resistivity gradient observed on the log. Resistivity gradient is obtained by dividing the change in resistivity, R_t, over the transition zone by the distance over which the change occurs and by the resistivity at 100% S_w, R_o, of the reservoir rock.

Approximations for fluid densities, as a function of water salinity and oil gravity, are given in the tables. For more information see the reference, Tixier (June 1949).

OIL	GRAVITY
1 Atmosphere	60°F (16°C)
°API	g/cm^3
15	0.966
20	0.934
25	0.904
30	0.876
40	0.825
50	0.780

FIGURE 5.3
Permeability from the resistivity gradient. (After Schlumberger, 1988.)

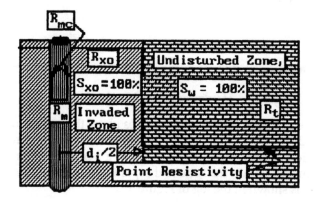

FIGURE 5.4
A simple step model.

The following assumptions must be made:

1. There must be no permeability barriers from the top to the bottom of the zone of interest.
2. R_w is uniform from the top to the bottom of the zone.
3. The zone has a moderate to high permeability.
4. The zone is shale-free.
5. The zone has been undisturbed long enough that separation by migration has taken place to the extent that even the residual hydrocarbon has migrated out of the lower part of the zone.

If the above assumptions are valid (and the fifth might not be), then there is a finite probability that the saturations, S_w and S_{xo}, of the lowest portion of the zone are at or very near 100%. If this is the case, then, in the lowest portion of the zone, $R_t = R_o$. If R_o and $R_{xo,s}$ are not obtainable from the lowest portion of the zone, they may be calculated from a porosity log or from the core analysis. This has been done in the log of Figure 5.2. In this case, the density values furnished the information to calculate the R_o curve. This is done routinely and in the field. In this log, the curve verifies that S_w is indeed at or nearly at 100%.

A typical pattern of the R_t curve is evident in Figure 5.2. If this is the case, several things can be observed:

1. The R_t vs. ϕ (Pickett) plot can be successfully used to determine the values of S_w (or S_{xo}) and R_w.
2. The values of R_t at the lowest part of the zone are apparently equal to R_o and can be used to approximate R_o and be used for calculation of S_w.
3. If the porosity of the zone is relatively uniform, then even without the R_t vs. ϕ plot, the values of R_t from the lowest part of the zone can be used to estimate S_w when used with the R_t values in the upper part of the zone (if the R_o curve is not available).
4. The resistivity gradient and the S_w values can be used to estimate the permeability of the zone, assuming it is relatively uniform. Figure 5.3 shows a simple adaptation from a plot from the SWC Chart Book. This type chart can be used for solving this problem for permeability of the zone to water.

5.5.1 Estimation of Permeability from the Resistivity Gradient

The resistivity gradient method for estimating permeability makes use of the differences of density between the waters of the pore space and the hydrocarbon. It also makes use of the great differences of resistivity of the

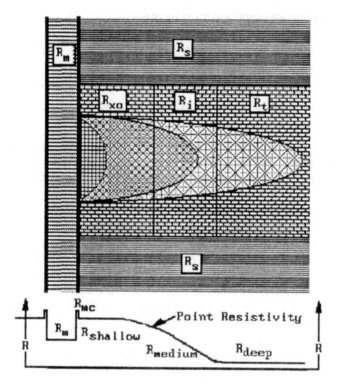

FIGURE 5.5
A representation of the investigation of the shallow, medium, and deep investigation resistivity devices.

pore water and the hydrocarbon. This was also covered in Chapter 5 of *Introduction to Geophysical Formation Evaluation*, a companion volume to this text.

Figure 5.3, was adapted from the Schlumberger Well Services, Inc. *Log Interpretation Charts, 1988*. To use this method, follow the instructions for the log, Figure 5.2 or, better yet, in the Schlumberger chart book, chart K-1. Since Figure 5.3 is greatly simplified, please refer to the chart, in the book, before attempting any quantitative use of it. This chart compares the oil "gravity" (API gravity), an inverse function of the oil density, with the resistivity gradient, *a*, to estimate the permeability of the zone. The density difference, $\Delta\rho$, of the hydrocarbon, across the gradient zone, is often used.

Lynch (1962) suggested that

$$m = 1 + \left(\frac{2\log\dfrac{L_e}{L_b}}{\log\phi} \right) \qquad (5.15)$$

where L_e is the effective length of the pore space, L_b is its straight line length and ϕ is its fractional volume and the pore water ($\sim$0.9 to 1.3 g/cc), is,

$$\Delta\rho = \rho_w - \rho_h \qquad (5.16)$$

This is presented on the vertical scale of Figure 5.3. This difference of density will, after a period of time, cause the hydrocarbon fluids to migrate upward in the formation zone, if the vertical permeability is moderate to large. This will cause a separation of the fluids.

The horizontal scale of Figure 5.3 is the resistivity gradient, a, of the hydrocarbon-bearing permeable zone. The separation of the formation fluids will be reflected in the resistivity curves because of their great differences in resistivity ($\Delta R = 10^{10}$ to 10^{12} Ωm). The resistivity gradient, a, as a function of depth, D, is

$$a = \frac{\Delta R}{\Delta D} \frac{1}{R_o} \qquad (5.17)$$

The probable permeability is shown in the family of curves within the body of the diagram. Tables in Figure 5.3 show typical densities for water and oils.

Since the determination of permeability is uncertain, this method and any others available should be evaluated and compared. Certainly, the value from the core analysis, if it is available, should be accepted as the most probably reliable.

5.6 Determination of the Diameter of Invasion

If we had systems which would measure the zone in which we are interested and no other, and if we had a sharply defined, simple, step interface between the invading fluid and the native fluid, and if the diameter of invasion was always evident, then determinations of saturation would present no real problem. This is shown in Figure 5.4. These simple situations are never present, except in simplifying simulations and models.

We have seen that each of these measurements can be influenced by the characteristics of the invaded zone and of the undisturbed zone. If the diameter of invasion is deeper than expected, the calculated value of R_t may be in error. If the diameter of invasion is shallower than expected, then the calculated value of R_{xo} may be in error. If the contrast between R_t and R_{xo} is great, then the calculation of R_t may be affected. If the invasion/undisturbed interface is complex, then all resistivity measurements are likely to be in error.

TABLE 5.3

Common combinations of resistivity devices

System	Use	Qualifications
IES	R_t&R_{xo}	Best in low R_t/R_m Geometry of Normal is unstable
DIL&R_{xo}	R_t&d_i	Best in low R_t/R_m
DLL&R_{xo}	R_t&d_i	Best in high R_t/R_m

We find, in actual practice, that two curves of different horizontal depths of investigation will not usually give a satisfactory solution. In order to get a valid idea of the diameter and resistivity of the zone of invasion, we will need more than two curves of different depths of investigation. In some experiments, as many as 13 measurements, each at a different depth of investigation, have been used. In order to get an idea of the depth or diameter of investigation, at least one curve of a device of medium depth of investigation has been found to be needed. This curve is usually the IM, IMPH, 5FF40, LLM, N_{64}, or one of several others. Table 5.3 shows some of the common combinations.

So far, not much has been said about the medium depth curve, except to say that it is used to determine the diameter of invasion. At first, the medium investigation depth curve was used qualitatively, if it was used at all. If the value of the medium curve read near the value of the deep investigation curve, the invasion was said to be shallow. If the medium curve read near the shallow curve, the invasion was said to be deep. You can picture this in Figure 5.5 and 5.6. R_Z is the mean resistivity value of the zone where the mixing of the formation water with the encroaching invading fluid takes place.

The major difficulty encountered is that of determining the actual diameter of the invasion. Figures 2.10 and 3.15 show the pseudo-geometrical factors for a number of focused devices. If the diameter of invasion is known or can be estimated, then this type of chart can be used to find the value of the geometrical factors, G_i and, thus, G_t, as explained in Chapters 2 and 3.

There are many problems with this method. It tells nothing quantitative about the invaded zone. It neither confirms its diameter nor determines its resistivity. It does not answer questions about the actual geometry of the invaded zone. There are problems, also, when the contrast between R_{xo} and R_t is very great or very small. If the contrast is great, the calculated value of R_t is uncertain. If the contrast is very small, the diameter of invasion is in doubt.

Using a sharp interface model, if the invasion diameter is known, as shown in Figure 5.7, two measurements are satisfactory. These would be a shallow investigation measurement for R_{xo} and deep one, which ignores the invaded zone, for R_t. Note, in Figures 5.7 and in most of the others to follow, R_{xo} is usually pictured as higher, more resistive, than R_t by a factor of 4 (the ratio of $R_{xo}/R_t \approx 4$). Actually, this ratio can easily be any value from

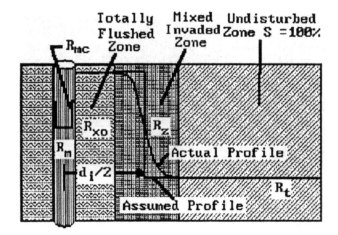

FIGURE 5.6
A more realistic step model.

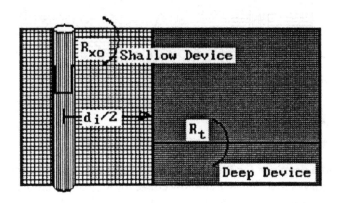

FIGURE 5.7
$R_{mf}/R_w > 1$.

much less than 1 to much greater than 4. Figure 5.8 shows the other situation.

The real boundary looks similar to that shown in Figure 5.9. It can, however, be satisfactorily approximated by the simplified model shown in Figure 5.6. Realize that even the interfaces between the flushed zone (R_{xo}) and the mixed zone (R_Z), and between the mixed zone (R_Z) and the undisturbed zone (R_t) are not sharp, as pictured in Figure 5.8, but are diffuse to some unknown and variable degree. The model is simplified for analysis purposes by assuming sharp apparent or virtual interfaces (the depth and resistivity values of R_Z) and a linear transition from the R_{xo}/R_Z interface to the R_z/R_t interface.

If we know the diameter of invasion and it is sharp, as shown in Figure 5.8 or 5.9, then there is no real problem. Two curves, a shallow one

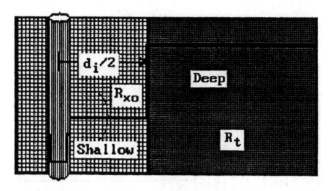

FIGURE 5.8
$R_{mf}/R_w < 1$.

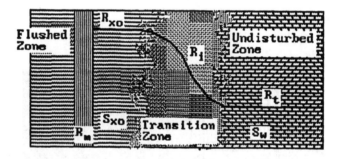

FIGURE 5.9
A representation of the actual zonation of the invaded formation.

for R_{xo} and a deep one for R_t will suffice. Usually, though the situation will require 3 curves to define it; a shallow curve for R_{xo} (essentially), a medium depth curve for the diameter of invasion (mostly), and a deep depth of investigation curve to define R_t. This can be estimated as a step model by picking the halfway point and defining it as the effective diameter of invasion. This can be done only with three or more curves. It turns out that three curves are enough for the vast majority of situations. This is shown in Figure 5.10. Figure 5.11 illustrates the simplifying approximation used in most log analyses.

Occasionally one finds a situation in a hydrocarbon-bearing zone where the invading fluid displaces the salty formation water and oil, and the oil will have a higher relative permeability than the water. In this case, the salt water will "pile up" between the invading fluid and the undisturbed fluid, or near the undisturbed fluid. This "annulus" condition can also be detected with three curves. You can see how this situation would disturb the reading of the LN, or a deep focused resistivity log. Such a situation is pictured in Figure 5.11. Figure 5.12 shows how to use these three curves to determine saturation.

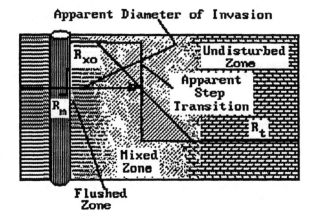

FIGURE 5.10
A step approximation of the zonation of the invasion.

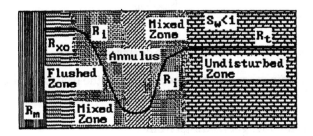

FIGURE 5.11
The annulus formed under some invasion conditions.

5.6.1   Reading "Tornado" Charts

Several logging contractors (Atlas Wireline, Schlumberger Well Services, and several others) have published charts giving solutions to these three curve problems for popular combinations of modern measurements. The methods are similar, although the charts may appear slightly different. These are the charts for determining the diameter of invasion (some of the recent charts solve for the depth of invasion, L_i. This depth is a radius.) The charts are popularly called "tornado" charts for reasons which will be obvious to anyone who has lived in the central part of the U.S. The Phasor Induction Log information, at this stage, is handled in the same manner at the information from the DIL and the DLL suites.

Refer to log interpretation chart book of the company which ran the logs you are examining. The principles are the same for all of these charts. Figure 5.13 shows a page from the *SWC Chart Book* explaining how to use the invasion correction charts ("tornado" charts). The purpose of these

---

SATURATION DETERMINATION IN CLEAN FORMATIONS

Either of the chart-derived values of R_t and R_{xo}/R_t can be used to find values for S_w. One value, which is designated as S_{wA} (S_w-Archie) is found using the Archie saturation formula with the R_t value and known values of F_R and R_w. An alternate value of S_w, designated as S_{wR} (S_w-Ratio), is found using R_{xo}/R_t with R_{mf}/R_w as in Chart Sw-2.

If S_{wA} and S_{wR} are equal, the assumption of a step contact invasion profile is indicated to be correct; and all values found (S_w, R_t, R_o, d_i) are considered good.

If $S_{wA} > S_{wR}$, either invasion is very shallow or a transition type of invasion profile is indicated; and S_{wA} is considered a good value for S_w.

If $S_{wA} < S_{wR}$, an annulus type of invasion profile may be indicated. In this case, a more accurate value of S_w may be estimated, using the relation: The correction factor, $(S_{wA}/S_{wR})^{1/4}$ can be found from the scale below.

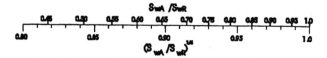

FIGURE 5.12
Chart showing how to use the three curves to determine saturation.

charts is to obtain more nearly correct values of the true formation resistivity, R_t, the invasion diameter, d_i, and the ratio of R_{xo}/R_t.

1. From the log, get readings from
   a. ILD (or 6FF40, LLD, LL7, LL3, 6FF28, etc.) = R_{ID}
   b. ILM (or 5FF40, LLM, SFL, etc.) = R_{IM}
   c. LL8 (or MLL, PL, MFSL, N_{16}, etc.) = R_{LL8}
2. Correct the log values for borehole and adjacent bed effects.
3. Determine the ratio for the horizontal axis (i.e., R_{IM}/R_{ID}), enter it, and extend a line upward. This may be different on other charts.
4. Determine the ratio for the vertical axis (i.e., R_{LL8}/R_{ID}), enter it, and extend a line to the right. This may be different on other charts.
5. Read the values at the intersection of the two lines.
6. The value of d_i is given by the dashed vertical lines. Interpolate, if necessary.
7. If the point is to the left of the left most line, then $d_i < 15$ in (<0.38 m).
8. If the point is to the right of the last line, then $d_i > 120$ in (>3.04 m).

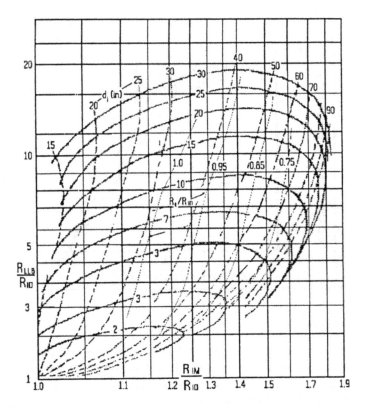

FIGURE 5.13
A "Tornado" diagram for determining the probable depth of invasion and R_t. Please note that this drawing is for illustration purposes, *only*. It was adapted from a Schlumberger Chart Book, *but is not accurate*. For actual use, refer to the latest Chart Book of Schlumberger Well Services, Inc.

9. The value of R_t/R_{ID} is given by the irregular, sloped lines, labelled 1.0 to 0.80, left to right. Interpolate, if necessary.

10. If the point is to the left of $R_t/R_{ID} = 1.0$, then $R_t/R_{ID} = 1$.

11. If the point is to the right of $R_t/R_{ID} = 0.75$, then extrapolate or estimate.

14. Multiply R_t/R_{ID} by R_{ID} to get R_t.

15. Find the ratio of R_{xo}/R_t from the horizontal solid lines. Interpolate, if necessary.

16. If the point falls above $R_{xo}/R_t = 30$, or below 2, then extrapolate.

17. R_{xo} may be found by multiplying the ratio by R_t.

These values of R_t, R_{xo}, and d_i are now corrected for the usual distorting effects of the borehole, invasion, and formation. They may be used for calculating saturation values, values of m and n, R_w, and other parameters.

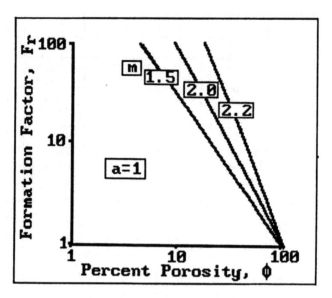

FIGURE 5.14
Values of "m", as a function of F_R.

In addition, the saturation values which were determined from the "tornado" charts can be compared with those determined by the Archie relationships. Figures 5.12 nd 5.14 show some estimations which can improve the accuracy of the determinations.

5.7 R_t vs. ϕ Crossplot — The Pickett Plot

Richard Pickett (1975) suggested using a plot of the Archie equation for porosity to determine the probable value of some of the unknowns and for making corrections when working with resistivity logs. The method is a valuable illustration of the resistivity relationships.

Archie's porosity equation states

$$R_o = R_w \phi^{-m} \qquad (5.18)$$

The Archie Saturation equation is

$$R_o = R_t S_w^n \qquad (5.19)$$

Combine these, solve for R_t, and take the logarithm:

$$Log(R_t) = \log(R_w) + n\log(S_w) - m\log(\phi) \qquad (5.20)$$

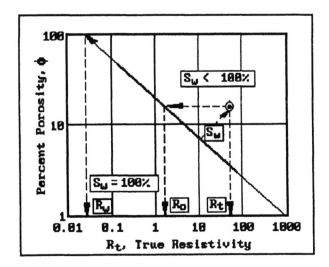

FIGURE 5.15
A crossplot of R_t vs. ϕ.

Equation 5.20 is a linear equation of the slope intercept form. Plotting it will result in a family of curves which are straight lines on a log/log grid. The y-axis is $\log(R_t)$, the x-axis is $\log(\phi)$, the y-intercept is $\log(R_w) + n \log(S_w)$ and the slope is $-m$. Iso-saturation lines are parallel to the 100% saturation line. See Figure 5.15.

The value of m is the negative slope value and may be determined by using values from one of the parallel saturation lines (i.e., the $S_w = 100\%$, R_o line):

$$-m = \frac{\Delta y}{\Delta x} = \frac{\log R_{t1} - \log R_{t2}}{\log \phi_1 - \log \phi_2} \tag{5.21}$$

If S_w is set equal to 1 (100%), then

$$R_t = R_o \tag{5.22}$$

and the $\log(S_w)$ term becomes equal to zero. Under these conditions, the trend will define values of R_o. The equation is now Archie's porosity equation.

If S_w and ϕ are both equal to 1, then both of their terms become zero and

$$R_t = R_o = R_w \tag{5.23}$$

In this way, the value of R_w for the system may be found.

If the value of S_w is less than 1 at any value of ϕ, the R_t value will be displaced at a constant porosity, ϕ, along the y or R_t direction, by a value of

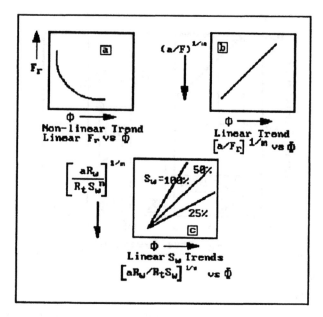

FIGURE 5.16
The process of obtaining the non-linear resistivity vs. porosity plot.

$$S_w = \left(\frac{R_o}{R_t}\right)^{\frac{1}{n}}$$ (5.24)

Since both R_t and R_o are now known, the value of S_w may be calculated for any value of n.

5.7.1 The Nonlinear (Hingle) Crossplot

A similar plot is the nonlinear. This diagram also plots the value of R_t against porosity, ϕ, or a function of porosity. The value of this plot is that one can use it to begin to identify the rock matrix type, in addition to estimating a good value of water saturation, S_w.

If the formation factor, F_r, is plotted against porosity, ϕ, the trend is nonlinear. See Figure 5.16a. This response can be made linear by using values of

$$\left(\frac{a}{F_r}\right)^{\frac{1}{m}}$$ (5.25a)

for the y-axis, in place of F_r. This is shown in Figure 5.16b. The saturation values can also be incorporated by using a value which is also a function of F_r and m, as well as saturation, S_w:

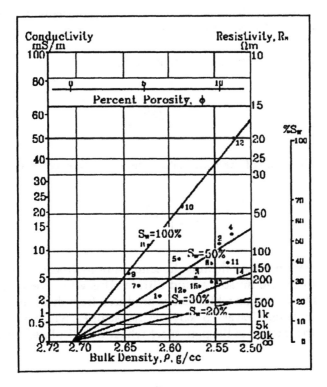

FIGURE 5.17
An example of a plot of formation resistivity against bulk density and porosity on a non-linear or Hingle plot.

$$\left(\frac{aR_w}{R_t S_w^n}\right)^{\frac{1}{m}}$$

(5.25b)

This is shown in Figure 5.16c. The y-axis is scaled in values of R_t in a negative direction (increasing downward). The grid is a function of "m". Trends of S_w values will be straight lines. All of the S_w trend lines will converge at $R_t = \infty$. At a value of $R_t = \infty$, the porosity is zero and the function of porosity will define the character of the solid rock matrix. A commercial sample of this is shown in Figure 5.17.

5.8 Moveable Hydrocarbon Method

The Moveable Hydrocarbon Method compares the amount of hydrocarbon (S_h) in the undisturbed zone with that remaining in the flushed zone

(S_{or}). The difference between the two values represents the amount of hydrocarbon which could be moved out of the reservoir with a water flushing.

The value of S_w, from Archie's relationships, is,

$$S_w = \left(\frac{F_R R_w}{R_t}\right)^{\frac{1}{n}} \tag{5.26a}$$

Similarly, the value of the water saturation in the flushed zone is,

$$S_{xo} = \left(\frac{F_R R_{mf}}{R_{xo}}\right)^{\frac{1}{n}} \tag{5.26b}$$

The difference between them is the amount of hydrocarbon which was flushed out by the invading mud filtrate. This will give a clue to the amount of hydrocarbon which can be produced from that zone.

Since Equations 5.26a and 5.26b have the factor F_R, which is ϕ^{-m}, the difference is the fraction of the pore space which was displaced by the invasion. Thus,

$$S_{h,d} = S_{xo} - S_w = \left(\frac{R_{mf}}{\phi^m R_{xo}}\right)^{\frac{1}{n}} - \left(\frac{R_w}{\phi^m R_t}\right)^{\frac{1}{n}} \tag{5.27}$$

If it is assumed that $m = n$, then relationship can be simplified by multiplying the difference by ϕ:

$$S_{h,d} = \left(S_{xo} - S_w\right)\phi = \left(\frac{R_{mf}}{R_{xo}}\right)^{\frac{1}{n}} - \left(\frac{R_w}{R_t}\right)^{\frac{1}{n}} \tag{5.28}$$

Equation 5.28 describes the moveable hydrocarbon as a fraction of the bulk volume, *not* a fraction of the porosity. Note that the bulk volume relationship does not require knowledge of the porosity for solution.

The Moveable Hydrocarbon Method assumes that the porosities and rock compositions of the deep zone are the same as in the flushed zone. In a uniform sand, this is a reasonable assumption.

5.9   The $F_{R/a}$ vs. F_R Method

As we previously saw in Equation 5.1, the value of F_R, the Formation Fac-

$$F_R = \frac{R_o}{R_w} \tag{5.1}$$

If we calculate the value of F_R (and call it $F_{R/a}$) from the log values, using R_t, instead of R_o, the value of $F_{R/a}$ will only equal F_R where S_w is 100%. The ratio of $F_{R/a}/F_R$ will be equal to 1.0 when S_w equals 100%. For any other value of $F_{R/a}$, the ratio will be greater than 1.0. Thus, if the ratio is plotted on the log, at the correct depth, it will tend to point out possible production zones.

5.10 The $R_{w,a}$ vs. R_w Method

A procedure, similar to the $F_{R/a}$ Method can be done with the value for the water resistivity, R_w. The relation for R_w is,

$$R_w = R_o \phi^m \tag{5.29}$$

If, however, we calculate $R_{w,a}$ using log values of R_t or R_a, instead of R_o, then $R_{w,a}$ will equal R_w only if S_w is 100%. Otherwise, the ratio of $R_{w,a}/R_w$ will be greater than 1.0 and indicate the possible presence of hydrocarbon.

5.11   Non-Hydrocarbon Usage of Resistance and Resistivity

The resistivity logs are not as important in non-hydrocarbon work as they are in the petroleum industry. They are, however, useful and frequently used. Both quantitative and qualitative curves are run. The three curve suites of the hydrocarbon industry are seldom run because invasion is not usually the same important process that it is in non-hydrocarbon work.

The single-point resistance curve, SGL, is frequently run because of its high resolution and because non-hydrocarbon boreholes are frequently small diameter. The resistance curve is used because resistivity is not always needed. Also, the electrode size is often changed to suit the need of a particular environment.

The single-point curve (SGL) is a very useful curve to qualitatively determine the stratigraphy of the zone being examined. It is a shallow investigation method and has a high resolution (typically about 5 in to 10 in (13 to 26 cm) diameter). It is an excellent curve for detecting depositional sequences. In small grained sands, a fining upward sequence will often result in a gradually lower resistivity upward. This may indicate a lowering of the depositional energy, with time. Sometimes, even seasonal variations can be detected. Coarsening upward and periodic flooding can also be seen. These effects are enhanced when increasing or decreasing shaliness adds to the effect. The depositional effect apparently occurs because of the change of the electrical characteristics of the sand grains as the grains become smaller and there are more exposed unsatisfied crystalline edges.

The SGL can be used quantitatively, if care is taken. The length of the SGL electrode is analogous to the spacing of the multi-electrode resistivity curves. The longer the electrode is, the deeper is its lateral depth of investigation. The shorter it is, the higher its resolution is. This is, however, usually a shallow lateral depth of investigation curve (the exception is the focused electrode family). It is very sensitive to the size and resistivity of the borehole. Chapter 9 in *Introduction to Geophysical Formation Evaluation*, explains the depth of investigation calculation and the need for restricting the use of the SGL to small diameter holes (Hallenburg, 1997). Further, like all unfocused systems, it performs best when the formation resistivity is lower than the drilling mud resistivity. The relationship for the SGL is,

$$R_{SGL} = 4\pi \frac{L_e}{\ln \frac{2L_e}{d_e}} \frac{E_e}{I} \tag{5.30}$$

where
   L_e = the length of the electrode, in meters
   d_e = the diameter of the electrode, in meters
   E_e = the potential read from the electrode, in volts
   I = the survey current, in amperes

For stratigraphic purposes, the SGL has several advantages over other, more widely used arrays. Because it is quite sensitive to borehole resistivity and diameter, its response is limited at higher formation resistivities. This means that a single log scale, if set judicially, will keep the curve on scale throughout the log.

The character of the SGL curve, on the log printout, can be changed, easily, to suit special conditions. If the electrode is lengthened, the recorded curve will have a smoother appearance. The depth of investigation will also be slightly deeper. If the electrode length is shortened, the curve will show more stratigraphic detail. Of course, for any quantitative purposes, the length of the electrode, at the time of logging, must be recorded on the log heading.

The limiting factor of the SGL response is the increasing influence of the borehole resistivity, as the formation resistivity increases. One can picture the log value as the conductivity of two resistors in parallel, the borehole and the formation. As the conductivity of the formation resistor approaches zero, the borehole conductivity becomes the controlling factor.

In recent years, and particularly with the advent of digital data processing, the focusing electrode systems have begun to supplant the SGL as the standard resistivity curve. It is useful because of the extremely large resistivity ranges often needed in mineral and other hydrocarbon work. The SGL is not useful in mineral environments when R_t is greater than ten times R_m. The focusing electrode systems, however, have been designed for coal work to read quantitatively with $R_t > 3000R_m$. Also, in the lower pressure environments of non-hydrocarbon holes, the focusing electrode arrays can be designed onto the housing of many of the probes.

Of course, the determination of porosity and locating the top of the water table can be accomplished in small diameter holes with the SGL. The use of the 3-electrode focusing system greatly aids these functions. The mineral type 3-electrode focusing system is excellent for use in coal exploration and evaluation. The resistivity of lignites is largely a function of their water content (which can exceed 45%) and its salinity. In bituminous coals, the resistivity is a lesser function of the moisture content, but is a factor of the ash and carbon (heat) content. A system designed for coal usage must be used, however. Petroleum systems usually do not have enough range nor linearity. More will be said about these things, later, in the cross plotting discussion.

6

Natural Gamma Radiation

6.1 Introduction

Natural gamma radioactive events in the earth originate in the spontaneous break-up or decay of naturally unstable (radioactive) nuclei. When this happens, gamma photons are usually emitted. The emitted gamma photons undergo successive scattering, losing energy in each, until they are finally captured in photoelectric reactions. Since the photons travel at the speed of light, their energies are evident in their frequencies, ν (or wavelength, λ). Loss of energy results in a longer wavelength (a decrease in frequency).

The gamma ray spectrum of a formation consists of peaks of the emitted gamma rays at energies characteristic of each reaction and the element and a background of lower energy, scattered gamma ray photons. A gamma ray spectrum made with a sodium iodide, thallium activated detector [NaI(Tl)] and a multichannel analyzer (MCA) is shown in Figure 6.1. The background will build up at the lower energies because of the accumulation of the contributions from all of the higher energy events. At some low energy, the gamma photons will undergo photoelectric reaction with the ambient atoms and will be absorbed. The background will decrease sharply below this energy.

The gamma ray response at the detector is a function of the radioactive material volume concentration and the specific activities of the radioactive materials in the formation. It is also an inverse function of the formation bulk density.

Natural gamma ray emitting elements originate primarily from the radioactive decay of three parent sources:

1. From daughter products of the radioactive decay of uranium-238, $_{92}U^{238}$

2. From daughter products of the radioactive decay of thorium-232, $_{90}Th^{232}$

3. From potassium-40, $_{19}K^{40}$

The primary gamma emitting daughters of $_{92}U^{238}$ are bismuth-214, $_{83}Bi^{214}$, and lead-214, $_{82}Pb^{214}$. The primary gamma emitting daughter of $_{90}Th^{232}$ is thallium-208, $_{81}Tl^{208}$. The amount and character of the radiation is

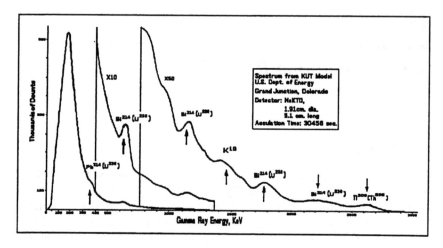

FIGURE 6.1
A gamma ray spectrum with a NaI detector.

a function of the type of the material and its history. Therefore, the range
of intensities and energies is large.

Figure 6.2 shows the decay series of uranium, thorium, and potassium.

6.2 Radiation from Formation Materials

Figure 6.3 shows typical relative amounts of natural gamma emission
from some of the common geologic materials.

6.2.1 Clays

Clays are the highest emitters, of the naturally-occurring, common, geological materials, because of their ability to bind metallic ions. These
metallic ions include uranium and thorium ions. Also, many clays are
potassium compounds, which contain $_{19}K^{40}$. The radiation from $_{19}K^{40}$, however, often amounts to less than 20% of the total, because of the differing
compositions of the various clays.

Clays are the alteration products of such potassium and sodium compounds as feldspar and other similar materials. The clays in shales may be
potassium or sodium compounds and are usually mixtures. They are the
primary component of all shales. Since a fixed fraction of the potassium is
$_{19}K^{40}$, the clay emissions from $_{19}K^{40}$ may or may not be present for a specific
clay. In reality, the $_{19}K^{40}$ gamma ray emission in a clay is usually about 20%
of the total gamma ray emission and is somewhat uniform for any one
clay-type shale mixture. Its amount *can* vary widely, however, because of

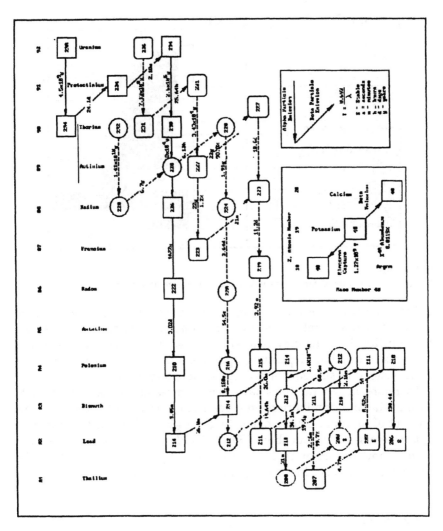

FIGURE 6.2
The decay series of uranium, thorium, and potassium.

the varying relative potassium amounts from clay to clay and, thus, from shale to shale.

Thorium compounds usually have limited solubilities in water at lower temperatures, such as near-surface temperatures. Most of the thorium in a clay comes from the source rock as particulate matter. Thus, the amount of thorium and its daughters is somewhat uniform from one clay to another. Thorium compounds are transported in solution at higher temperatures. Anomalous amounts of thorium and its daughters may appear in fractures, fault planes, and such communicating passages.

Many uranium compounds are very soluble in water when they are oxidized (they can be as soluble as 20,000 ppm) and very insoluble when

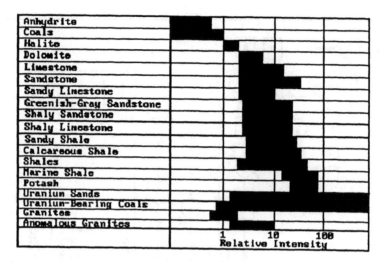

FIGURE 6.3
The relative amounts of gamma radiation from some formation materials.

reduced: perhaps as low as 10^{-15} ppm. In many cases, especially when the contact is of long duration, the uranium atoms can migrate or diffuse from the solution into the clay. Therefore, clays may have large or small amounts of uranium depending upon the source of the clay, its depositional history, its subsequent history, and its present state.

Therefore, in general, clays have some gamma ray emission from $_{19}K^{40}$, a somewhat uniform thorium emission from $_{81}Tl^{208}$, and a widely variable amount of uranium emission, mostly from $_{83}Bi^{214}$. Their normal range of gamma emission is from about 75 to 150 APIg units. They may, however, have anomalous amounts of radioactive components, especially in the presence, or presence in the past, of uriniferous solutions.

6.2.2   Sands

Many of the basic sand minerals, such as quartz, have little or no natural gamma radiation. Others, such as granite, may be radioactive. Sands usually have some amount of K-feldspar, traces or substantial amounts of clay, and may have some granites (containing mica). Thus, most sands will have a low $_{19}K^{40}$ gamma emission. Solutions of uranium compounds can easily move through sands because of the usual high sand permeability. If the environment is reducing, uranium may be deposited. Petroleum products, organic trash or humate, bacteria, and metal sulfide materials can cause such reducing conditions. Hot geothermal solutions may carry both uranium and thorium compounds and deposit them at lower temperatures. Thus, sands may have very low to very high radiation levels. Normally, their levels are low. Quartz sands usually have levels of 15 to 30 APIg units. Granitic sand levels will be 25 to 150 APIg units. Normal (non-anomalous) granitic sands will be more radioactive than normal quartz sands because the granites have higher K-feldspar and mica contents. Granites are also believed to be one of the major source rocks for sedimentary uranium deposits. Weathering of granites allow leaching of uranium compounds and its subsequent transport through permeable sand zones. Uraniferous sands may show many thousands APIg units radiation.

6.2.3 Carbonates

Limestones are usually deposited in environments which are low in potassium, uranium, and thorium compounds and their daughter products. Besides, permeability is often low and restricted, compared to sands. Furthermore, during crystallization and dolomitization, foreign metallic ions may be expelled from the crystalline structure. Thus, in general, carbonates have low radiation levels; from 5 to 25 APIg units. Note, though, that carbonates have high-pH environments associated with them. This can cause the precipitation of uranium, radium, and bismuth compounds in or on carbonates.

6.2.4 Igneous and Metamorphic Materials

Like the sediments, the igneous and metamorphic formation materials have wide ranges of natural radioactivity. Granites, quartz, and carbonates have already been mentioned in conjunction with sedimentary deposits. Those remarks apply equally well to the "hard rock" environments.

Granite and the granite-like rocks generally show substantial amounts of gamma radiation. These rocks are probably one of the source rocks for sedimentary uranium deposits. In igneous form, they are mined for uranium in Canada and Greece. In addition to uranium, they contain mica minerals and other potassium compounds. The normal radiation level of the hard rock granites is about that of the shales; from about 70 to about 150 APIg units. On occasion, they can show much higher radiation levels. They are subject to weathering into sedimentary materials.

Feldspathic minerals are widespread in hard rock environments. These minerals can be either potassium or sodium compounds. The potassium compounds, of course, contain $_{19}K^{40}$, which is a gamma emitter. These alter into clay minerals.

Quartz normally has a low level of or no gamma radiation. This probably is because of the tendency to expel foreign atoms during the crystallization process. Quartz, however, is slightly soluble in slightly high pH water. Therefore, quartz intrusions in non-radioactive rocks may have anomalous radiation because of the uranium and thorium in the solution that carried the quartz. Quartz intrusions are a major source of uranium in some hard rock deposits.

Of course, there are many minerals which are naturally radioactive. The most common ones are the uranium, thorium, and potassium minerals. These are part of many rock systems, either by solution while molten or by intrusion through fracture and fault systems.

6.2.5 Fractures and Faults

Fractures and faults often conduct water. Therefore, they frequently have associated anomalous radiation because of the mineral-bearing solutions moving through them. This is one method of detecting the presence of fractures, faults, and microfractures. Also, if the radiation is primarily from uranium daughters ($_{83}Bi^{214}$), then the solutions were probable relatively cold and oxidized (surface solutions). Hot hydrothermal solutions, on the other hand, will often have thorium compounds, as well as uranium compounds in them.

Figure 6.4 is a diagram from Garrels and Christ (1965) showing the relationships of pH and Eh of various environments. Figure 6.5 shows the relationships between uranium compounds and the iron compounds frequently associated with them. Eh is the potential of a half-cell referred to a standard hydrogen half-cell. The potential of the hydrogen half-cell is taken as zero at any temperature. The value of Eh is a function of the redox state of the material in question.

6.3 Gross Count or Total Count Gamma Ray Systems (GCGR)

6.3.1 Ranges of Detection

Natural gamma rays exist in a wide range of energies because of their initial relatively high energies and subsequent energy degradation by scattering through the formation material. Therefore, a gross count system is frequently used for general geophysical logging. It is sensitive to most of the gamma ray spectrum; until photoelectric reactions reduce the gamma ray field at the low energy end (this is the sharp drop-off at the left end of the spectrum shown in Figure 6.1). Therefore, its counting rates are high and statistical variations in the counting rates are less of a problem than with many of the other types of gamma ray measurements. Gross count systems are simpler and have fewer stability problems than spectral systems. Logging speeds can be faster and statistical variations smaller. The available information content is limited, however. We are not usually able to distinguish the emissions of one element from that of another with the GCGR.

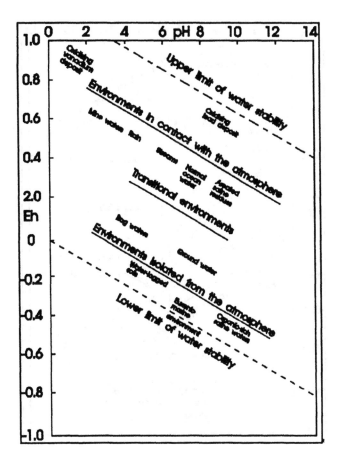

FIGURE 6.4
Natural environment redox positions. (After Garrels and Christ.)

6.3.2 Equilibrium

Radioactive elements will decay by the emission of one or more of its nuclear particles. These may be one or more of several particles: a beta (a negative electron), a positron (a positive electron), a proton (a hydrogen-1 nucleus), an alpha (a helium-4 nucleus), a neutron or other mechanisms. These mechanisms are usually accompanied by the emission of electromagnetic radiation.

As a radioactive element decays the parent element, such as $_{92}U^{238}$, decreases in amount. It changes into its daughter element. As the parent decreases, the daughter element increases. Unless the daughter is stable, however, it will also decay to its daughter element. None of these elements will likely have the same halflife (activity). This process will continue until a nonradioactive, stable element is reached. The stable element will

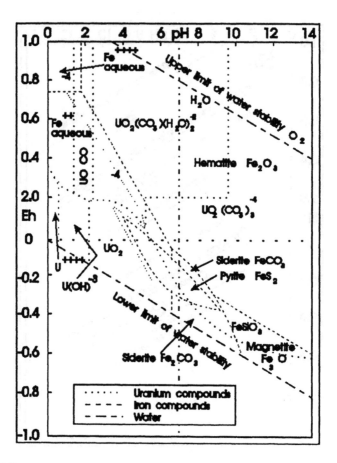

FIGURE 6.5
Stability relationships of uranium and iron compounds in water. (After Garrels and Christ.)

increase as the parent and all of the daughter elements transmute toward the end, stable element. If the system remains undisturbed, all of the daughter elements and the parent element will eventually exist in pseudo-stable ratios of amounts determined by their respective rates of accumulation and decay; their representative halflife (activity) ratios.

Each element, parent and daughters, will decay at a rate determined by its particular activity, λ. This activity is indicated by its halflife. The equation showing the amount of the first daughter element produced from the parent by time, t, is,

$$N_1 = N_0 e^{-\lambda_1 t_1} \tag{6.1a}$$

where N_0 is the original number of atoms of the parent element and N_1 is the number of atoms of the first daughter element at any time, t, in seconds. The daughter will, in turn, decay:

$$N_2 = N_1 e^{-\lambda_2 t_2} \qquad (6.1b)$$

Therefore, at any time, t_n, the number of atoms of the nth element will be

$$N_n = N_{n-1} e^{\Sigma(-\lambda_n t_n)} \qquad (6.2)$$

At some time, if nothing disrupts the sequence, the decay/growth action, the rate of decay of each daughter will equal its rate of augmentation. The series will then exist in a pseudo-stable state. This is termed a state or condition of equilibrium. That is, each daughter product amount is relatively constant with respect to all others except the parent and the end element. If some process or event disturbs it, it will be in disequilibrium. This will remain the state until the chain is disrupted or until the parent can no longer maintain the first daughter in relative constant amount.

This concept of equilibrium (and disequilibrium) is particularly important in some uses of gamma ray logging (both GCGR and spectrometric logging). $_{92}U^{238}$ is a weak emitter of gamma rays (it emits only a small number of gamma photons of low energy). The major gamma ray emitter of the $_{92}U^{238}$ series is $_{83}Bi^{214}$, which is eleven steps removed from $_{92}U^{238}$. Thus, when we detect gamma rays with the GCGR system, the majority of them come from $_{83}Bi^{214}$, not from the element uranium, as usually assumed.

If the series is in equilibrium, the amount of gamma radiation can be equated with the amount of uranium present through calibration techniques. If it is not in equilibrium, the "disequilibrium factor" must be known before the amount of uranium in the zone can be evaluated by means of gamma radiation techniques.

Uranium can easily be transported in solution when its compounds are oxidized. Thus, a uranium deposit can establish a chain of daughters and then be wholly or partially removed by solution from the deposit. This is termed a "low" disequilibrium factor (<1) because the amount of uranium is lower than would be indicated by the amounts of its daughters. Note that if the disequilibrium factor is low, the relative amounts of the daughter products will be decreasing. If it is high, they will be increasing.

The second daughter of $_{92}U^{238}$ is protactinium-234, $_{91}Pa^{234}$, which occurs within 24 hours of $_{92}U^{238}$. Thus, it is an important indicator of the presence of uranium. Refer to Figure 6.2.

The sixth daughter of $_{92}U^{238}$ is radium-226, $_{88}Ra^{226}$. This element forms complexes with iron. It appears to be a major constituent of the scale that occasionally forms on the outside of well casing. It is a gamma ray emitter and will sometimes show up on cased hole logs, especially at casing collars.

The seventh daughter of $_{92}U^{238}$ is radon-222, $_{86}Rn^{222}$. This is a noble gas, a gamma ray emitter with a halflife of about 4 days. Because it is a gas, it can escape from the formation and result in an absence of further daughter elements. It is also 100% soluble in formation gases. Mixed with these gases, it can enter the open hole, be dispersed in the drilling fluid, and temporarily raise the background radiation level of that hole. If the bubbles are large enough, it is possible to observe them (with the gamma ray system), migrating upward in the borehole.

The ninth and tenth daughters are lead-214, $_{82}Pb^{214}$, and bismuth-214, $_{83}Bi^{214}$. These are the strongest gamma ray emitters in the $_{92}U^{238}$ series. The detection window for uranium, in fixed-window type spectral equipment, is usually set at one of the emission zones of $_{83}Bi^{214}$, 1.76 MeV. The actual amount of gamma radiation, in this case, is that of $_{83}Bi^{214}$, and not $_{92}U^{238}$, of course. The value *must* be corrected for disequilibrium before it can actually represent the uranium amount. This is never done with petroleum logs nor with surface measurements. It is done routinely in uranium exploration and development.

Most of the bismuth emission lines will affect the potassium line (1.41 MeV) and, thus, will have to be stripped from the $_{19}K^{40}$ window. The thorium ($_{81}Tl^{208}$) emission line will also affect the uranium window. However, thorium ($_{90}Th^{234}$) is usually not found in large quantities in sedimentary deposits. Therefore, it is usually not a serious problem.

A new or recent deposit of uranium may not have been in place long enough to come to equilibrium with its products. In that case, the factor will be high (>1). $_{92}U^{238}$ requires about a million years to establish equilibrium under stable conditions. If any of the daughter elements, such as radon-222, $_{86}Rn^{222}$, is removed, the rest of the chain cannot come to equilibrium with the amount of $_{92}U^{238}$. Again, the factor will be high because the amount of uranium will be high compared to the amounts of its daughters.

Disequilibrium, D, can be closely approximated by

$$D = \frac{eU_3O_8, \ actual}{eU_3 \ radiometric} \tag{6.3a}$$

or

$$D = \frac{eU_3O_8, \ core}{eU_3O_8, \ radiometric} \tag{6.3b}$$

where eU_3O_8 is the hypothetical amount of U_3O_8 which would contain the same amount of uranium as the compound in question.

Much the same phenomena will be evident with the other isotopes of uranium and thorium. With thorium, however, the modes of transport are quite different. Disequilibrium is not as severe a problem with thorium compounds as it is with uranium compounds. The first daughter of $_{19}K^{40}$ is $_{40}Ar^{18}$ or $_{40}Ca^{20}$, both of which are stable.

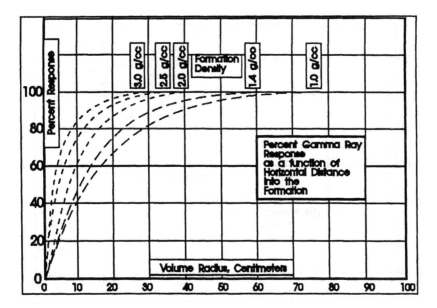

FIGURE 6.6
Percent gamma ray response as a function of distance into the formation.

6.3.3   Volume of Investigation and Borehole Corrections

The diameter of the sensitive volume of investigation of a gross count natural gamma ray system depends upon the density of the material surrounding the detector. It is about 1.6 ft or 1/2 m radius and is approximately spheroid for 98% of its response in sand and shales and other materials (about 2.4 g/cc density). See Figures 6.6 and 6.7.

Since the gamma ray log, as with other logs, is recorded from the borehole, a correction must be made to the recorded gamma ray value because of the radiation lost from formation material missing where the borehole is and by absorption by the drilling fluid. This absorption is a function of the size of the borehole and the density of its contained fluid. The borehole normally contributes no useful information and no radiation to the log. Empirical correction charts, such as those shown as Figure 6.8, are generally used to make this correction. The correction is a ratio of the correct value to the recorded value. That is, it is a correction factor.

The borehole correction charts are determined with models especially constructed for that purpose. These models are usually uranium ore (with, perhaps, some thorium and potassium compounds), stabilized with concrete. The model will often contain several boreholes of different sizes and with a slight slant. The logging tool is placed in each hole, in turn. A gamma ray reading is made with the sonde eccentered and again with it

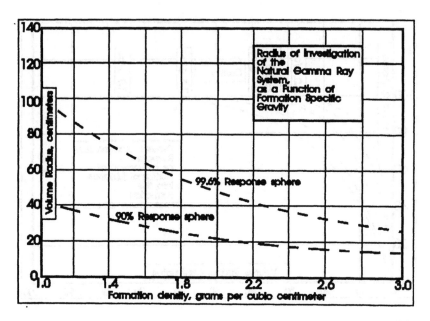

FIGURE 6.7
Radius of investigation of the gamma ray, as a function of formation density.

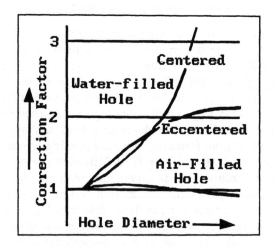

FIGURE 6.8
A hypothetical comparison the gamma ray response curves, as a function of borehole size and fluid.

centered and with the hole filled with air and again with it filled with water or drilling fluid. Long counts are made (a minute or more long) to reduce the statistical error. A complete set of correction factor curves can be generated from the readings in three or four holes.

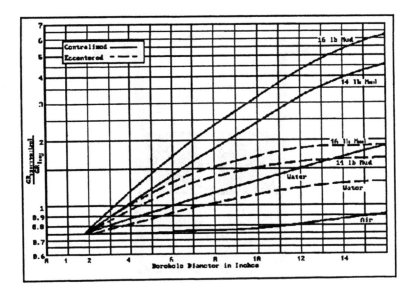

FIGURE 6.9
Gamma ray borehole correction for hole size and mud weight for a 1 11/16 in instrument.

There is a marked difference in the response curve shapes in the case where the sonde is eccentered and where it is centered (see Figure 6.8). When the borehole is the same diameter as the sonde, there is no correction to make. That is, the correction factor is 1.0. Note that the correction in an air-filled hole changes slightly with hole diameter because of increased air absorption of the gamma rays as the hole diameter increases. Some of the logging contractor's charts put in a fixed correction to the logged value so that no additional correction need be made in usual hole diameters and muds for the centered or the eccentered cases. This is seen on the charts in the chart books published by the various logging contractors, one of which is Figure 6.9.

As the hole diameter increases and approaches infinity, the cylindrical borehole wall configuration approaches a flat plane. In the eccentered case, the sonde is against the wall and slightly less than one half of the sensitive volume of the detection is in the formation and slightly more than one half is in the borehole. Therefore, the correction factor is asymptotic to a value of slightly greater than 2.

In the centered case, the sonde is farther from the wall as the hole diameter increases. At an infinite diameter, the sonde is an infinite distance from the wall and the correction factor has asymptotically approached infinity.

6.3.4 Calibrations

Petroleum usage of the gamma ray curve does not demand the high degree of accuracy and stability that non-hydrocarbon uses do. A calibration model

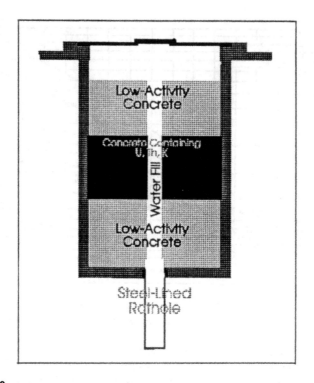

FIGURE 6.10

A typical petroleum-type gamma ray calibration model. Petroleum industry standardization assigns an arbitrary value to the contrast between the center and the two other zones. Each zone is at least 8 ft (2.5 m) thick.

was installed by the American Petroleum Institute at the University of Houston, in Houston, Texas. The standard calibration of petroleum gamma ray recording equipment is done in such a model. The active zone is uranium ore, thorium ore, and muscovite whose quantities were adjusted to give a gamma ray spectral response twice that of a "typical mid-continent shale". It is mechanically stabilized with concrete. By definition, this model has a radiation level, at its center, of 200 APIg units. Figure 6.10 shows the principle of such a model.

Uranium or potassium minerals can be located and assayed quantitatively with a useful degree of accuracy, if the gamma ray system is stable and its response is calibrated. The response is accurate if the calibration is frequent, accurate, and realistic. Figure 6.11 outlines one of the many mineral type models, this one at the U.S. Department of Energy installation in Grand Junction, Colorado.

Mineral, engineering, environmental, and scientific gamma ray logging equipment are calibrated much more precisely than that needed for petroleum work. The complete logging system is run in models similar to that shown in Figure 6.11. The area under the gamma ray curve of the log of the model is proportional to the grade of the radioactive mineral in the model,

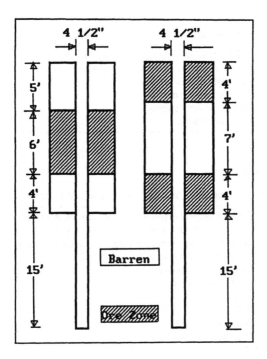

FIGURE 6.11
Two of the U.S.D.O.E. calibration models.

because the uranium content is in equilibrium with its daughters. Appropriate corrections must be made for borehole size and fluid content. The counting rate must also be corrected for the coincidence loses at high counting rates. A "K-factor" is determined for that system and for that calibration. The "K-factor" has the units of %mineral, by volume, per count per second. In very high or very low density materials, a correction must also be made for the effect of the bulk density (effective volume of investigation).

The relationship between the APIg unit and the K-factor must be defined for each system. This is because the APIg unit is a formation parameter, while the K-factor is a system sensitivity. One APIg unit is approximately equal to 20 ppm eU_3O_8.

6.3.5 Uses

The GCGR systems have several uses in petroleum work. The first important use of the gamma ray measurement was as a cased hole substitute for the *SP* curve. This was possible because both of the curves respond to the clay content of the zone. Of course, the gamma ray has little to do with the salinity of the formation and cannot be used for determining R_w. The GCGR system is one of the best correlation tools available. It is almost

TABLE 6.1

Uses of the gamma ray log in the petroleum industry

1. Differentiation between sands (or carbonates) and shales
2. Volumetric estimation of shale or clay content
3. Correlation between logs
4. Marker detection
5. Bed thickness determination
6. Fracture location

TABLE 6.2

Non-hydrocarbon uses of the gamma ray curve

1. Differentiation between sands (or carbonates) and shales
2. Volumetric estimation of shale or clay content
3. Determination of concentrations of radioactive minerals
   (ore grade determination and value)
4. Correlation between logs
5. Marker detection
6. Coal location
7. Bed thickness determination
8. Fracture location
9. Coal quality estimation

independent of borehole conditions. It will log satisfactorily in air, water drilling mud, or oil, in a cased hole or open, with any type of casing and/or cement. It can easily be built into or combined with other logging systems. It is essentially a point detector system. It has a stable and predictable response in most environments. Table 6.1 shows some of the uses of the GCGR in the petroleum industry. Table 6.2 is a similar listing of some of the non-hydrocarbon uses of the GCGR. A typical sedimentary uranium log is shown in Figure 6.12.

6.3.5.1 Volume of Clay/Shale Estimation, V_{sh}

The gamma ray curve is frequently used to estimate the volume of shale in the formation, V_{sh}. Since the response of the gamma ray system is proportional to the amount of radioactive material in the formation, the curve amplitude is a factor of the (radioactive) clay or shale content of a zone. It can be used to estimate the probable shale or clay content in a sand or carbonate. It is not exactly proportional because of formation density/absorption considerations, but a close approximation can be made.

Since a "clean" sand or carbonate (clay/shale-free) usually has some radioactivity level, γ_s, of its own, its radiation level must be taken as the zero amount of shale. A pure shale (radiation γ_{sh}) is, obviously, 100% shale. Therefore, the amount of the deflection, γ, which is above that of the sand

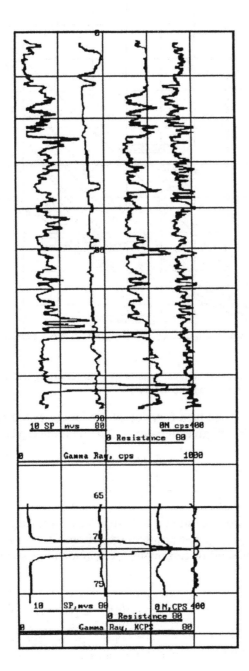

FIGURE 6.12
A typical sedimentary gamma ray log.

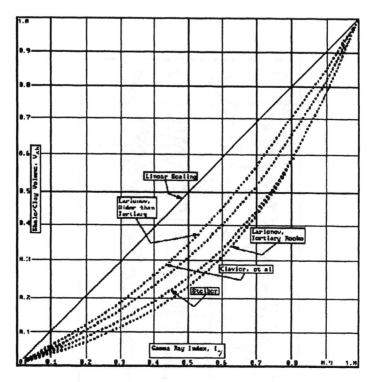

FIGURE 6.13

The chart for estimation of V_{sh} from the gamma ray value (Courtesy of Atlas Wireline Services Division of Western Atlas International, Inc.)

(or carbonate), γ_s, will be a function of the amount of shale in the shaly sand $(\gamma - \gamma_s)$. This will give an index, I_γ. The relationship is,

$$I_\gamma = \frac{\gamma - \gamma_s}{\gamma_{sh} - \gamma_s} \tag{6.4}$$

Several investigators, as shown in Figure 6.13, have determined that the volume of shale, V_{sh}, can be approximated from GCGR readings. In a Tertiary or younger sand or carbonate, the relationships suggested by Larionov (1969) is

$$V_{sh} = 0.083 \left(2^{2.71 I_\gamma} - 1 \right) \tag{6.5}$$

or, in older rocks by

$$V_{sh} = 0.33 \left(2^{2 I_\gamma} - 1 \right) \tag{6.6}$$

Clavier, et al. (1977):

$$V_{sh} = 1.7 - \left[3.38 - \left(I_\gamma + 0.7 \right)^2 \right]^{1/2} \qquad (6.7)$$

Stieber (1970), for South Louisiana Miocene and Pliocene:

$$V_{sh} = \frac{I_\gamma}{3.0 - 2.0 I_\gamma} \qquad (6.8)$$

Stieber, variations:

$$V_{sh} = \frac{I_\gamma}{2.0 - I_\gamma} \qquad (6.9)$$

$$V_{sh} = \frac{I_\gamma}{4.0 - 3.0 I_\gamma} \qquad (6.10)$$

Some of these relationships are also plotted in Figure 6.13.

Figure 6.14 shows a typical petroleum type gamma ray log which illustrates picking of values for the calculation of V_{sh}. Table 6.3 illustrates how to use this method. A clean, Louisiana Gulf Coast sand is evident in the lower portion of the log (actually at 10,280 ft to 10,300 ft) because of the low resistivity and low gamma ray readings. Its average radiation level, γ_s, is about 23 APIg units. The zone immediately above this sand (from 10,160 ft to 10,200 ft) looks like a good, solid, clean shale, because of its resistivity and gamma ray responses. Its radiation level, γ_{sh}, is about 103 APIg units. Both of these readings are typical of this region in this type formation. The radiation levels and the shale volumes, according to the Larionov calculation, are shown in Table 6.3.

6.3.5.2 Correlation

Correlation is the main use of the gamma ray curve in petroleum usage. This system will log in any type of hole and mud. The *SP* and gamma ray curves resemble each other enough that zones can usually be located, either in the open hole or through the casing, with great precision. Therefore, an *SP* curve is run in the open hole. Then, before and after the hole is cased, the gamma ray log is used, with other logs, to locate zones, sands, reservoirs for perforating and production. A casing collar locator is usually run with the cased hole gamma ray, so that the positions of the casing collars will be known during perforation.

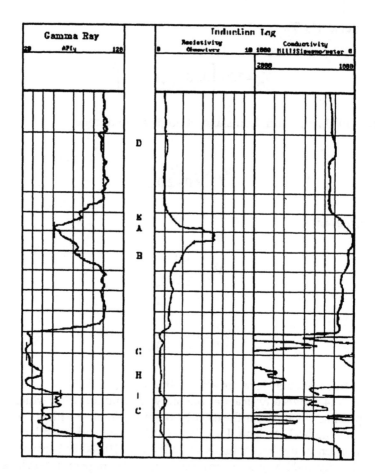

FIGURE 6.14
A typical petroleum log through a sand/shale sequence.

TABLE 6.3

Use of the gamma ray log to calculate V_{sh}

Zone	γ	I_γ	$V_{sh, Tertiary}$	$V_{sh, older}$
A	51	0.360	0.126	0.214
B	72	0.610	0.314	0.439
C	40	0.213	0.060	0.113
D	99	1.000	1.000	1.000
E	77	0.707	0.426	0.549
G	24	0.000	0.000	0.000
H	29	0.067	0.016	0.032
I	59	0.467	0.192	0.300

The calculation for V_{sh}, tertiary, zone A, is,
$I_\gamma = (51 - 24)/(100 - 24) = 0.360\ (14 - 4)$

6.3.5.3 Lithology

In sediments, the gamma ray curve is useful as a lithology curve. This is the first lithology curve we have examined in this text. The gamma ray curve can often distinguish between shales, sands, carbonates, and coals, because of their typically differing radiation levels. Particular shales can often be identified over wide areas by their individually characteristic radiation levels. Coals show radiation levels of 0 to about 10 APIg units. Coals containing large amounts of shale and/or extensive fracturing containing radioactive materials are the exception to this. Carbonates show about 10 to 25 APIg units level, sands 20 to 35 APIg units, shales 75 to 150 APIg units.

Any of the formation materials, however, can contain anomalous amounts of radioactive material and show a much higher radiation level. Sands and shales with levels of many thousands of APIg units are not unusual in uranium deposits. Hard rock materials frequently show anomalous radiation levels because of the residue from fluid movements through fracture systems or because of intrusions by radioactive mineral-bearing materials. If only the gamma ray curve is examined, granitic sands can easily be mistaken for shales.

6.4 Deadtime

Deadtime is that period of time, after the initiation of a pulse, that a system is insensitive to any other signal. If the counting rate in a zone is high, or if the deadtime of the detector and system is long, or both, a statistical probability correction must be made for the number of events per second lost to overlapping count pulses.

The event of a gamma ray photon passing through the detector is marked by a pulse in the detector. It is a light pulse in the detector crystal and converted to an electrical pulse in the photomultiplier (for example). it is then processed by the remainder of equipment. The pulse, whatever its form, has a finite length which is determined by the duration of the event and by the optical and electrical characteristics of the system. If other pulses occur while the first pulse is being processed, the subsequent pulses will be ignored. The area under the pulse envelope is proportional to the power of the pulse. In modern equipment, the pulse duration may be anything from much less than one microsecond at the output of the preamplifier of the detector to more than 200 microseconds. The longer a pulse (deadtime) is, the more subsequent pulses will be lost. The shape of the pulse is similar to that shown in Figure 6.15.

Even if a pulse is short to begin with, it will be spread out as it is detected, amplified, clipped, transmitted, or otherwise processed. The resulting width of the pulse will depend upon the inherent time constants

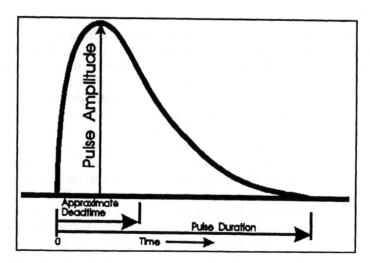

FIGURE 6.15
The approximate shape of a pulse in the gamma ray circuit.

of the various circuits through which the pulse will pass, particularly the longest of the individual system time constants. The pulse length is one item by which to judge the quality of a system. The worst offender is the logging cable. With analog equipment, which transmits unprocessed pulses up the logging cable, a 5 microsecond long pulse may be lengthened by a factor of 2 to 5 for every thousand feet of cable through which it travels. This is because of the distributed series resistance and parallel capacitance of the cable. A typical mineral logging cable, 5000 ft (1500 m) long, will have a time constant of about 28 microseconds. No matter how short the length of the driving pulse is, the pulse, at the surface, with analog transmission will have a length of at least 28 microseconds. Modern digital transmission takes care of this problem by encoding the signal in a low frequency pulse code which represents, unambiguously, the count rate for that time interval. With digital transmissions, the repetition rate is limited by the deadtime of the pulse, and thus, the character of the cable. This latter is not as serious, however, as the analog problem.

Since radioactive events and the pulses in radioactive instruments occur randomly, there is *always* a finite chance that one pulse will occur so close to the one before it that the second one cannot be detected by the system. This problem becomes more severe as the counting rate increases. From the initiation of a pulse (this is one gamma ray count), the system will not process another pulse until most of the first pulse has passed through the system. This is the "deadtime". Deadtime is roughly the time for the initiation of the pulse until the amplitude has dropped to about 1/3 of its peak value. Any events which occur during the deadtime (during which the system cannot handle another pulse) will be lost. Since the pulses occur in random sequence, the probability of this occurring is

$$P + \frac{\Delta N}{N} = \mu n \qquad (6.11)$$

where

    N = the true rate of events
    ΔN = the counting rate error due to coincidence
    μ = the deadtime, or effective pulse width in seconds
    n = the recorded counting rate

The counting rate error, ΔN, is

$$\Delta N = \frac{\mu n^2}{1 - \mu n} \qquad (6.12)$$

and the probable corrected counting rate is

$$N = \frac{n}{1 - \mu n} \qquad (6.13)$$

In general, a deadtime loss correction must be less than a factor of 2. If the correction is greater, there is a greater probability of it being wrong than there is of it being correct.

6.5 Bed Boundary Effects

The thickness of a bed can be determined accurately with a gamma ray system because the gamma ray detector is essentially a point detector. A gamma ray detector, in a barren zone, will detect a radioactive zone before it arrives at the bed boundary between the two zones. This is because the system has a finite volume of investigation. If the practical volume of investigation is assumed to be that from which 98% of the signal originates, the volume in a sand will be a spheroid whose radius is about 1/2 m (20 in). At a distance of 1/2 m from the bed boundary of a radioactive bed, the detector will begin to "see" the radioactive material. The actual distance will depend upon many things, such as the density of the formation, the thickness of the sonde housing, the borehole diameter and fluid density. The distance for a given system will depend mostly upon the density of the formation material, however.

The response shape of the gamma ray system (the log trace), as it moves toward and across the bed boundary will be approximately sinusoidal. For a thick bed, the deflection, D, at any point, will be

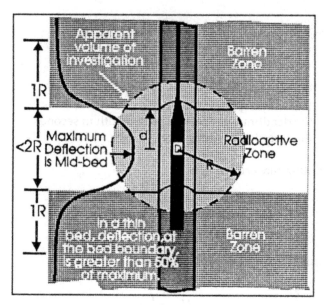

FIGURE 6.16
Gamma ray response in a thin bed.

$$D = \frac{\pi}{3}k\left(2R^3 + 3dR^2 - d^3\right) \qquad (6.14)$$

where

d = the distance from the detector to the bed
     boundary
k = the constant of sensitivity
R = the radius of investigation

This is illustrated in Figure 6.16.

If the second derivative of Equation 6.14 is taken, it will indicate the inflection point of the response curve across the bed boundary. At the bed boundary, the distance, d, is equal to zero. At that point, the second derivative becomes zero as it passes from positive to negative, or negative to positive. Thus, the bed boundary will always be indicated by the inflection point of the curve.

The maximum deflection will occur when all of the sensitive volume is within the radioactive bed and Equation 6.14 becomes

$$D_{max} = 2\frac{\pi}{3}R^3k \qquad (6.15)$$

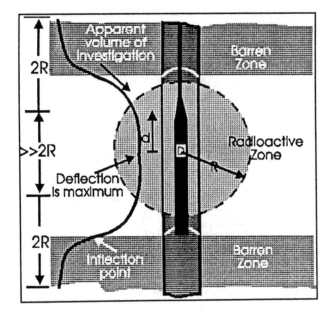

FIGURE 6.17
Gamma ray response in a thick bed.

6.6 Thin Beds

If the bed is thin (less than 2R in thickness), Then the maximum deflection
will be less than that in a thick bed of the same radioactivity. This is shown
in Figure 6.17. In this case, the actual deflection, D, will be

$$D = C_1 D_1 + C_2 D_2 + C_m D_m \qquad (6.16)$$

where C_1 and C_2 are the fractional volumes of the sphere of investigation
in the upper and lower adjacent beds, time the sine of the relative distance
of the detector from the maximum deflection point. C_m is the fractional volume of the sphere of investigation in the radioactive bed. D_1, D_2, and D_m
are the deflections these zones, if they were thick beds.

The actual deflection in a thin bed, if the adjacent beds are not radioactive, is

$$D_{max} = \frac{16R^3(d-1)+1}{z(12R-z)} \qquad (6.17)$$

where z is the thickness of the zone.

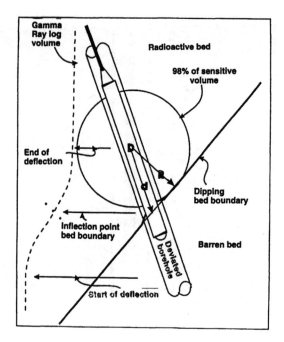

FIGURE 6.18
A deviated borehole through a dipping bed.

6.7 Dipping Beds and Slant Holes

The assumption is usually made that the borehole (and, thus, the sonde) passes through the bed normal to its top and bottom surfaces. This is almost never true. Figure 6.18 depicts a steeply dipping bed with a vertical borehole passing through it. The measure point of the logging detector is indicated at "D" (zero).

If the bed is horizontal (dip angle $\beta = 0$) and the borehole is normal to the bed (deviation angle $(90 - a) = 0$), the detector, D, will respond to the bed at a distance R, before the bed. If the bed is dipping ($\beta > 0$) and the borehole is deviated $(90 - a) > 0$), then the detector will respond to the bed at a borehole distance of $(R + A)$. The value of A is

$$A = R\cos(\alpha - \beta) - R \tag{6.18}$$

where
   R = the normal radius of investigation
   A = the increased borehole distance, greater than R, due to a dipping bed

α = the borehole deviation angle from *horizontal* (logged deviation angles are measured from the vertical and are $(90 - a)$

β = the dip angle of the bed boundary, from horizontal

Thus, when a log is run in a deviated hole and/or a dipping bed, the total transition of the gamma ray curve from barren to radioactive and reverse is spread out to more than the radius of investigation. If an expanded depth scale is used, this makes a possible way to detect a deviated borehole and/or a dipping formation bed.

6.8 Grade Calculations

Since the response of the gamma ray system is proportional to the volume amount of the radioactive material in the formation material, the area under the gamma ray curve deflection, as it goes through a radioactive bed, is proportional to the volumetric amount of the radioactive material in the zone:

$$GT = kAF \qquad (6.19)$$

where

G = the uranium or potash grade, usually in percent,

T = the thickness of the zone from inflection point to inflection point,

k = a conversion constant with units of $eU_3O_8\%$ per count per second or eK_2O per count per second. The value of k is usually determined by measuring the response of the system in a known radioactive environment, such as the calibration models in Grand Junction, Colorado; Ottawa, Canada; Adelaide, New South Wales; and many other places.

A = the area under the curve from background to background in usual units of counts per second depth units,

F = the product of all of the correction factors.

Note that the actual bed thickness is T, the distance from inflection point to inflection point. The radioactivity, on the other hand, is measured from background to background. This latter, will be greater than the bed thickness by one sensitive diameter.

The average grade of the zone is the sum of the grade thickness products from background to background, divided by the true bed thickness:

$$G_{ave} = \frac{GT}{T} \qquad (6.19a)$$

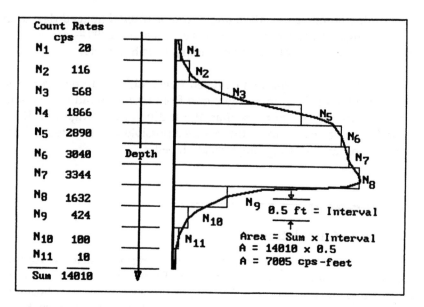

FIGURE 6.19
Determination of mineral grade.

$$A = w\Sigma N_j \qquad\qquad (6.19b)$$

where

w = the uniform width of each depth increment,

N = the increment height in corrected counts per second,

n = the total number of increments from beginning to end of the anomaly, and the summations are from j = 1 to j = n (from background to background).

The area, A, under the curve is determined by using a manual integration process, such as has been used by civil engineers for a century. Starting at the background value, the counting rates are noted at regular intervals (such as every half foot, every foot, or every 10 centimeters). The sum of these counting rates, multiplied by the interval used, is the area under the curve. This method, in essence, breaks the area under the curve into a series of rectangles, each of whose width is the uniform width interval and whose length is the counting rate, at that point. The smaller the width interval, the greater is the accuracy of representation of the area. This is illustrated in Figure 6.19.

The rectangles shown in the Figure 6.19 must be as narrow as possible in order to represent the shape of the curve accurately. Note that in the mathematical process of integration, the rectangles are made infinitesimally thin. This of course is impracticable in a visual process. They must be thin, however. Typically, the widths, with visual selection, are depths of 0.5 foot

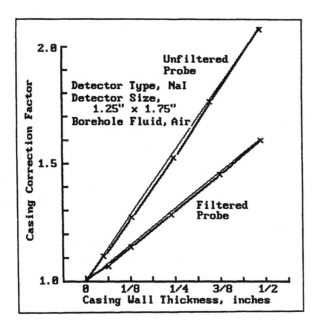

FIGURE 6.20
Response curves of filtered gamma ray systems.

or 10 cm. In computer processing, any width interval may be used. In fact, if the width interval is entered separately for each term, the value of "w" can vary from term to term.

The tabulation for the determination of the grade of a zone starts at the point of the anomaly where the gamma ray curve is beginning to be influenced by the bed of interest. This is about 1.7 ft or 1/2 m before the bed boundary in a sand or shale. Note the amplitude (counting rate) of the curve at every interval (i.e., every 0.5 ft or 10 cm). The counting rates must be individually corrected for deadtime effects. The method was described in Equation 6.13 of this text and chapter. The sum of the counting rates times the width of the intervals will be the area value under the gamma ray anomaly. The area units will be cps-feet or cps-meters. Any deflection dimension may be substituted for the cps.

Correction factors are hole size correction, disequilibrium, casing factors, etc. all of which must be determined for the particular type of system being used and the particular location being logged. The area of the anomaly may be multiplied by the product of all of the correction factors. This, then, is the corrected area of the anomaly. Some of these are illustrated in Figures 6.20, 6.21, and 6.22.

The same process as shown in Figure 6.19 can be applied to thin beds, as shown in Figure 6.23. If the notation of counting rates is started one sensitive volume radius (approximately 1.5 ft or 1/2 m in 2.4 g/cc sediments; less in higher density rocks and greater in lower density materials) before

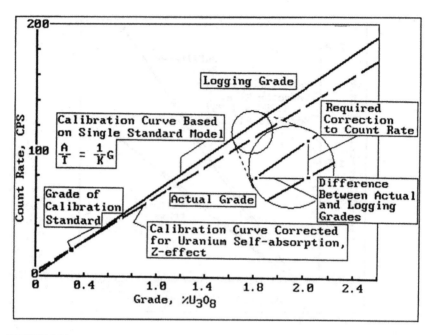

FIGURE 6.21
The Z-effect correction in calibration.

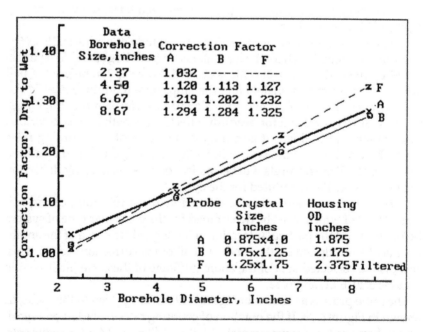

FIGURE 6.22
Effect of detector type on the response of the gamma ray system.

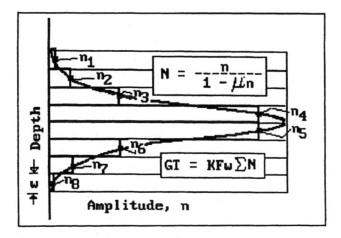

FIGURE 6.23
Grade determination in a thin bed.

the bed boundary (the inflection point of the curve) and continued to the same distance past the second bed boundary, even a thin bed determination will give correct results, without further thin bed corrections.

Improvement can be made in the accuracy of the determination by taking into account the losses due to photoelectric absorption in the bed and in the housing of the sonde. An alternate to this is to restrict the measurement to high energy (>100keV gamma rays). This is illustrated in Figure 6.20. The popular term for this method is "filtering". It can be done by shielding the detector or by electronic amplitude filtering. Figure 6.19 shows the effect of the photoelectric absorption (Z-effect) upon the casing correction, where the photoelectric absorption can be severe.

Once the curve area is accurately determined, the area need only be multiplied by the conversion factor or "K-factor" (Equation 6.15). This factor will have been determined at the time of calibration for the particular system and mineral being investigated. The K-factor converts the instrument response (i.e., counts per second) to environmental parameters. The common factor for uranium use is $\%U_3O_8$ per count per second. Any suitable parameters may be used. For example, K_2O or % clay may be used instead of the uranium value. Millimeters or inches deflection may be used instead of counts per second. The result of this operation will be the "grade-thickness product" (*GT*). The bed thickness can be determined by measuring from one inflection point of the curve of the anomaly to the other. If the GT product is divided by the thickness, the grade of the radioactive mineral will be determined (Equation 6.15a). All of these operations can be easily programmed into a dedicated or general purpose computer or a programmable calculator.

If the bed is a thin one, the deflection at the center of the bed will not be full for the grade in the bed. This is because the sensitive volume of the detector never fully leaves one of the adjoining beds. If the adjoining beds are barren, then the center deflection of the thin bed will be too low. Of course, if the adjoining bed or beds are higher grade than the thin bed, the center deflection of the thin bed will be too high. The grade of the thin bed can be determined by the same method just described, if one is careful to include *all* of the deflection, from background to background. The area under the curve *still* reflects the volume of radiating material in the sensitive volume. This is illustrated in Figure 6.23.

An alternate method used by the U.S. Department of Energy and others is the GAMLOG program or modifications of it. The GAMLOG program is a computer program which examines the anomaly in 1/2 foot intervals. It calculates the tentative grade for each 1/2 foot interval independently. See Figure 6.24. The intervals and their total responses are added in depth sequence. The shape and amplitude of the resulting curve is then compared with the original log. If the agreement is good, the sum of all of the overlapping counting rates at each depth in the hole should be equal to the recorded value on the log. If it is not, then an adjustment is made to each grade value and a new curve calculated and compared. This process is repeated until a satisfactory fit is obtained. The 1/2 foot interval grades are then published.

6.9   Fracture Detection

Frequently, when a solution moves through a fracture system or a permeable plane of any kind, it carries with it uranium compounds in trace amounts. This is especially true if the Eh state of the solution is oxidized. If the solution is hot, it may have, in addition, thorium compounds. Some of these may be left in the alteration products or on the surfaces of the rocks through which the solutions are moving. These radioactive deposits will show up on the gamma ray log as anomalies in otherwise barren or relatively barren zones.

6.10   Tracers

Tracers are used for many different purposes. Perhaps the movement of fluids up or down the borehole is to be detected or traced. Water loss into the formation or zones that are making fluid can easily be detected. There are many. Flow within a formation zone can also be determined by injecting the tracer solution into the formation in one borehole and recording the time interval in one or more other downstream boreholes.

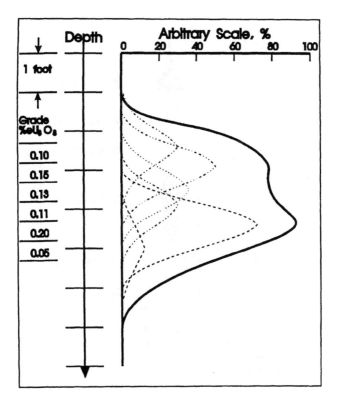

FIGURE 6.24
The U.S.D.O.E. GAMLOG program.

Radioactive compounds and solutions are available with a wide range of half-life and chemical characteristics. These can be selected for the particular application at hand. Special sondes with various kinds of remotely controlled injectors are available from many of the contractors (see Figure 6.25). These can be used to release or inject pulses of known amounts of fluid at precise locations and times. Timing of the pulse movement by gamma ray detectors strategically placed will give the desired information.

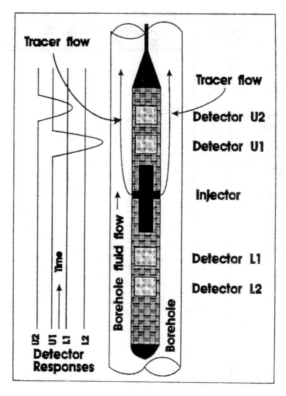

FIGURE 6.25
A diagram of a gamma ray tracer sonde.

7

Gamma Ray Spectroscopy

7.1 Introduction

Most of the gamma radiation associated with sedimentary formation materials is from $_{19}K^{40}$ and the active daughter products of $_{92}U^{238}$ and $_{90}Th^{232}$. Thus, useful data can be obtained by determining from which of the emitting elements the radiation originated. There is no provision for doing this with any reliability with the gross counting systems. The spectral gamma ray systems take care of this problem. Gamma ray spectrometric methods measure the counting rates in one or more narrow frequency (energy) bands or "windows" of gamma radiation. There are many forms available. They may be single energy window systems, three window systems (KUT), or multiple window systems (full spectrum).

The gamma ray spectrograph is new to the petroleum industry, but has been widely used in the mineral industry for many years in airborne, surface, and subsurface applications. Spectral methods are used for both wide area and detailed exploration. They are used for quantitative analysis and for various monitoring purposes.

7.2 Chemical and Geological Implications

In the petroleum business, the KUT system is used to analyze fluid movement through fractures and fault planes, identify clay types, determine clay or shale content, determine the degree of dolomitization, identify anomalous permeable zones, and locate possible petroleum deposits. It is valuable for determining shale zones and zones which merely appear to be shale, but are not.

Most (but not all) of the gamma ray spectrometric methods, in present use, detect one or more of the strongest gamma emitters daughter elements of the three major (geologically) radioactive series (see Figure 6.2). The radioactive series are the uranium-238, the thorium-232, and the potassium series. The isotopes usually detected are bismuth-214 ($_{83}Bi^{214}$), potassium-40 ($_{19}K^{40}$), and thorium-232 ($_{90}Th^{232}$). These are used to represent the volume of the parent uranium ($_{92}U^{238}$), thorium ($_{90}Th^{232}$), and potassium ($_{19}K^{40}$), present in the formation material.

There are, of course, many things which must be taken into consideration with the assumption that these three daughter isotopes actually tell the amount of the parent uranium, potassium, and thorium which is present. Some of the newer, and many of the experimental systems additionally detect other isotopes than those listed above. Also, the gamma ray spectrograph instrument is often the detecting component of many of the induced gamma radiation systems, which are gaining in usage. These will be discussed in a later chapter. There are many types of gamma ray systems available. One must pick the system which suits his need and which suits the phenomenon which he is trying to read.

Presentations may be of one or more windows in log form, logs of ratios of one or more pairs of windows, partial or full spectra at selected points, cross-plots, and/or isorad, isolith, thickness, grade, grade-thickness products, and location maps.

7.3 System Types

The various types of spectrometric systems available will variously measure a discrete energy channel, several channels, or a full spectrum. The quality of the system (figure of merit) will depend upon the window width and stability, the detector efficiency, the detector resolution, the maximum system pulse length (deadtime). The system must be chosen to suit the application and the end result. Too sophisticated a system is economically as bad as a poor system.

7.3.1 Detectors

Several types of detectors are in use. The most common one is thallium activated sodium iodide [NaI(Tl)]. These crystals are deliquescent (They absorb moisture from the air and decompose) and must be sealed. They are frequently inadequately sealed in older equipment. The color of the crystal, as well as the optical coupling between the crystal and photomultiplier, should be routinely checked. The crystal should be water-clear. A faulty crystal will be colored yellow and rapidly lose efficiency. The optical coupling must be bubble-free and use a water-clear, heat resisting, optical compound, made for that purpose. A reflector is usually incorporated on the inside of the protective shell to increase the detection efficiency. Newer equipment often uses detectors which have the crystal and photomultiplier sealed in one unit. These are much more rugged and trouble-free.

Recently, bismuth thallate has been used as a scintillation crystal because of its high density. This makes it more efficient than a similar sized NaI(Tl) crystal. Some systems use plastic scintillators, which have the advantages of a shorter light decay time and that they do not have to be hermetically sealed and are mechanically rugged. They have a lower efficiency partially due to their low density. Cesium iodide crystals {CsI(Tl) and Cs(Na)} also do not have to be sealed. They have a higher efficiency than the sodium iodide crystals because of their higher average atomic number, compared to NaI(Tl).

Several systems are commercially available which use intrinsic (ultra-pure) germanium detectors. These must be operated at cryogenic temperatures (about 77°K) to reduce the thermal noise to a practical level. They have about a 50x better resolution than the NaI(Tl) detectors. Cryogenic temperatures cannot be maintained downhole by use of liquid gas, such as liquid nitrogen because of pressure/temperature problems. Low temperatures are maintained for up to 8 hours by use of melting propane in canisters which are frozen in liquid nitrogen before use.

7.3.2 Single-Window Systems

Windows are usually achieved by the use of pulse amplitude discriminators to define the upper and lower window edges. Some gamma ray spectrographic systems, particularly the older models, examine a single channel or energy window. The channel frequently has adjustable upper and lower energy levels. The operator can compromise between the best resolution and the best probable error. Some older equipment had a provision for a motor drive for the window to scan the whole spectrum. Single window systems now are mostly confined to alarms, switches, and other dedicated instruments.

7.3.3 KUT Systems

The KUT (potassium, uranium, thorium) system is the most common of the downhole field spectrometric systems. It detects three different channels of gamma ray energies. The K channel is centered on the 1.459 MeV $_{19}K^{40}$ emission. A second, the U channel, is centered on the 1.764 MeV $_{83}Bi^{214}$ emission. The third, the T channel, is centered on the 2.615 MeV emission $_{81}Tl^{208}$. Figure 7.1 shows three spectra, each with a K, U, or T channel illuminated. In addition, a gross count or total count curve is also usually recorded. The window width depends upon the needed counting rate and is one of the criteria of quality of the system. The narrower the window is the lower will be its counting rate, but the higher will be its resolution. An early, analog, KUT log is shown as Figure 7.2.

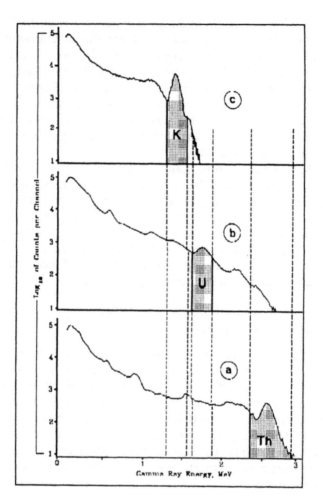

FIGURE 7.1
The three channels of a KUT system.

If the equilibrium situation between the daughter products and the parent minerals is known, the $_{81}Tl^{208}$ curve will indicate the thorium content and the $_{83}Bi^{214}$ curve will indicate the uranium content of the formation material, without equilibrium problems, after appropriate corrections have been made. The $_{19}K^{40}$ curve will indicate the potassium content. The potassium channel needs no equilibrium correction. This is shown in Figure 6.2 of Chapter 6.

7.3.4 MCA Systems

Many modern systems incorporate multichannel analyzers (MCA) of 256 or more channels. Some of these MCA instruments are used downhole.

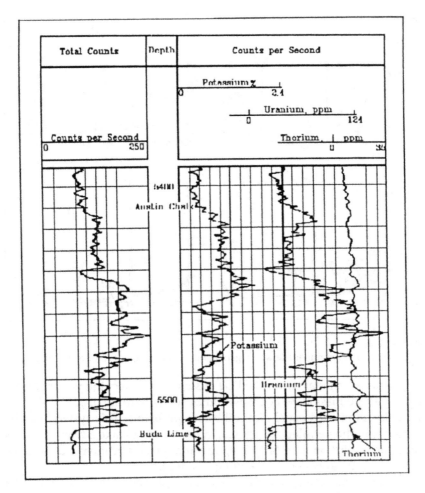

FIGURE 7.2
An early KUT downhole log (Courtesy of Atlas Logging Services Division of Western Atlas International, Inc.).

One system (PGT from Princeton Gamma Tech) allows additional selection of any single channel or group of channels to be plotted in log form, besides the individual full-256 channel spectra. The output of the PGT system controls the logging winch speed to achieve a constant probable error. This system employs a high-resolution germanium detector.

At this time, this author is unaware if the high-resolution full-spectrum system is offered on a regular commercial basis or not. However, several of the major logging contractors are working on such devices. This system is very complex and expensive. It has, however, about 50 times the spectral resolution to gamma rays than the usual sodium iodide detector has. See Figure 7.3. It uses an ultrapure (intrinsic) germanium detector in a low temperature (77°K) environment.

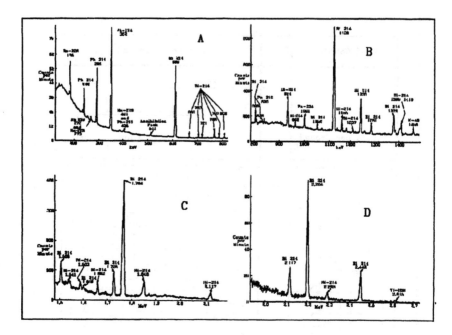

FIGURE 7.3
The gamma ray spectrum made with a germanium detector (after Princeton Gamma Tech).

The major original use of the PGT system is to detect the protactinium-234 ($_{91}Pa^{234}$) emission, which is within 24 hours half-life of $_{92}U^{238}$. Thus, it virtually eliminates usual disequilibrium problems. This system may be available in the petroleum business, in the future, to take advantage of its superior resolution. The technical, operational, and economic difficulties of the system, however, are delaying its adoption.

If you examine the spectrum in Figure 7.3b, at 1.001 MeV you will find a peak for $_{91}Pa^{234}$. This isotope exists within 24 hours of $_{92}U^{238}$. Thus, it is a high-probability indicator of the presence of $_{92}U^{238}$, regardless of the disequilibrium situation (unless it has been removed within the last 24 hours). Now, refer to Figure 6.1 (Chapter 6), to a spectrum recorded with a NaI(Tl) detector. It is evident that the $_{91}Pa^{234}$ line has been completely obscured, because of the poorer resolution of the NaI detector, by the overlap of the $_{83}Bi^{214}$ lines at 0.934, 0.960, and 1.052 MeV energies. Much the same problem can be seen with the lines of $_{88}Ra^{226}$, $_{86}Rn^{219}$, $_{86}Rn^{223}$, $_{82}Pb^{211}$, and $_{84}Po^{218}$.

7.3.5 Monitoring Systems

There are many gamma ray and X-ray monitoring systems. Many of these have little place in this discussion. There are several, however, which deserve mention.

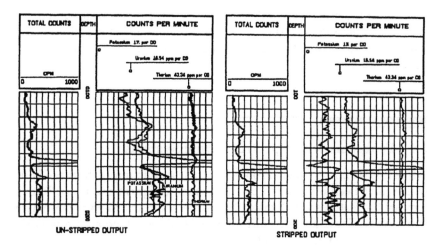

FIGURE 7.4
An example of a log portion before and after stripping. (Courtesy of Western Atlas Logging Services Division of Western Atlas International, Inc.)

The hand-held survey meters of the gross count variety have their counterpart in spectrographic systems. They are not as frequently used as the gross count systems, however, because the necessary portability demands small size and light weight.

One of the exceptional hand-held units is in the X-ray spectrograph family. These systems are about the size and weight of the larger hand-held gross count units. Some of these units employ a filtering system which makes use of the absorption of X-rays up to the wavelength of the K-edge of the filter's electrons. They use a low energy source of gamma rays to excite X-ray emission of the material (rock) under study. The instrument pulses resulting from the detection of the X-ray emission are counted for a fixed length of time. The filter is then changed to one which has a longer wavelength absorption edge and the resulting pulses are subtracted from the first reading and for the same length of time. The difference between the two readings is a function of the amount of material in the sample for which the filters were chosen.

The airborne gamma ray spectrometric systems are large and frequently very elaborate. The space and carrying capacity of large aircraft allow not only large crystals, but also the computer assembling and analyzing systems. The system usually detects the radon radiation, also, which is identified as the atmospheric background. In addition, quite often a magnetic system records the magnetic field. The aircraft must fly precise patterns. Thus, high-quality navigation equipment must be on board. The intensity information is usually presented in map form. These systems have the advantage of being able to explore tremendous areas quickly and accurately.

Many of the ocean exploration vessels test the ocean water and atmospheric air for radiation. These systems closely resemble the airborne systems. They are frequently combined with water salinity and temperature determinations.

7.4   Problems with Spectrographic Systems

While this section is titled "Problems", this is not to imply these are faults which have been overlooked by the system designers and users. Rather, they are the features which make them different from ordinary gamma ray equipment. These are the additional difficulties which accompany the greater detail and more complex system.

The most common and serious problem with spectrometric equipment, compared to the gross count systems, is that the counting rates of practical detector types and sizes are frequently so low that it is difficult to obtain reliable statistical samples in reasonable amounts of time. Occasionally it is necessary to resort to stationary readings. One commercial system automatically varies the logging speed to obtain a preset statistical probable error.

Airborne systems are the least restricted in detector size. Airborne systems with 1000 in³ (16,400 cc) volume thallium activated sodium iodide NaCl(Tl) crystal detectors are not uncommon. The largest that the author has seen was 3000 in³ (49,000 cc). Surface and car-borne detectors are usually about 100 to 1500 cc volume. Downhole oilfield crystals are often about 200–400 cc. Mineral-type crystals are 1.6 to about 30 cc volume.

Large crystal sizes are difficult to grow. Also, they are mechanically and thermally delicate. A partial solution to this problem is the use multiple, smaller crystals. This has become popular recently because it can also incorporate solutions to other problems, such as "noise" reduction.

System stability is an instrumentation problem. Instability is usually the result of analog sensitivity drift with temperature and aging. Airborne systems sometimes have thermostatically controlled heaters for the crystal. This helps to alleviate thermal drift and shock. The usual system uses a crystal which is doped with a small amount of low energy, mono-energetic, radioactive material to give a single reference energy line of known intensity. Some use a separate source, in contact with the detector. Some modern systems use a natural, common peak in the spectrum to monitor for stability. One system uses an LED in the detector to give a constant count rate. The constant counting rate of that line is sensed and the resulting signal is used to adjust the power supply to keep the sensitivity of the detector constant.

Other problems include circuit thermal and electronic noise. Electrical leakage of glass insulators and windows, especially at high-voltage gradients and high temperatures, is an example of this. "Shot" noise becomes a problem if a photomultiplier cathode is operated much more negative than the potential of the crystal case. This noise has much the same shape as the signal pulses. Also, glass becomes more conductive at higher temperatures, with a similar effect. These problems are usually overcome by using dual detectors and requiring a coincidence of events (a pulse from one event occurring simultaneously in both detectors) for acceptance. Also, usual practices are to operate the detector at as low and constant a temperature as possible. A germanium detector is operated at a temperature of about 77°K to achieve this thermal noise reduction.

The small size of the detectors presents another problem in downhole logging equipment. This not only limits the count rate, but it allows some of the photons to pass though the detector without expending all of their energy. Those that pass through, of course, retain an unknown energy, since only part of the energy of such photons shows up in the detector. This problem is taken care of by making the detectors as large and as dense as possible. It is difficult, in borehole equipment, however, to increase the size of the detector substantially, as its diameter is limited by the inside diameter of the sonde housing. Increasing a single crystal volume by increasing its length has limits because of the less-than-perfect transparency of the crystalline material. In downhole instrument design, the length limit is taken, approximately, to be less than four times the crystal diameter. Some systems use multiple detectors which are electrically in parallel and mounted serially in the housing. The problem is the easy one of taking the positions into consideration. This increases counting rates and thus reduces probable error.

The Compton background buildup tends to hide emission lines. Again, see Figure 6.1 (Chapter 6) and compare it with Figure 7.3. Some of the gamma rays of any energy will be degraded in energy by Compton scattering. Degraded radiation from $_{81}Tl^{208}$ (2.615 MeV) appears all down the spectrum. This results in a background which must be removed from the channel count before any quantitative value may be assigned to it. This is done by "stripping". The $_{83}Bi^{214}$ channel (at 1.764 MeV), used for $_{92}U^{238}$, must have the contributions of all higher energy contributions removed or stripped from it. The $_{19}K^{40}$ channel (at 1.459 MeV) must have all of the higher energy contributions from uranium daughters, thorium daughters, and any other higher energy emissions stripped from it before it can be used to determine the potassium content accurately. Figure 7.4 shows a piece of a log before and after stripping. Note that the thorium curve is hardly affected, the uranium curve is moderately affected, and the potassium curve, which is the lowest energy of the three, is greatly affected.

Stripping factors or ratios are calculated by determining the ratio of the 1.764 MeV peak ($_{92}U^{238}$) amplitude to the 2.615 MeV peak as the amount of thorium is changed. The same measurement is made with the $_{19}K_{40}$ peak amplitude at 1.459 MeV and the $_{91}Tl^{208}$ peak amplitude. Then, the

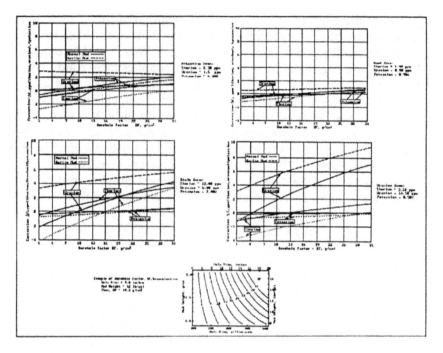

FIGURE 7.5
The borehole correction charts for the spectral gamma ray log (after Schlumberger, 1988).

measurement of the $_{19}K^{40}$ peak is compared to the $_{92}U^{238}$ peak amplitude, as the uranium amount is changed. There will be at least two stripping values determined for $_{19}K^{40}$ and one for $_{92}U^{238}$. Some suppliers use more complex stripping procedures, but the principle remains essentially the same.

Since each channel, of the several of a gamma ray spectrographic system, operates at a different energy level, each has a different absorption coefficient. Therefore, the absorption-type corrections, such as the borehole correction, should be made separately for each channel. This is illustrated in the borehole correction charts (Figure 7.5). These curves include the stripping corrections.

The Compton buildup also limits the effective dynamic range of any analyzer system. The lower energy limit of the detector range is also limited by the mass of the detector environment. This includes the instrument housing, the borehole material and the formation material.

7.5 Major Mineral Descriptions

At this time, the natural gamma ray spectrographic systems are mostly designed to measure the presence and quantities of uranium, thorium, and

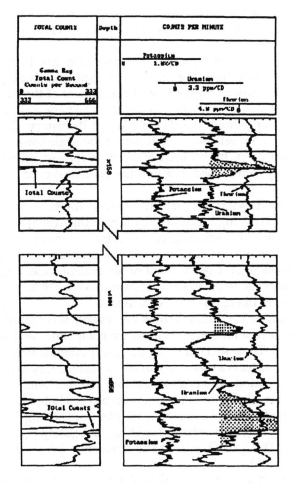

FIGURE 7.6
Shale-free Zones detected by the KUT ((Courtesy of Western Atlas Logging Services Division of Western Atlas International, Inc.)

potassium. The properties of these minerals and their daughter product minerals are worthy of brief mention.

7.5.1 Uraniferous Mineral Systems

Uranium has been known for more than 200 years and used for more than 2000 years. Glass colored by uranium and dating from 79 A.D. has been found. Uranium was recognized by Klaproth in 1789. It has a density of 18.95 g/cc. It can have a valence of 2, 3, 4, 5, or 6. Its compounds are found in many of waters of the "baths" in Europe and in the "therapeutic" muds of Africa. Its major use, at this time, is as a source of electrical power, from the fission of the $_{98}U^{235}$ isotope. Depleted uranium (the $_{92}U^{238}$ left after the

$_{98}U^{235}$ has been used) is made into rotors for gyro instrument and into pendulum weights, because of its high density. Small amounts are still used for coloring glass.

Uranium is found in many minerals in trace amounts. This is especially true of the clays in a shale. The uranium content (and its daughter products), probably account for the largest component of the gamma radiation from most shales. This uranium content can vary widely because of the physical properties of the clay/shale and its history. Thus, the uranium content is not reliable for estimating the shale content of a shaly sand. This also affects the estimate of the shale content when the reading of the GCGR system is used. The thorium content should be used when it is available.

Uranium is a heavy metal (density 19g/cc). Therefore, uranium compounds are often distinguished by their higher-than-usual densities. They frequently have a characteristic yellow color, "uranium yellow." Many uranium compounds are soluble when the uranium is in the oxidized state. This is usually the pentavalent or hexavalent state. They are often very insoluble when in the lower valence (reduced) states. Exploration in sediments makes use of this last characteristic.

Uranium can form extremely complex and variable compounds. It is frequently associated with iron and vanadium compounds and also in complex with them. The association with iron compounds is frequent enough that a magnetometer is a useful tool in many sedimentary deposits. Many vanadium compounds also contain uranium and thorium.

The major ore compounds for uranium are pitchblende, UO_2 (up to 76.7% uranium by weight); carnotite, $(K, Na, Ca, Cu, Pb)_2(UO_2)_2[VO_4]_2 \cdot 3H_2O$ (up to 54.6%); uranaphane, $Ca(UO_2)_2[SiO_4]_2 \cdot 3H_2O$ (52.4 to 55.7%); uranitite $I(U, Th)O_2 mUO_3 \cdot nPbO$; and bannerite, $(U, C_a, Fe, Y, Th)(Ti, Fe)_2O_6$, (7 to 27.5%).

The source rocks for uranium deposits are crustal rocks (average 2.7 ppm uranium by weight), granite (average 4.8 ppm), some basalt (average 0.6 ppm), rhyolite (5 ppm), shales (average 3.2 ppm), andesite (2 ppm), and diorite (2 ppm). The uranium content in seawater is low (~0.003 ppm), but is higher than the average in fresh water by a factor of 3 or more.

Uranium isotopes exist in natural deposits in definite proportions. The bulk of the uranium is $_{92}U^{238}$. It constitutes, at present, 99.28% of all uranium. $_{98}U^{235}$ is the fissionable isotope and amounts to 0.71%. Some Africa deposits, however, have much lower $_{98}U^{235}$ percentages than normal. It is believes that they were deposited at higher than a critical mass and underwent fission. The isotope, $_{98}U^{234}$ is a daughter of $_{92}U^{238}$ and amounts to 0.006%. Minable concentrations occur from about 0.20% (by weight) to (rarely) 16%. It is produced from both igneous and sedimentary deposits.

Uranium isotopes have long decay series of many radioactive daughter products, some of them gamma emitters. See Figure 6.2. The presence of uranium is assumed from the amount of the daughter, $_{83}Bi^{214}$. There are several things, however, which must be considered in this assumption.

Uranium minerals can be transported in particulate form. Some uranium minerals are highly soluble and easily transported, in solution, when they are in an oxidized state. They can be precipitated by decreases in temperature and/or decreases in oxidation or alkalinity. Therefore, there is a finite probability that the $_{92}U^{238}$ has been present short enough time that it is not in equilibrium with its daughters ($<10^6$ years) or has been wholly or partially removed by solution, leaving its daughters behind (this, too will result in disequilibrium).

The second daughter of $_{92}U^{238}$ is protactinium-234, $_{91}Pa^{234}$, which occurs within 24 hours of $_{92}U^{238}$. Thus, it is an important indicator of the presence of uranium. Refer to Chapter 6 for a more detailed description of the decay mechanism of uranium, especially as it influences the interpretation of the gamma ray logs.

Table 7.1 and 7.2 shows some of the important uranium minerals (also, the thorium and potassium). Note, though, that these are not necessarily the actual uranium and thorium contents, but rather the log readings of the uranium and thorium channels. These readings assume equilibrium.

Precipitation of uranium compounds is a function of the pH value and the redox state (Eh) of the local environment. These states are affected by the presence and chemical reactions of dissolved oxygen in the solutions (Eh), carbon dioxide (pH), hydrogen sulfide (Eh and pH), organic material (Eh and pH), sulfide mineral (Eh), anaerobic bacteria (Eh), carbonates (pH), and many other things. Refer to the diagrams of Garrels and Christ (1965) for an example, Figures 6.4 and 6.5, in the previous chapter.

The precipitated uranium (and its daughter products) may be found in an environment through which solutions move. This will include permeable sands and carbonates, fracture systems, fault planes, and in permeable clastic, argillaceous and igneous rocks. Uranium may be absorbed by ferric hydroxide, which can co-precipitate with calcium. Radioactive compounds, especially radium, can co-precipitate and be transported with barium compounds. Hard rock uranium deposits will be found in intrusive minerals which are either uranium minerals or contain dissolved uranium minerals.

Russian literature points out correlations of occurrence and amounts of radium with water salinity, geologic age, and gas content of solutions. Waters in contact with hydrocarbon reservoirs show increases in radium content. Uranium and radium compounds are known to be soluble in hydrocarbons. Radium content was found to be higher in waters containing Cl^-, and Ca^{++} ions. The amounts were found to be lower in waters containing HCO^- and Na^+ ions. Waters containing radium compounds are frequently low in sulfate content.

TABLE 7.1

Various minerals and their radioactive mineral content

Minerals	Potassium (%)	Uranium (ppm)	Thorium (ppm)
Accessory Minerals			
Allenite		30–700	500–5000
Apatite		5–150	20–150
Epidote		20–50	50–500
Monzanite		500–3000	$2.520(\times10^4)$
Sphene		100–700	100–600
Xenotime		$500–3.4\times10^4$	low
Zircon		300–3000	100–2500
Andesite (average)	1.7	0.8	1.9
Andesite, Oregon	2.9	2.0	2.0
Basalt			
Alkali basalt	0.61	0.99	4.6
Plateau basalt	0.61	0.53	1.96
Alkali olivine basalt	<1.4	<1.4	3.9
Tholeiites (orogene)	<0.6	<0.25	<0.05
(nonorogene)	<1.3	<0.05	<2.0
Basalt, Oregon	1.7	1.7	6.8
Carbonates			
Range	0.0–2.0	0.1–9.0	0.1–7.0
Average	(0.3)	2.2	1.7
Calcite, chalk, limestone, dolomite (all pure)	<0.1	<1.0	<0.5
Dolomite (clean, West Texas)	0.1–0.3	1.5–10	<2.0
Limestone (clean), Florida	<0.4	2.0	1.5
Texas, Cretaceous trend	<0.3	1.1–15	<2.0
Hunton Lime, Oklahoma	<0.2	<1.0	<1.5
West Texas	<0.3	<1.5	<1.5
Clay Minerals			
Bauxite		3–30	10–130
Glauconite	5.08–5.30		
Bentonite	<0.5	1–20	6–50
Montmorillinite	0.16	2–5	14–24
Kaolinite	0.42	1.5–3	6–19
Illite	4.5	1.5	
Mica, Biotite	6.7–8.3		<0.01
Muscovite	7.9–9.8		<0.01
Diabase, Virginia	<1.0	<1.0	2.4
Diorite	1.1	2.0	8.5
Dunite, Washington	<0.02	<0.1	<0.2
Feldspars, Plagioclase	0.54		<0.1
Orthoclase	11.8–14.0		<0.1
Microcline	10.9		<0.1
Gabbro (mafic igneous)	0.46–0.58	0.84–0.9	2.7–3.85
Granite (silicic igneous)	2.75–4.26	3.6–4.7	2.7–3.85
Rhode Island	4.5–5	4.2	25–52
New Hampshire	3.5–5	12–16	50–62
Precambrian, Okla., Minn., Colo., Texas	2–6	3.2–4.6	14–27
Granodiorite	2–2.5	2.6	9.3–11
Colorado, Idaho	5.5	2–2.5	11.0–12.1

TABLE 7.1 (continued)

Various minerals and their radioactive mineral content

Minerals	Potassium (%)	Uranium (ppm)	Thorium (ppm)
Oil shales, Colorado	<4.0	<500	1–30
Peridotite	0.2	0.01	0.05
Phosphates		100–350	1–5
Rhyolite	4.2	5	
Sandstones, range	0.7–3.8	0.2–0.6	0.7–2.0
Average	1.1	0.5	1.7
Silica, quartz, quartzite	<0.15	<0.4	<0.2
Beach sands, Gulf Coast	<1.2	0.84	2.8
Atlantic Coast, Florida, North Carolina	0.37	3.97	11.27
New Jersey, Massachusetts	0.3	0.8	2.07
Shales, Common, range	1.6–4.2	1.5–5.5	8–18
Average	2.7	3.7	12.0
200 samples	2.0	6.0	12.0
Schist (biotite)		2.4–4.7	13–25
Syenitee	2.7	2500	1300
Tuff (feldspathic)	2.04	5.96	1.56

7.5.2 Thorium Minerals

Thorium is a heavy, moderately reactive metal. Its density is 11.7 g/cc. It is a surprisingly uniform trace element in shales. It is a much more reliable indicator of the shale amount (by gamma ray measurement) than is uranium. There has been little study of the methods of transport and deposit of thorium compounds. Like uranium, thorium is fissionable and is used in some power reactors. The reactors in Canada will use uranium or thorium/uranium mixtures. It is 3 to 4 times more abundant in sediments, granites, and crustal rocks than uranium.

A major form of transport of thorium minerals seems to be in a particulate form by water and air. Thus the presence of the thorium in clays and shales is probably primarily syngenetic. Some of its minerals are soluble in hot water, but there has not been much study of this. Thorium-bearing water deposits some of its mineral content in fractures. The author presumes this to be mostly from hot solutions.

The principle thorium minerals are monazite, $(Ce, La, Y, Di)PO_4$, (2 to 24%); thorianite, ThO_2 (0–14%); uraninite, UO_2 or U_3O_8 (0–14%); thorite, $ThSiO_4$ (average 71.4%); huttonite, $ThSiO_4$ (81%); thorogummite, $Th(SiO_4)_{1-x}(OH)_{4-x}$, where x < 0.25. Monzanite, thorite, and thorogummite are the main ore minerals. Notice that frequently thorium occurs in the uranium and rare earth minerals. These are shown in Table 7.2. The source rocks for thorium are crustal rocks (average 9.6 ppm), granite (average 17 ppm), basalt (average 2.2 ppm), shale (average 11 ppm), andesite (low), dacite (4 ppm), and sea water (average 4.6 ppm).

TABLE 7.2

Physical characteristics of some of the thorium minerals

Mineral	Composition	ThO$_2$ Content, %
Thorium Minerals		
Cheralite	(Th,Ca,Ce)(PO$_4$SiO$_4$)	30, variable
Huttonite	ThSiO$_4$	81.5 (ideal)
Pilbarite	ThO$_2$.UO$_3$.PbO•2SiO$_2$.4H$_2$O	31, variable
Thorianite	ThO$_2$	Isomorphous series to UO$_2$
Thorite[2]	ThSiO^4	25–63; 81.5 ideal
Thorogummite[2]	Th(SiO$_4$)$_4$.x(OH)$_{4x}$, x<.25	24–58 or more
Thorium-bearing Minerals		
Allanite	(Ca,Ce,Th)$_2$(Al,Fe,Mg)$_3$Si$_3$O$_{12}$(OH)	0–~3
Bastnaesite	(Ce,La)Co$_3$F	<1
Betafite	~(U,Ca)(Nb,Ta,Ti)$_3$O$_9$.nH$_2$O	0–~1
Brannerite	~(U,Ca,Fe,Th,Y)$_3$Ti$_5$O$_{16}$	0–~12
Euxenite	(Y,Ca,Ce,U,Th)(Nb,Ta,Ti)$_2$O$_5$	0–~5
Eschynite	(Ce,Ca,Fe,Th)(Ti,Nb)$_2$O$_6$	0–~17
Fergusonite	(Y,Er,Ce,U,Th)(Nb,Ta,Ti)O$_4$	0–~5
Monazite[1]	(Ce,Y,La,Th)PO$_4$	0–~30[3]
Samarskite	(Y,Er,Ce,U,Fe,Th)(Nb,Ta)$_2$O$_6$	0–~4
Thucholite	A hydrocarbon mixture containing U, Th, and rare earth elements.	
Uraninite	UO$_2$ w/Ce,Y,Pb,or Th,etc.	0–14
Yttrocrasite	~(Y,Th,U,Ca)$_2$(Ti,Fe,W)$_4$O$_4$	7–9
Zircon	ZrSiO$_4$	Usually <1

[1]   The most important ore of thorium. Deposits in Brazil, India, USSR, Scandinavia, South Africa, and U.S.A.

[2]   A potential thorium ore mineral.

[3]   Usually 4–12.

Figure 6.2 shows the decay series of thorium. $_{90}$Th^{232} is a poor gamma ray emitter. Therefore, the major thorium mineral gamma emission does not come from the thorium, but rather, from its daughters. The daughter which emits the greatest amount of gamma radiation, in the case of $_{90}$Th^{232}, is thallium-208, $_{81}$Tl^{208}. The emission energy is 2.61 MeV. This is the highest of the lines detected with the KUT systems. Thus, the stripping process is minimal. The parent element and its daughter products produce many peaks of emitted energy. The emissions of gamma energy of some of the higher uranium, thorium, and potassium daughter emitters are shown in Table 7.3. The best for the thorium series is $_{81}$Tl^{208}.

Disequilibrium is not as great a problem with thorium deposits as it is with uranium. One reason is that the thorium compounds are not as readily transported as are uranium compounds. A second reason is that the radon gas interval in the thorium decay chain involves $_{86}$Rn^{220} instead of $_{86}$Rn^{222}. $_{86}$Rn^{220} has a half-life of about 55 seconds. Thus, the chance of it migrating away before decaying is less than that of $_{86}$Rn^{222} by a factor of 1.65×10^{-4}. The chance is still there, however. Thus, if gamma ray systems are used for detection and evaluation of the uranium and thorium contents,

TABLE 7.3

The spectral composition of the major emitters

Nuclide	Gamma Ray Energy MeV	Number of Photons per Disintegration in Equilibrium Mixture
$_{82}\text{Pb}^{214}$	0.242	0.11
	0.295	0.19
	0.352	0.38
$_{83}\text{Bi}^{214}$	0.609	0.47
		0.769
	0.05	
	1.120	0.17
	1.238	0.06
	1.379	0.05
	1.764	0.16
	2.204	0.05
$_{89}\text{Ac}^{228}$	0.336	0.11
	0.790	0.06
	0.908	0.25
	0.964	0.20
	1.587	0.12
$_{82}\text{Pb}^{212}$	0.239	0.40
$_{83}\text{Bi}^{212}$	0.729	0.06
	0.83	0.06
$_{81}\text{Tl}^{208}$	0.511	0.11
	0.583	0.28
	2.614	0.35
$_{19}\text{K}^{40}$	1.46	0.11

there is a chance that the measurement will be in error because of radon leakage or transport of the parent element compounds to or away from the deposit.

The measurement of thorium emission is actually the emissions of $_{81}\text{Tl}^{208}$. In order for the measurements to be correct, the formation material must be in secular equilibrium or the degree of disequilibrium must be known.

7.6 Miscellaneous Effects

See also the discussions in Chapter 6.

Quartz-bearing solutions frequently contain uranium and thorium in trace amounts. These co-precipitate with the quartz. It is likely that some uranium and thorium compounds are soluble in molten forms of many mineral, such as quartz and basalt. This would account for their presence in batholiths, dikes, and intrusions.

TABLE 7.4

Gamma ray data for granitic rock groups

Classification	K (%)	U (ppm)	Th (ppm)	Th/U	Th/U Average
Rocky Mountains	3.3	2.9	15.7	6.9	5.4
Canadian Shield	2.5	1.8	11.0	5.9	6.1
Pre-Cambrian	3.2	2.6	14.0	5.6	5.4
Post-Cambrian	2.2	3.0	8.6	3.3	2.9
U.S. West Coast	1.8	3.1	8.5	2.9	2.7

TABLE 7.5

Physical properties of some potassium minerals

Mineral	Specific Gravity	ρ_{ma} g/cc	Δ_{ma} μs/ft	ϕ_N %	γ API	K_2O %
Anhydrite	2.96	2.97	50	0	0	0
Gypsum	2.32	2.35	52	49	0	0
Halite	2.16	2.03	67	4	0	0
Trona	2.12	2.10	65	40	0	0
Trachydrite		1.66	94	62.5	0	0
Sylvite	1.98	1.86	74	0	500	63
Carnallite	1.61	1.57	78	65	200	17
Langbeinite	2.83	2.82	52	0	275	22.6
Polyhalite	2.78	2.79	57.5	15	180	15.5
Kainite	2.13	2.12	—	45	225	18.9

The annulus effect, due to a saline water pile-up before the invading fluid in the invasion of a permeable zone and detected by resistivity methods (Chapter 10), appears to have a corollary in anomalies of radioactive salts. This appears to originate in a migration of radium salts and, perhaps other uranium daughters, out of the hydrocarbon.

Granitic sands and massive rocks are source rocks for a number of the radioactive sedimentary minerals. Also, granitic sand is sometimes petroliferous. The KUT systems will allow the identification of these and prevent costly confusion with shaly zones. Table 7.4 shows the radioisotope content of several granites.

The use of the gamma ray spectrograph in the petroleum business is growing rapidly. The KUT systems are excellent for recording the potassium content of the formation material (Table 7.5) and the Th/K ratio for use in determining the types of clays in the permeable zones. This is needed to determine the value of the cation exchange capacity (CEC) of the shales in shaly sands. If the clay can be identified, then the CEC value may be assigned. The characteristics of clays are shown in Table 7.6. This is a particularly useful method when spectrographic gamma ray data are

TABLE 7.6

Characteristics of some clays

Clay Minerals	Remarks	Density g/cc	Hydrogen Index	CEC meq/100g	Spectral Distribution		
					K%	U ppm	Th ppm
Chlorite	Low water absorptive properties. As coating and/or pore bridging in reservoir pore space. Small effect on resistivity measurement due to moderate surface area.	2.60–2.69	0.34	10–40			
Illite	No absorbed water. As pore lining and/or bridging in reservoir pore space. Reduce resistivity measurements. Moderate surface area.	2.64–2.69	0.12	10–40	4.5	1.5	<2.0
	Biotite: As pore bridging and/or thin mica seams as laminae in reservoir rock.	2.7–3.2	0.12		6.7–8.3		<0.01
	Muscovite: Drastic effect on vertical permeability.	2.76–3.0	0.13		7.9–9.8		<0.01
Kaolinite	Patchy Kaolinite as discrete particles in reservoir pore space. As migrating fines, creating formation damage. Small effect on resistivity measurements. Low surface area.	Theoretical 2.61 Literature 2.60–2.68 Frequently quoted 2.63	0.36	3–15	0.42	1.5–3.0	6–19
Smectite	Montmorillinite, montronite. As coating and/or pore bridging in reservoir pore space. Critical to physical and chemical formation damage. Large reducing effect on resistivity measurments. High surface area.	2.20–2.70	0.13	80–150	0.16	2.0–5.0	14–24
Bentonite:	Low iron smectite: 3.6% iron content:	2.53 2.74		<0.5	1–20	6–50	

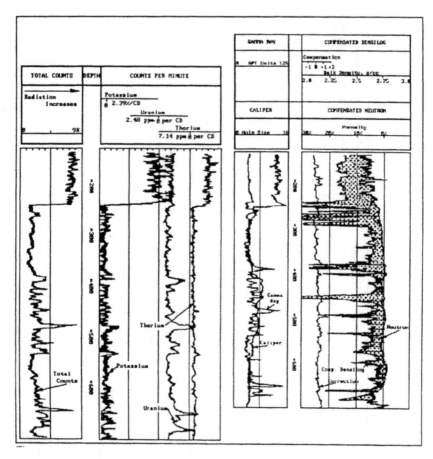

FIGURE 7.7
A log through a naturally fractured zone (Courtesy of Atlas Logging Services Division of Western Atlas International, Inc.).

cross-plotted with other data, such as density and the photoelectric cross section.

The four-curve log generated by the KUT system is used for many things. Since uranium compounds can be transported so readily in oxidized solution, the GCGR systems are not always reliable indicators of the actual situation. For example, a porous carbonate, such as a chalk, can absorb uranium-bearing solutions and resemble a shale on a GCGR log. On a KUT log, the high uranium indication, coupled with low thorium and potassium readings probably contra-indicates a shale (see Figure 7.7).

Fracture systems often have high uranium channel readings because uranium-bearing solutions have moved through them. Figure 7.7 shows a log through a naturally fractured zone. It also shows the difference between a

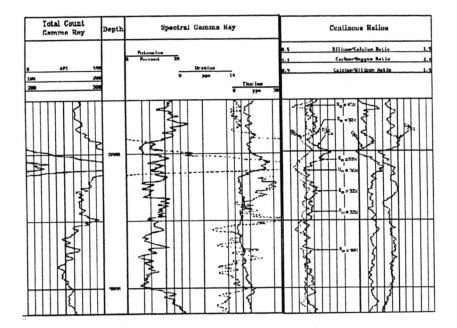

FIGURE 7.8
An example of uranium deposition because of the presence of hydrocarbon (Courtesy of Atlas Logging Services Division of Western Atlas International, Inc.).

shale and a fractured zone. Anomalous radiation may also be present in vugular porosity and in caves. High thorium readings in a fracture system may indicate that the solutions which had passed through the system may have been hydrothermal. Thorium indications in a uranium zone may indicate an igneous deposit. It will also, in a sediment, indicate contamination of the uranium mineral by thorium and possible hydrothermal deposition. Uranium indications around fracture systems may indicate the presence of hydrocarbon and especially, gas seepage (Figure 7.8). This is probably due to the reducing nature of hydrocarbons and organic material.

The occurrence, in shale-free sands, of radioactive, non-shale minerals can be determined with the use of the spectral gamma ray log. The gamma ray log must be used with other logs and cores. Such materials as sands containing micaceous or feldspathic minerals (i.e., granitic sands) can be identified by their sand-like indications on the electric log plus their potassium or thorium response on the gamma ray log.

Radioactive scaling can occur in several forms. Many times it is obvious. Occasionally, however, it can be quite confusing. The log of Figure 7.9 is an illustration of one aspect of this problem.

Of course, the spectral gamma ray logs can be well used for mineral type environments. Figure 7.10 shows a log through an evaporite interval. The

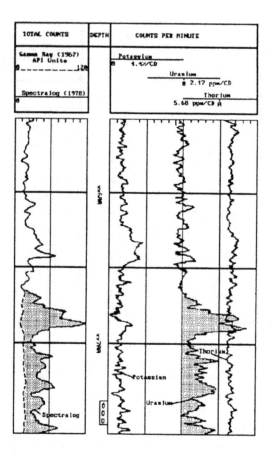

FIGURE 7.9
Scaling can include radioactive materials (Courtesy of Atlas Logging Services Division of Western Atlas International, Inc.).

potassium-rich evaporite is obvious. Figure 7.11 shows logs through a uranium-rich zone. It should be noted that the instruments used for these last two examples were made with petroleum logging systems and should not be used for quantitative mineral work. Petroleum gamma ray logging systems must only be used for qualitative analytical work. A well-calibrated mineral-type spectral system should be run to follow these logs.

The radioactive gamma ray logs, including the spectral logs, can be affected by the drilling environment. The effect of KCl mud is illustrated in Figure 7.12. In areas where there is an appreciable amount of natural uranium, the presence of radon gas can severely affect the log. Since radon is infinitely soluble in hydrocarbon gases and helium, these can enter the borehole and be dispersed in the mud during mud. It can also bubble up through the mud during logging.

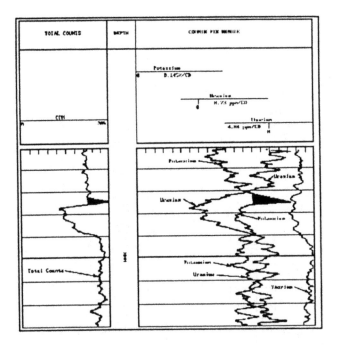

FIGURE 7.10
A KUT log through potassium evaporites. (Courtesy of Western Atlas Logging Services Division of Western Atlas International, Inc.).

7.7   Spectrometric Ratios

Spectrometric channel value ratios are often used because they tend to combine indications. These ratios are often used for cross plotting. For example, when plotted against the photoelectric cross section, Pe, a high ratio of Th/K in a shale tends to indicate a glauconite or an illite. A low ratio indicate a possible montmorillonite, chlorite, or kaolinite. See Figure 7.13. Ratios of channel values from the KUT system can be especially valuable for uranium deposit evaluation. The ratio of U/Th, for example, is a measure of the quality of the uranium deposit, in sediments. In igneous deposits, this ratio can determine how the uranium can be used, if produced. In sediments, the ratio is an indicator of the depositional history of the deposits.

7.7.1   Uses of Ratios

Figure 7.14 shows the correlation between the organic carbon content and the Th/U ratios in three specific formations. The following are some sample ratios in different general types of formation materials. The information is from Dresser Atlas.

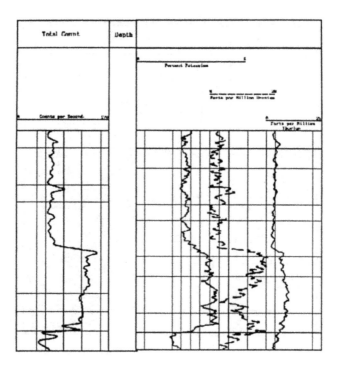

FIGURE 7.11
A uranium anomaly on a petroleum log (Courtesy of Atlas Logging Services Division of Western Atlas International, Inc.).

7.7.1.1   Uses of the Thorium/Uranium Ratio, Th/U

1. In sedimentary rocks, Th/U varies with depositional environment, i.e., continental, oxidizing environment, weathered soils, etc.

    Th/U < 7: marine deposits, gray and green shale, graywacke

    Th/U < 2: marine black shales, phosphates

2. In igneous rocks, high Th/U is indicative of oxidizing conditions by magma before crystallization and/or extensive leaching during the post-crystallization history.

3. Determining the source rock potential estimates of argillaceous sediments (shales):

    Identifying major geological unconformities

    Estimating the distance to ancient shore lines or location of rapid uplift during time of deposition

    Making stratigraphic correlations, transgressions vs. regression, oxidation vs. reduction regimes, etc.

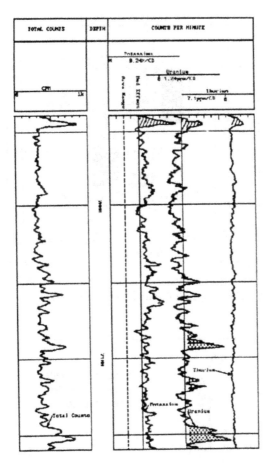

FIGURE 7.12
This shows the effect of radioactive material in the mud (Courtesy of Atlas Logging ine Services Division of Western Atlas International, Inc.).

7.7.1.2 Potential Uses of the Uranium/Potassium Ratio, U/K

1. Determining the source rock potential of argillaceous sediments
2. Making stratigraphic correlations
3. Identifying unconformities, diagenetic changes in argillaceous sediments, carbonates, etc.
4. Frequently may be used for making correlations with vugs and natural fracture systems in subsurface formations, including localized correlation with hydrocarbon shows on mud logs and core samples in both clastic and carbonate reservoirs.

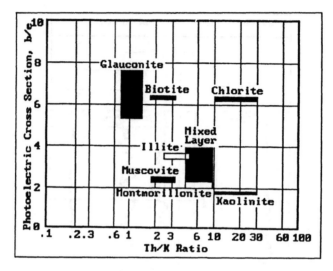

FIGURE 7.13
A crossplot of the Th/K ratio against Pe (after Schlumberger, 1988).

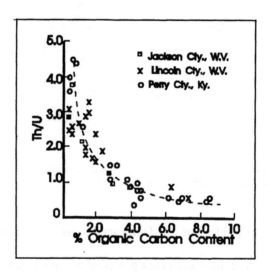

FIGURE 7.14
The Th/U ratio vs. organic carbon in three black shales. (Courtesy of Western Atlas Logging Services Division of Western Atlas International, Inc.)

7.7.1.3   Some Uses of the Thorium/Potassium Ratio, Th/K:

1. For recognition of rock types of different facies
2. Use for paleographic and paleoclimatic interpretation of facies characteristics

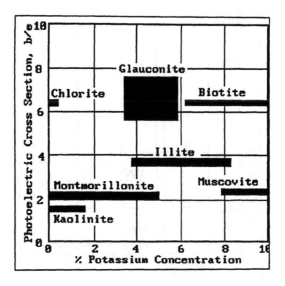

FIGURE 7.15
A crossplot of Pe vs. K% (after Schlumberger, 1988).

3. Locate and identify depositional environments, distance from ancient shore lines, etc.

4. Identify diagenetic changes of argillaceous sediments

5. Use to type clays

   Th/K increases from glauconite → muscovite → illite → mixed layer clays → kaolinite → chlorite → bauxite

6. Determine probable correlation with crystallinity of illite, average reflectance power, paramagnetic electronic resonance.

7.8 Cross-Plotting

The absolute values of the KUT log can also be used for cross plotting with other types of measurements. A $_{19}K^{40}$ curve can be plotted against the photoelectric cross section, Pe, to indicate clay type, as in Figure 7.15. The Th/K ratio can also be cross plotted with the Pe value for clay identification (Figure 7.13). Chlorite, glauconite, biotite, illite, muscovite, kaolinite, and montmorillonite can be identified. Figure 7.16 shows a crossplot of the K channel with the carbon/oxygen ratio.

A major method to determine clay type is the cross-plotting technique, shown in Figure 7.17. This figure shows the K-channel values plotted

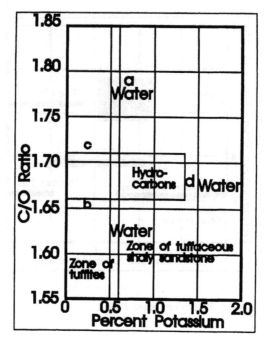

FIGURE 7.16
A crossplot of the C/O ratio vs. K%.

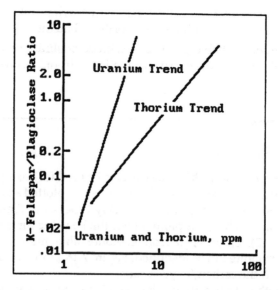

FIGURE 7.17
A crossplot of the U channel vs. the T channel.

against the T-channel values. Essentially, this is a two-dimensional presentation of the Th/K ratios. It shows how the various types of potassium-clay minerals can be distinguished from each other and from the non-clay minerals. The potassium content (or lack of it) and the thorium content are both attributes of the several clays and shales, but are independent of each other.

The U channel curve is useful in uranium exploration and development because the grade of uranium mineral can be known more accurately from it than from the GCGR curve. There are several reasons for this. The uranium curve is free from interference from the thorium content, when spectral systems are employed. Also, the $_{83}Bi^{214}$ peak is at 1.7 MeV, well above the photoelectric reaction range. Thus, the uranium calibration method for this curve is straightforward and simple. In low-grade deposits, no allowance for background counting rate need be made when using the U-channel values.

8

Scattered Gamma Ray Methods

8.1 Introduction

Scattered gamma ray systems use sources of non-formation gamma radiation and one or more gamma radiation detectors. The source is usually isotopic. The major scattered gamma ray methods in geophysical logging are the formation density systems, the high-resolution coal device, and the scattered gamma ray spectrographic methods. Other commercial uses of scattered gamma ray systems include determination of drilling mud density (and other fluids), metal-thickness gauges for rolling mills, location of other tubing strings in multiple-completion wells, annular cement quality and tops, and many other applications.

The oilfield density systems are probably inherently the most precise and accurate of the geophysical logging systems. This is one of the more important systems used in the mineral exploration and development, and in geotechnical engineering. It is on an equal footing, in the petroleum business, with the neutron porosity methods and lesser only to the resistivity and acoustic methods, in value. The geophysical density logging methods in the petroleum business are quantitative. In the mineral, engineering, and scientific geophysical logging fields most of them are quantitative. These methods include formation density logs, compensated formation density logs, spectrographic scattered gamma ray methods, high-resolution coal systems, and fluid density devices.

Scattered gamma ray logging methods are used in the petroleum business mostly to determine porosity, rock type, and to calculate R_o. The methods, however, have many other uses. Among these are identification of formation mineral type, assessment of coal quality, locating fractures, identifying the fracture type, determining mud density and the densities of other fluids, and locating gassy and shaly zones. A short spacing, soft radiation (<90 keV) device is used to accurately (~±1 cm) determine the depths and thicknesses of coal beds and partings. The formation density log is used in Western Australia to locate the altered-unaltered interfaces where gold deposits occur.

TABLE 8.1

Uses of scattered gamma ray methods

1.	Formation bulk density determinations
2.	Lithology determinations
3.	Effective porosity determinations
4.	Corrected porosity determinations
5.	Porosity type determinations
6.	Cross plotting with R_t for S_w, m and R_w
7.	Calculating R_o
8.	Gas and shale detection
9.	In combination with the gamma spectrograph and/or the photoelectric absorption response, Pe, to determine lithology, esp. rock type, and clay type
10.	Used in combination with gravity systems for detail
11.	Used in combination with the acoustic methods for mechanical parameters
12.	Used in combination with other types of measurements to determine rock type, independent of porosity
13.	Determining the volume and type of clay or shale in a sand
14.	Determining coal quality
15.	Determining concrete and fill quality
16.	Estimating water salinity

8.2 Formation Density Logging

During the course of this chapter, we will discuss:

1. Electron density, ρ_e, (see Equation 8.7a) which is the density actually read by the system

2. Apparent density, ρ_a, which is the density recorded on the log, or that at any stage before ρ_b

3. Bulk density, ρ_b. This is the actual ratio of the mass of the material to that of an equal volume of water. ρ_b is obtained by making Z/A, mudcake and borehole corrections to ρ_a

Formation density logging systems all put moderately high energy (500 keV to 1.3 MeV) gamma rays into the formation. The sources, at this time, are all isotopic. Electronic sources, however, are being investigated. The photons which are scattered back to the logging tool detector are counted. The flux density at the detector is a function of the average number of electrons within the effective sensitive volume of formation material. Density logging determines the probable bulk density, ρ_b, of the formation material. It does this by measuring the electron density, ρ_e, of the formation material, which is a function of the bulk density. The method can be quite accurate, if the corrections are performed properly. See Table 8.1.

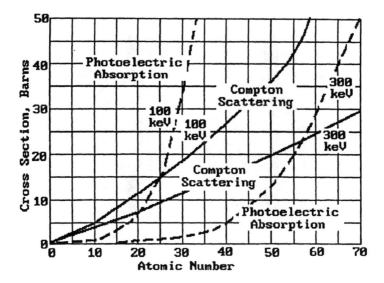

FIGURE 8.1
Relationship between atomic number and the Compton scattering and photoelectric reactions for two energies.

8.3 Source Energy Requirements

The photon source energy for density measurements must be low enough that the formation materials present a reasonable cross section to the incident gamma photons. Since most of the elements found in rock formations are composed of the first twenty elements, the response to these elements should, ideally, be linear or nearly linear. Figure 8.1 illustrates this relationship. This diagram indicates that the Compton reaction with 300 keV or higher photons is desirable. The photoelectric absorption, since it is quite nonlinear (a function of $Z^{3.6}$), should be minimized.

The half-life of the source must be long enough to minimize corrections for source strength (this is of lesser importance with modern computer and calibration techniques than with the older, analog equipment). However, for quantitative measurements, the source activity and strength must be known and relatively constant.

The source half-life must be short enough to have a high activity so that the resultant physical size will be small, to approximate a point source. $_{27}Co^{60}$, with a half life of approximately five years, will decrease in emission by about 13% a year. $_{55}Cs^{137}$, with a half-life of about 37 years, will only decrease about 2% in this length of time. Of course, with modern digital equipment, the serial number and date of initial strength measurement

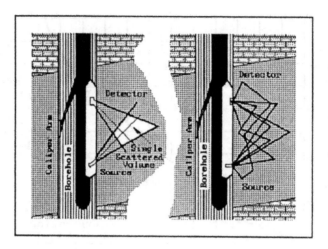

FIGURE 8.2
Two types of source orifice designs.

can be programmed into the computer, which can then make a correction for the decrease of source strength, for any type of source. The relative strength of the source can be determined during calibration and a normalizing factor can be applied, if needed. Both $_{55}Cs^{137}$ and $_{27}Co^{60}$ satisfy this requirement. $_{88}Ra^{226}$ has been used as a source of gamma rays, especially in older equipment. The major reason for this is that, at that time, no license was required the use the radium source. $_{88}Ra^{226}$ has a half-life of 1600 years. Thus, the change of source strength in a year is negligible.

The source energy is chosen to keep the singly scattered photons within the center portion of the Compton energy range. The injected photons scatter from the orbital electrons of the formation atoms. The higher the source energy is, the more personnel shielding becomes a problem. The highest energy of $_{27}Co^{60}$ approaches the upper limit of this consideration, with an energy of 1.3 MeV.

8.4   Operation

Gamma rays are collimated (not focused) from a sidewall tool into the formation. See Figure 8.2. Compton scattering takes place mostly from the orbital electrons in the formation material (in a well-designed system). The amount of scattering is a function of the number of electrons present within the measurement volume of the tool.

Scattering and absorption of gamma rays involve the orbital electrons of an atom. The probability is that scattering will be away from the detector. See Figure 8.3. The path from the source to the detector is not a straight line

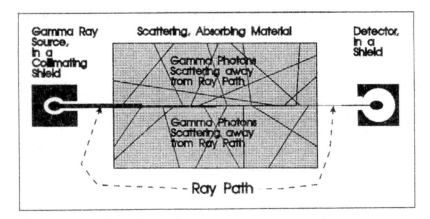

FIGURE 8.3
Schematic of gamma ray scattering measurement.

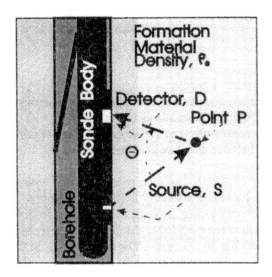

FIGURE 8.4
Hypothetical path of a single-scattered gamma photon.

in most density tools. The gamma ray paths which end at the detector are triangular with an average of approximately a 90° apex for the single scattered rays. Refer to Figure 8.4. These single scatterings are all Compton events. The more dense the formation material is, the more orbital electrons it will have. There will be a high probability that the gamma rays will be scattered away from the path from the source to the detector. Therefore, number of the scattered photons which arrive at the detector will be an inverse function of the number of orbital electrons within the sensitive volume of the system.

The passage of a gamma ray through any medium follows the rule,

$$I = I_0 e^{-\mu \rho x}$$ (8.1)

where

I_o = the original intensity
I = the intensity at the detector
μ = the mass absorption coefficient
ρ = the bulk density of the material
x = the distance traversed

See Figure 8.3.

An approximation of the intensity, I, of the scattered gamma rays at the detector can be determined, using the attenuation equation, Equation 8.1.

A beam of gamma rays from the source, S, of Figure 8.4, will travel to point P, and be scattered, as it travels. The intensity, I, at point P, will be,

$$I_p = I_s e^{-\mu_{SP} \rho_e (SP)}$$ (8.2)

where

I_S = the source intensity
μ_{SP} = the mass attenuation coefficient for the energy of the gamma ray beam on the path SP to the point P
ρ_e = the average electron density along the path SP
(SP) = the distance along path SP, in centimeters

Point P can be considered to be a remote source within the formation material. It will have the intensity I_p and its energy will be that of the gamma scattered once along the path PD, at an angle of θ to the path SP. The beam will follow the path PD to the detector and lose intensity by scattering, until it reaches the detector. The intensity, I_d, at the detector wall will be,

$$I_d = I_p e^{-\mu_{PD} \rho_e (PD)}$$ (8.2a)

But, since the intensity at point P depends upon the source strength and the action along the path SP of length (SP), then,

$$I_d = I_s e^{-\mu_{SP} \rho_e (SP)} e^{-\mu_{PD} \rho_e (PD)}$$ (8.2b)

And, if the mass attenuation coefficients $\mu_1 = \mu_2 = \mu_{(SPD)}$,

$$I_d = I_s e^{-\mu_{(SPD)} \rho_e [(SP)+(PD)]}$$ (8.2c)

where $\mu_{(SPD)}$ is the mass attenuation coefficient of the average energy of the scattered gamma rays on the path $_{SPD}$, of length $(SP) + (PD)$, in centimeters.

Proper tool design and source strength minimizes the effect of the natural formation radiation, except when the formation is anomalously high in natural radiation. Usual practice is to use a high source strength and a low-sensitivity detector. This will tend to "swamp out" the natural gamma rays. The amount of radiation scattered from the borehole direction can be quite large. And, it dilutes the desirable density signal. Proper back-shielding and collimation will reduce these unwanted components to a negligible value. Thus, with a well designed, large diameter tool, borehole effects are small. With proper design, they can be small, also, with a small diameter, mineral type sonde. Much more care must be taken with the smaller diameter mineral tools, however.

There has been some recent work by Australian investigators in the use of dual spacings in a non-sidewall tool to allow compensation for borehole effects. The author has never used these devices, but has only read of them. This design uses two detectors at different spacings (similar to the principle of the mudcake compensated density system) to compensate for the presence of the borehole effects and the natural gamma radiation. This allows very small source strengths to be used. The U.S. Department of Energy successfully investigated the use of two different spacings, in a density tool, to compensate for natural formation radiation in uranium beds.

The final result is a signal which is an inverse function of the average electron density of the material within the sensitive volume of the detector/source assembly. The relationship is,

$$\rho_e = c + d \log_{10} a \qquad (8.4)$$

where

    c = tool design constant
    d = tool design constant
    a = the signal at the detector

The density measurement, with commercial tools, can be very precise. The author has made measurements of density (in a good limestone test model with smooth walls), with a commercial system, which were repeated to ±0.05% and were accurate within ±0.1%.

The geometry of the gamma photon transport approximates an ideal situation when the source and detector are at the interface between an absorbing sink (the tool body and the borehole) and the formation. The detector response, J (the gamma flux at the detector), (Tittman, 1986) then is

$$J = \frac{M}{4\pi r^3}\left(1 + \frac{r}{L}\right)e^{-\frac{r}{L}} \qquad (8.5)$$

where
- L = the diffusion length
- r = the source to detector spacing
- M = the dipole moment of the source

The number of electrons, n_e, in a sample of an element, is

$$n_e = N_0\left(\frac{Z}{A}\right)\rho_b \qquad (8.6)$$

where
- N_0 = 6.03×10^{23} per mole, Avogadro's number
- Z = the atomic number of the element
- A = the atomic mass of the element
- ρ_b = the bulk density of the formation material, in g/cm^3

There are three methods commonly in use to correct for or take into account the difference between the electron density and the bulk density:

1. One can do all of the reduction calculations in electron densities. If that is done, the density answer is in apparent density (electron) units,
2. One can convert all densities to bulk densities by calculating and correcting all values for the individual Z/A ratios,
3. The density logs of the major oilfield contractors assume an average Z/A ratio equal to 0.5 (a good assumption in most sediments) and use an average correction which is valid in water- and/or oil-filled sediments:

$$\rho_a = 1.0704\rho_e - 0.1883 \qquad (8.7)$$

where

$$\rho_e = \rho_b 2\frac{Z}{A} \qquad (8.7a)$$

or

$$\rho_e = \rho_b\left(\frac{\Sigma Z}{molecular\,mass}\right) \qquad (8.7b)$$

The relationship, Equation 8.7 is correct or nearly correct as long as the logging is done in porous sediments which are filled with water and/or hydrocarbons (with a Z/A ratio of 0.5 to 0.7) and in matrices whose average

Z/A ratio is close to 0.5. The assumption of a matrix Z/A ratio of 0.5 is good as long as the sediments contain only elements 2 through 20, in quantity.

If the logging is done in any other environment (i.e., hard rock or any other zone which lacks hydrogenous material (i.e., water and/or hydrocarbon) and/or where large quantities of elements heavier than $Z = 20$ are present) and/or where $Z/A \neq 0.5$, the relationships will not give correct values. In that case, the full set of corrections, must be made; that is, the Z/A correction and the mudcake and borehole corrections.

The ratio of Z/A, in most sedimentary matrix materials, closely approximates 0.5. The major exception is in hydrogenous materials. Hydrogen has a Z/A ratio of 1.0. Thus, water, with two hydrogen atoms and one oxygen atom, has an electron density of 1.111 g/cm^3 at one atmosphere pressure and 4°C, instead of the 1.000 g/cm^3 bulk density of pure water that is listed in the tables. The use of Equation 8.7 effectively takes care of the effect of the water in the pore spaces. The assumption made in Equation 8.7 is very nearly correct in most sediments. The major oilfield contractors now use this correction. Older logs, however, may not have had this correction applied. When using a contractor's density log, however, it is wise to find out which system is being used.

The Z/A ratio of any material is the sum of the partial ratios of the component molecules. The molecular weight, A_m, of a molecule consisting of n elements, is,

$$A_m = A_1 N_1 + A_2 N_2 + \ldots + A_n N_n \tag{8.8}$$

where N is the number of atoms per molecule. For example, calcium carbonate, $CaCO_3$, has a molecular weight of

$$A_{CaCO3} = 40.081(1) + 12.011(1) + 16.000(3) = 100.091 \tag{8.8a}$$

The mole fraction, M_n, of the nth element is,

$$M_n = \frac{N_n A_n}{A_m} \tag{8.9}$$

The mole fraction, for example, of oxygen in $CaCO_3$, is

$$M_O = \frac{3(16.000)}{100.091} = 0.480 \tag{8.9a}$$

The Z/A ratio for any molecule is,

$$\left(\frac{Z}{A}\right)_m = \sum_{j=1}^{n} M_j \left(\frac{Z}{A}\right)_j \tag{8.10}$$

Similarly, the ratio, Z/A, for any mixture of materials, is,

$$\frac{Z}{A} = W_1 \frac{Z_1}{A_1} + W_2 \frac{Z_2}{A_2} + \ldots + W_n \frac{Z_n}{A_n} \qquad (8.11)$$

where W is the relative mass of each elemental component of the molecule. Refer to Table 8.2, for bulk densities, apparent densities, and Z/A ratios for many materials.

8.4.1 Example — The Mole Fraction Method:

The Z/A ratio of calcium carbonate, $CaCO_3$, is,

1. The mole fraction of calcium in $CaCO_3$ is

$$M_{Ca} = \frac{1(40.08)}{100.091} = 0.400$$

2. The mole fraction of carbon in $CaCO_3$ is,

$$M_C = \frac{1(12.011)}{100.091} = 0.120$$

3. Thus, the Z/A ratio of $CaCO_3$ is,

$$\left(\frac{Z}{A}\right)_{CACO3} = 0.400\left(\frac{20}{40.08}\right) + 0.120\left(\frac{6}{12.011}\right) + 0.480\left(\frac{8}{15.9994}\right) = 0.500$$

8.4.2 Example — The Molecular Mass Method:

The Z/A ratio of water, H_2O, is (from Equation 8.2),

1. The molecular weight, M_{H2O}, of water, H_2O, is

$$I_p = I_s e^{-\mu_{SP} \rho_e (SP)}$$

2. The atomic number of hydrogen is 1. That of oxygen is 8. Therefore, the sum of the atomic numbers for water is:

$$2(1) + 8 = 10$$

TABLE 8.2

Rock and mineral densities and Z/A ratios

Material	Z/A Ratio	Matrix Density	Apparent Density*
Lead, Pb	0.3953	11.34	8.97
Uraninite, UO_2	0.4000	8.25(6.5–10.8)	6.60
Cinnabar, HgS	0.4143	8.1 (8.0–8.2)	6.71
Iron, Fe	0.4687	7.87	7.38
Galena, PbS	0.4093	7.5 (7.4–7.6)	6.14
Wulfenite, $PbMoO_4$	0.4187	6.9 (6.7–7.0)	5.78
Arsenopyrite, FeAsS	0.4605	6.1 (5.9–6.3)	5.62
Cobaltite, CoAsS	0.4517	6.1 (6.0–6.3)	5.51
Chalcosite, Cu_2S	0.4610	5.65(5.5–5.8)	5.21
Hemitite, Fe_2O_3	0.4787	5.26(4.9–5.3)	5.04
Magnetite, Fe_3O_4	0.4774	5.18(4.97–5.18)	4.95
Bornite, Cu_5FeS_4	0.4643	5.15(4.8–5.4)	4.78
Pyrite, FeS_2	0.4850	5.06(4.95–5.17)	4.91
Illmanite, $FeTiO_3$	0.4757	4.75(4.5–5.0)	4.52
Zircon, $ZrSiO_4$	0.4691	4.69(4.2–4.86)	4.40
Stibnite, Sb_2S_3	0.4436	4.57(4.52–4.62)	4.05
Pyrrhotite, Fe_5S_6	0.4812	4.55(4.58–4.64)	4.40
Barite, $BaSO_4$	0.4454	4.45(4.30–4.60)	3.96
Chromite, $FeCr_2O_4$	0.4753	4.45(4.30–4.60)	4.23
Rutile, TiO_2	0.4756	4.20(4.15–4.25)	3.80
Chalcopyrite, $CuFeS_2$	0.4751	4.2 (4.1–4.3)	3.99
Corundum, Al_2O_3	0.4904	4.02(3.95–4.10)	3.94
Carnotite, $K_2O.2UO_3.V_2O_5.2H_2O$	0.4350	4+	3.48
Rhodocrosite, $MnCO_3$	0.4793	4.0 (3.5–4.0)	3.84
Sphalerite, ZnS	0.4720	4.0 (3.9–4.1)	3.78
Siderite, Fe_2CO_3	0.4797	3.88(3.0–3.88)	3.72
Limonite, $2Fe_2O_3.3H_2O$	0.4897	3.8 (3.51–4.0)	3.72
Dunite,	0.4978	3.3 (3.24–3.74)	3.29
Olivine, $(Mg,Fe)_2SiO_4$	0.4892	3.3 (3.27–3.37)	3.23
Magnesite, $MgCO_3$	0.4992	3.1 (3.0–3.2)	3.1
Norite	0.4970	2.98(2.72–3.02)	2.97
Diabase	0.4954	2.98(2.96–3.05)	2.95
Gabbro	0.4938	2.98(2.85–3.12)	2.94
Anhydrite, $CaSO_4$	0.4995	2.95(2.89–3.05)	2.95
Aragonite, $CaCO_3$	0.4995	2.94(2.85–2.84)	2.94
Muscovite, $KAl_2(AlSi_3)O_{10}(OH)_2$	0.4966	2.93(2.76–3.1)	2.91
Biotite, $H_2K(Mg,Fe)_3Al(SiO_4)_3$	0.4900	2.90(2.65–3.1)	2.84
Dolomite, $CaMg(CO_3)_2$	0.4994	2.85(2.80–2.99)	2.85
Illite, $KAl_5Si_7O_{20}(OH)_4$	0.4954	2.84(2.60–3.0)	2.81
Diorite	0.4964	2.84(2.72–2.96)	2.82
Langbeinite, $K_2Mg_2(SO_4)_3$	0.4961	2.83	2.61
Polyhalite, $2CaSO_4.MgSO_4.K_2SO_4.2HO$	0.5013	2.78	2.79
Synite	0.4971	2.76(2.63–2.90	2.74
Granodiorite	0.4963	2.72(2.67–2.78)	2.696
Chlorite, $(Mg,Al,Fe)_{12}(Si,Al)_8O_{20}(OH)_{16}$	0.5056	2.71(2.60–3.22)	2.74
Calcite, $CaCO_3$	0.4996	2.71(2.71–2.72)	2.71
Aluminum, Al	0.4818	2.70	2.60

TABLE 8.2 (continued)

Rock and mineral densities and Z/A ratios

Material	Z/A Ratio	Matrix Density	Apparent Density*
Plagioclase Feldspar, $xNaAlSi_2O_8, yCaAl_2Si_2O_8$	0.4925	2.69(2.62–2.76)	2.65
Limestone	0.5000	2.69(2.66–2.74)	2.69
Granite	0.4969	2.67(2.52–2.81)	2.65
Quartz, SiO_2	0.4993	2.65(2.65–2.66)	2.65
Sandstone	0.4990	2.65(2.59–2.84)	2.655
Kaolinite, $(OH)_8Al_4Si_4O_{10}$	0.5103	2.63(2.40–2.68)	2.68
Albite, $NaAlSi_3O_8$	0.4885	2.62(2.61–2.65)	2.56
Orthoclase Feldspar, $KAlSi_3O_8$	0.4958	2.57(2.55–2.63)	2.55
Kieserite, $MgSO_4.H_2O$	0.4724	2.57	2.43
Concrete**		2.35(1.98–2.35)	
Montmorillonite, $(OH)_4Si_8Al_4O_{20}.nH_2O$	0.5009	2.35(2.00–3.00)	2.35
Gypsum, $CaSO_4.2H_2O$	0.5111	2.32(2.30–2.35)	2.37
Glauconite, $KMg(FeAl)(SiO_3)_6.3H_2O$	0.4998	2.30(2.20–2.80)	2.30
Graphite, C	0.4995	2.22(2.09–2.23)	2.22
Serpentine, $Mg_3Si_2)_5(OH)_4$	0.5062	2.20	2.23
Halite, NaCl	0.4799	2.16(2.13–2.16)	2.07
Nahcolite, $NaHCO_3$	0.4905	2.20	2.16
Kainite, $MgSO_4.KCl.3H_2O$	0.5140	2.13(2.1–2.13)	2.19
Trona, $Na_2CO_3HNaCO_3.2H_2O$	0.5043	2.12(2.11–2.15)	2.14
Sulfur, orthorhombic, S	0.4990	2.07(2.05–2.09)	2.07
Potash, $KCO_3.2H_2O$	0.5049	2.04	2.06
Sylvite, KCl	0.4829	1.99(1.97–1.99)	1.92
Cement**		1.99	
Sulfur, monoclinic, S	0.4990	1.96	1.96
Kernite, $Na_2B_4O_7.4H_2O$	0.5026	1.91	1.92
Magnesium, Mg	0.4975	1.74	1.76
Carnalite, $KMgCl_3.6H_2O$	0.5095	1.61(1.60–1.61)	1.64
Coal, anthracite**	0.5134	1.60(1.32–1.80)	1.64
Coal, bituminous**	0.5201	1.35(1.15–1.7)	1.40
Coal, lignite**		1.10(1.1–1.5)	1.16
Water, (3.0×10^5 ppm NaCl)	0.5325	1.219	1.298
Water, (2.5×10^5 ppm NaCl)	0.5363	1.1825	1.268
Water, (2.0×10^5 ppm NaCl)	0.5401	1.146	1.238
Water, (1.5×10^5 ppm NaCl)	0.5438	1.109	1.206
Water, (1.0×10^5 ppm NaCl)	0.5476	1.073	1.175
Water, (5.0×10^4 ppm NaCl)	0.5513	1.0365	1.143
Water, (3.0×10^4 ppm NaCl)	0.5528	1.022	1.130
Water, pure, STP	0.5551	1.00	1.11
Hydrocarbon oil, nCH_2, 10°API	0.5703	1.00	1.14
Hydrocarbon oil, nCH_2, 30°API		0.88	1.00
Hydrocarbon oil, nCH_2, 40°API		0.85	0.97
Hydrocarbon oil, nCH_2, 50°API		0.78	0.85
N-Octane, C_8H_{18}, 70°API	0.5778	0.70	0.81
N-Pentane, C_5H_{12}, STP	0.5823	0.626	0.733
N-Pentane, C_5H_{12}, 200°F, 7k psi		0.603	0.702
N-Hexane, C_6H_{14}, STP	0.5803	0.659	0.765
N-Hexane, C_6H_{14}, 200°F, 7k psi		0.628	0.739

TABLE 8.2 (continued)

Rock and mineral densities and Z/A ratios

Material	Z/A Ratio	Matrix Density	Apparent Density*
N-Heptane, C_7H_{16}, STP	0.5778	0.684	0.790
N-Heptane, C_7H_{16}, 200°F, 7k psi		0.657	0.759
N-Octane, C_8H_{18}, STP	0.5778	0.703	0.812
N-Octane, C_8H_{18}, 200°F, 7k psi		0.673	0.778
N-Nonane, C_9H_{20}, STP	0.5768	0.718	0.828
N-Nonane, C_9H_{20}, 200°F, 7k psi		0.686	0.791
N-Decane, $C_{10}H_{22}$, STP	0.5763	0.730	0.841
N-Decane, $C_{10}H_{22}$, 200°F, 7k psi		0.701	0.808
N-Undecane, $C_{11}H_{24}$, STP	0.5759	0.740	0.852
N-Undecane, $C_{11}H_{24}$, 200°F, 7k psi		0.713	0.821
Methane, CH_4, STP	0.5703	6.77×10^{-4}	7.6×10^{-4}
Methane, CH_4, 200°F, 7k psi		0.2189	0.2497
Ethane, C_2H_6, STP	0.5986	1.269×10^{-3}	1.5×10^{-4}
Ethane, C_2H_6, 200°F, 7k psi		0.4104	0.4913
Propane, C_3H_8, STP	0.5896	1.86×10^{-3}	2.2×10^{-3}
N-Butane, C_4H_{10}, STP	0.5850	2.46×10^{-3}	2.9×10^{-3}
Helium, He, STP	0.4997	1.7×10^{-4}	1.7×10^{-4}
Carbon Dioxide, CO_2	0.4999	1.858×10^{-3}	1.857×10^{-3}
Nitrogen, N_2, STP	0.4998	1.182×10^{-3}	1.185×10^{-3}
Oxygen, O_2, STP	0.5000	1.350×10^{-3}	1.350×10^{-3}
Hydrogen Sulfide, H_2S, STP	0.5281	1.438×10^{-3}	1.1519×10^{-3}
Air (dry), STP	0.4997	1.224×10^{-3}	1.223×10^{-3}
Natural Gas, average, STP**	0.5735	7.726×10^{-4}	8.86×10^{-4}
Natural Gas, 200°F, 2k psi**		0.252	0.289

* Based on an assumed Z/A ratio = 0.5
** For particular, typical samples

$$Z_{H2O} = N_H Z_H + N_O Z_O = 2(1) + (1)8 = 10$$

3. Therefore, the Z/A ratio of water, $(Z/A)_{H2O}$, is

$$\left(\frac{Z}{A}\right)_{H_2O} = \frac{2(A_H) + A_O}{M_{H_2O}} = \frac{10}{18.016} = 0.555$$

4. The density of water, ρ_{H2O}, then is,

$$\rho_{H_2O} = \frac{\rho_e}{2\frac{Z}{A}} = \frac{2(1.11 gl\ cm^3)}{2(0.555)} = 1.000\ gl\ cm^3$$

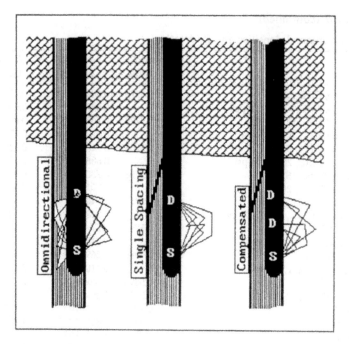

FIGURE 8.5
Three general types of density sondes.

8.5 Downhole Tool Types

There are several forms of downhole density or scattered gamma ray tools and systems which are in use today. Each system is designed for a particular field of investigation. Also, several different types of downhole tools might be interchangeable on a system, depending upon the use to which the system is to be put. See Figure 8.5.

8.5.1 Omnidirectional Density Systems

Omnidirectional, single spacing probes (see Figure 8.6) have been used in the mineral industry, especially for coal. Because of the adverse borehole effects, the measurements from this type of tool could not be quantified and are no longer used. Investigators in Australia have apparently successfully been experimenting with an omnidirectional borehole compensated density system.

Many of the early density systems were omnidirectional; that is, a natural gamma ray tool with the addition of a short bottom sub, containing an isotopic gamma ray source. The usual spacing was 10 in (25 cm) to 20 in

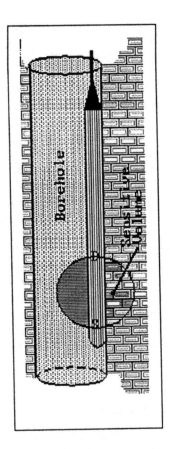

FIGURE 8.6
Schematic of the omnidirectional density measurement volume.

(50 cm). Many times the spacing was changeable in the field (a record of the spacing was seldom kept). The systems were not deliberately eccentered, but were not centered. Also, they were not shielded from the borehole. Normally a caliper log was not run. Thus, the signal contained large, unknown components from the borehole, drilling mud, mudcake, and wall rugosity. If a caliper log was made, it was a separate run. The result was that the logs were not normally useable for quantitative assessments.

The most frequent use of the omnidirectional systems was in the mineral industry for learning the location, identification, and thickness of coal beds. The coals are evident on the combination logs (natural gamma ray, omnidirectional density, and single-point resistance) by their high resistivity (5000 Ωm to >10,000 Ωm for good quality bituminous coals), low density (1.1 to 1.3 g/cc for bituminous coals), and low gamma radiation (usually <5 API units). This combination gives a distinctive signature in lignite and bituminous coals. Also, the bed boundaries and shale contamination are sharply noticeable. The signature is more difficult to identify in the case of the higher grades of coals; the anthracite or "hard" coal. These coals can have high densities (sometimes as high as 1.8 g/cm^3), and low resistivities. The resistivity is low because conductance begins to take place through the carbon in the harder coals. The resistivities can be as low as 10^{-3} ohms.

The omnidirectional systems are seldom used now. They could, however, find application in fields in low-budget determinations where the parameters are well known and where the only parameter needed might be the stratification, depth, and thickness of a thick coal. This, of course, is the purpose of the "coal tool."

The *coal tool* is a modification of the omnidirectional probe. It was designed to accurately determine coal bed boundaries and partings with a safe, fast, and economical system. It does this superbly. The downhole tool is usually a small diameter (0.75 in to 1.5 in diameter; 1.9 to 3.8 cm diameter) probe. The detector is a simple scintillation gamma ray detector. The source-to-detector spacing is short (0.5 in to 1 in; 1.3 to 2.5 cm). The source is often $_{93}Am^{242}$, which has an easy-to-shield, low energy (49 keV) emission, although $_{55}Cs^{137}$ (660 keV) is being used by one contractor. The log is responsive to coals, because of their usual low density (0.9 to 1.3 g/cm^3). Boundaries and partings may be located to within ±1 in (±2.5 cm). The absolute accuracy of the location depends upon the accuracy of the depth measuring system. The relative accuracy is excellent. Irregular and rugose holes can degrade the accuracy.

8.5.2 Single-Spacing Sidewall Systems

Most of the practical density tools, at this time, are sidewall tools. See Figures 8.2 and 8.4. That is, the investigation is directed into the formation material and the tool is forcibly eccentered in the borehole. The backup arm (eccentering device) is frequently used to operate a borehole caliper device.

Early density systems used only a single detector with a single source. Thus, they were quite sensitive to the thickness and density of the mudcake. Corrections were tedious but necessary. These systems are still popular in non-sedimentary application because of the usual lack of mudcake and because of their lower cost and complexity. Many of the early systems did not have adequate shielding of the source and detector to isolate them from the borehole. This was especially true of the small diameter systems, where diameter was severely limited. The reason, besides small diameter, was that lead was used as a shielding material (it is relatively dense: 11.34 g/cm^3, inexpensive, easily obtained, and easily handled). It does not, however, provide adequate shielding from borehole scattering. The result was that these systems required large environmental (borehole) corrections. In addition, many of the early systems used $_{27}Co^{60}$ for a gamma ray source. The gamma energy of this source is 1.17 MeV. Both instrumental and personnel shielding are more difficult with this high energy.

Later systems use tungsten as a shielding material. Tungsten has a density of 19.3 g/cm³. Therefore, it is almost 3 times more effective as a gamma ray shield. Shielding was also interspersed between the crystal and the photomultiplier of the detector. In addition however, the $_{55}Cs^{137}$ source, with an output of 0.662 MeV photons was adopted. The single scattered gamma rays are near 300 keV, with this source. The net result is that the system is shielded from the borehole about 9 times better than with some of the small diameter, early, inexpensive tools. This lower energy source also increases the density response contrast. The shielding problem will be discussed later in this chapter.

Some of the early systems used radium-mixture sources. These sources emit a very wide range of low energy gamma. The maximum energy of any significant emission is 190 keV. The results are uncertain for quantitative work because of the mixture. Photoelectric absorption is very high when radium is used, because of the predominance of low energy emissions. When using old logs and old mineral logs especially, be careful of this possibility.

A single-spacing density system has the advantage of relative simplicity. And, for many applications, such as measuring the density of a competent coal or in hard rock formations (because of the lack of mudcake in both situations), the single spacing system is quite adequate. It should be considered wherever density logging is to be done in impermeable materials. A well designed, well shielded tool is imperative. Contrary to popular belief, the compensated systems *do not* handle rugose borehole wall and caves any better than the single spacing instruments.

If logging is to be done in permeable sediments, the presence of the usual mudcake can cause serious errors in the density readings. This effect, caused by the "standoff" of the tool from the wall of the hole, is shown in Figure 8.7. The mudcake occurs in the portion of the measurement volume which is the most sensitive. It is usually much lower density than the formation material.

It is vital that a simultaneous caliper be run in this environment, to allow for more accurate estimations of the mudcake thickness, borehole rugosity, and general curve quality. Usually, the caliper mechanism is built into modern tools and is measured simultaneously with the density. Note however, this particular caliper is not suitable for measurement of borehole diameters. It was designed to determine the probable mudcake thickness. It does this quite well. Except for these factors, the information of the single spacing systems should be handled in the same manner as that for the compensated systems.

8.5.3 Mudcake Compensated Density Systems

Two gamma ray detectors at different spacings usually are used to compensate for the effect of the mudcake on density measurements. The principle is

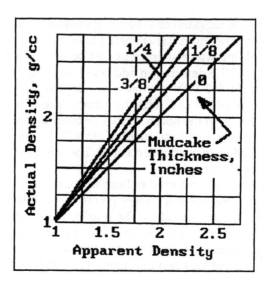

FIGURE 8.7
The effect of standoff (mudcake) upon a single spacing density array.

shown in Figure 8.8. Figure 8.9 shows the positioning of the system in the borehole. One spacing from the source to the detector is long (16 in to 24 in; 40 cm to 60 cm); about the same as the spacing of the single-detector petroleum tool. This is the deeper measurement (8 in to 12 in or 20 to 30 cm). The other measurement uses a shorter spacing (6 in to 10 in or 15 cm to 25 cm); about the same as the single spacing used in mineral work. The depth of measurement is about 3 to 5 in or 8 cm to 13 cm. The volume of the mudcake is proportionally greater in the short spacing measurement. Thus, the effect of the mud cake upon the measurement may be separated form that of the formation. This is done during the recording of the log, whether the system is analog or digital. The correction curve indicates the amount of correction that was applied to the log during recording. This, and the caliper appearance indicate the quality of the density values.

In the case of the mudcake compensated density system, the principle is that the mudcake forms a proportionally larger part of the signal at the short spacing detector than at the long spacing detector. The formation density affecting each detector is the same. Therefore, if the short spacing response is multiplied by a factor to make the mudcake component of the signals equal and one is subtracted from the other, the remainder is a function only of the formation density.

Referring to Figure 8.8, the signal at detector 2 (the long spacing detector), D_{long}, is

$$D_{long} = V_{mc,long}\rho_{mc} + V_{form,long}\rho_{form} \qquad (8.12)$$

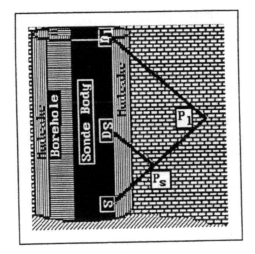

FIGURE 8.8
Approximate average photon paths for a dual detector system.

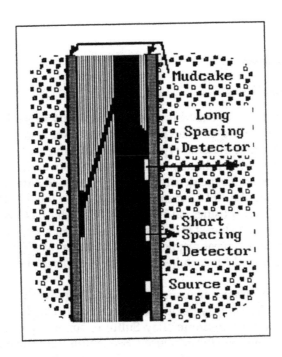

FIGURE 8.9
A compensated density array schematic.

At the short-spacing detector, D_{short},

$$D_{short} = V_{mc,\,short}\,\rho_{mc} + V_{form,\,short}\rho_{form} \qquad (8.13)$$

where

$V_{mc,\,long}$ = the volume amount of mudcake included in the sensitive volume of the long spacing detector

$V_{mc,\,short}$ = the volume amount of mudcake included in the sensitive volume of the short spacing detector

V_{form} = the volume amount of formation within the sensitive volume of the indicated detector

ρ = the bulk density of the indicated material

If

$$V_{mc,\,long} - FV_{mc,\,short} = 0 \qquad (8.14)$$

Then

$$D_{long} - FD_{short} = \rho_{form}\left(V_{form,\,long} - FV_{form,\,short}\right) = f\left(\rho_{form}\right) \qquad (8.15)$$

8.6 Calibrations

In order to obtain these measurements, the density system, whatever its type, must be calibrated. The primary calibration is done in a zone of known density and material type (Z/A ratio). Commonly, the same limestone models used to calibrate the neutron porosity and the acoustic systems are also used for the density systems. See the illustration in the next chapter showing the API Calibration Model. These models have precisely known densities, porosities, and compositions. After the density response has been measured at least three different densities, a response curve may be determined. A typical response curve is shown as Figure 8.10. Such a curve is determined for each instrument of each model before it is sent to the field.

Secondary calibrators are usually used locally, in the shop or even in the logging truck. These may be one or more blocks of acrylic resin, magnesium alloy, and/or aluminum alloy. Sulfur or some other material whose Z/A ratio is known may also be used. See Figure 8.11. The purpose of the secondary calibrator is to determine if the system has drifted or not. If it has drifted a small amount (which would be normal) a normalizing factor can be determined for the immediate measurements.

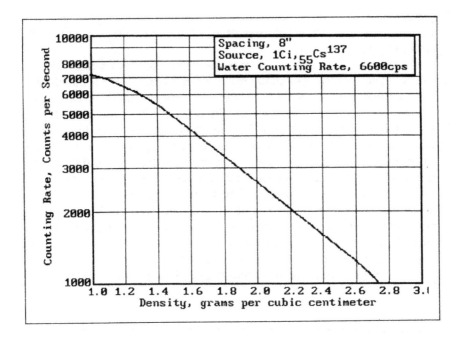

FIGURE 8.10
A typical density response curve.

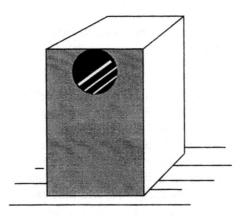

FIGURE 8.11
A typical calibrator block. These calibrator blocks are designed for secondary calibration use. This particular design is for a sidewall tool only. The blocks are often carried in the logging truck. The material may be any known, uniform solid. Typically, the blocks may be made of aluminum(2.69 g/cc), magnesium (1.8 g/cc), sulfur 2.0 g/cc), or acrylic resin (1.2 g/cc).

Modern digital equipment applies the normalizing factor to the data as it is being processed and before plotting. Older, analog logs must have the factor applied manually.

After logging, any density measurement must be corrected for environmental effects. These are usually restricted to a correction for the borehole effects; the thickness and density of the mudcake (if not taken care of by a compensation system), the borehole diameter and the type of fluid filling the borehole. With a well-designed and maintained system, these corrections will usually be small. The corrections are empirical and will depend upon the tool design, the source energy, the hole diameter, and the scattering properties of the borehole fluid. Copies of two of Schlumberger's charts are shown as Figure 8.12. These corrections must be made to the raw measured values before any other processing is done. Modern digital systems frequently apply these corrections as the log is being plotted. Be sure to check with the contractor to determine what has been done and what has not been done. Notice the larger amount of borehole correction needed for the smaller diameter tool (Figure 8.12b) in liquid-filled holes.

8.7   Interference by Natural Gamma Radiation

Since the detectors in a density system operate in the same energy range as natural formation radiation, they are sensitive to this natural radiation. This effect is negligible under normal logging circumstances, if it is minimized by good tool design. If, however, logging is to be done through known beds of high radiation, such as uranium beds, that portion of the log must be discarded.

Common practice involves several features:

1. The detector is shielded to a maximum extent by very high density material, such as tungsten. Often, a lead glass shield is placed between the detector and photomultiplier, to eliminate any radiation coming through the photomultiplier. The only exposure of the detector and source is toward the volume of the probable greatest amount of scattering from the formation material, of the source gamma photons. The borehole is excluded as much as possible.

2. The detectors are made relatively insensitive to minimize their response to the natural radiation. This is done by limiting the crystal size and photomultiplier gain. Also, where necessary, X-ray shielding is incorporated to eliminate response to the soft radiation caused by higher energy gamma photons.

3. The source strength is made a high as is considered safe, in order to compensate for the low detector sensitivity.

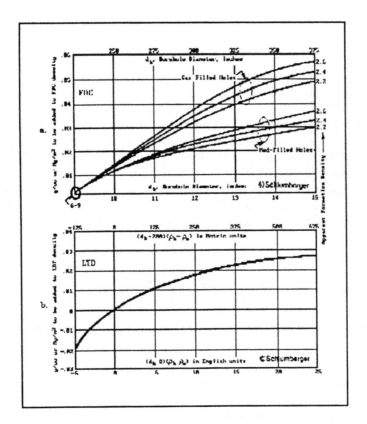

FIGURE 8.12
Density correction charts. Under some conditions Formation Density Log, and the Litho-Density Log need to be corrected for the borehole size effect. These charts provide those corrections. For the FDC log, enter the chart with the borehole diameter, go to the *Apparent Formation Density* (the FDC log reading); read, in the ordinate, the amount of density to be added to the log reading. For the LDT log, enter the chart abscissa with the product of the borehole diameter (less 10 in) and the density reading (less the mud density). Read, in the ordinate, the correction to be added to the density reading. (after Schlumberger, 1988).

Natural gamma interference becomes very important if the system is to be used in zones of anomalous radiation, such as uranium beds. The U.S. Department of Energy (personal communication, 1970) performed experiments in uranium beds with a single detector, variable source-to-detector spacing, small diameter, sidewall density system. They made multiple runs at different spacings and then performed manual compensations to eliminate the constant natural radiation. The results were quite encouraging. But, as far as this author is aware, it was never developed commercially. An Australian omnidirectional system claims to be insensitive to natural gamma rays. Therefore, a low strength, relatively safe source can be used.

8.8   Rock Type Identification

The matrix rock types may be tentatively identified by their characteristic densities. Limestone density, $\rho_{LS,ave} = 2.690$ g/cm^3 (range 2.660 to 2.740 g/cm^3 and $Z/A = 0.500$); sandstone density, $\rho_s = 2.655$ g/cm^3., etc. These are the matrix densities: ρ_{ma}. Many of the geological material densities are listed in Table 8.2. These values can also be found in the literature published by the major geophysical logging contractors, the CRC Handbook, and various geological and geophysical texts.

Frequently, the rock matrix is a mixture. For example, a carbonate often consists of a mixture of limestone and dolomite. The rock matrix density is the sum of the n fractional volume–density products of the matrix components:

$$\rho_{ma} = \sum_{j=1}^{N} V_j \rho_j \qquad (8.16)$$

where V_j is the relative or fractional volume of each component.

8.9   Porosity Calculations

Common practice when logging formation density is to assume a limestone matrix (even if another type matrix is present) and calculate and plot apparent limestone porosity, $\phi_{\rho,LS,a}$. The apparent limestone porosity scale on a log is a scaling and comparison device, unless the log is actually in a pure matrix of the type assumed. If the presumed limestone matrix is dolomite or partially dolomitized, the porosity will be too low or even negative. If this is evident, then the possibility of dolomitization must be considered.

The apparent density porosity, $\phi_{\rho,a}$, is often used with the apparent neutron porosity, $\phi_{N,a}$, to indicate and locate shale. Because the shale densities are very near those of quartz sands, there is little difference of the shale density from the sand value. Therefore, ϕ_ρ is very near the correct **effective** porosity, ϕ_{eff}. On the other hand, ϕ_N is nearly **absolute** porosity, ϕ_{abs}. Thus, it will read an apparent high porosity because of the bound water content of the shale. The presence of clay or shale may be detected and evaluated by comparing $\phi_{\rho,a}$ with $\phi_{N,a}$, using the correct matrix density for the non-shale portion of that zone (i.e., sandstone or limestone). Then, ϕ_ρ and ϕ_N will have the same values in a clean sand but ϕ_ρ will be lower than ϕ_N in a shaly sand or a shale.

The volume of shale, V_{sh}, may be determined approximately:

$$V_{sh} \approx \frac{\phi_N - \phi_p}{\phi_p} \qquad (8.17)$$

If the pore space is filled with water and/or oil and if the matrix is correctly identified, when ϕ_p and ϕ_N are calculated they will be identical and correct. If, however, the pore space contains gas, then the bulk density will be low and the resulting ϕ_p calculation will be too high because of the lower average density of the pore fluid. The gas, however, contains fewer hydrogen atoms per unit volume (compared to water or liquid oil). Thus, ϕ_N will be too low. This is a *gas log*.

The approximate porosity in the sand or limestone, in this case, will be

$$\phi \approx \frac{\phi_{p,a} + \phi_{N,a}}{2} \qquad (8.17a)$$

and, in the gas-bearing zone,

$$\phi = \frac{\rho_{ma} - \rho_{b,a} + \left(\rho_L - \rho_g\right)\left(\dfrac{R_w}{R_t}\right)^{\frac{1}{2}}}{\rho_{ma} - \rho_g} \qquad (8.17b)$$

where the subscript, L denotes the liquid phase and g the gas phase. The liquid saturation will be

$$S_L = \frac{\rho_{ma} - \rho_g}{\rho_L - \rho_g \dfrac{\rho_{ma} - \rho_{b,a}}{\left(\dfrac{R_w}{R_t}\right)^{\frac{1}{2}}}} \qquad (8.18)$$

The density system averages *all* densities of the materials within the sensitive volume of the device. Therefore, it determines primary porosity, ϕ_{pri} (i.e., intergranular, etc.), plus secondary porosity, ϕ_{sec} (i.e., fractured, etc.).

Salinity of water may be approximated with the density values:

$$\rho_w \approx 1 + 0.73\left(ppm \times 10^6\right) \qquad (8.19)$$

where ppm is the parts per million of total dissolved solids in the water.

Figure 8.13 is a portion of a petroleum-type compensated density log.

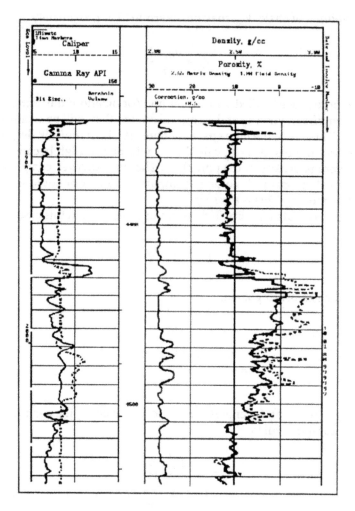

FIGURE 8.13
A portion of a petroleum-type density log.

8.10 Coal Analysis

Density logs have been plotted against the ash content of the coal, as determined from the proximate analysis to determine the probable ash content of unknown samples. Figure 8.14 shows an example of this.

This type of cross-plot is a very useful technique for many parameters other than the ash content. One must be careful, however, that the uncertainty of the determination is not too great. Note, on Figure 8.14, at a density of 1.5 g/cc, the trend would indicate an ash content of about 22%, in a

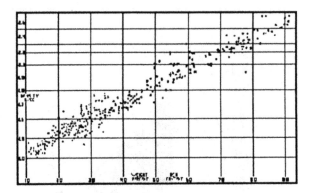

FIGURE 8.14
Coal ash content, as a function of density. Data from BPB Instruments. This is an example of a technique which can easily be used in a known area. Data were extracted from logs in one field of South African coals. One must take care, however, because this plot only compares ash content with density. The density is also affected by the amounts of several other components.

hypothetical sample. Notice, however, that the spread of actual measurements indicate that 1.5 g/cc could represent anything from about 15% to over 34%. Thus, the actual determination value would be 22% ±9.5%, for a spread or possible error of ±43%. The reason is clearly shown in Equation 8.20. The amounts of coal, moisture, and volatile can also vary. This uncertainty is intolerable for any quantitative purposes. It could be useful, however, for predicting possible trends. The coal properties can also be determined, in more detail, by cross plotting techniques. These methods will be further covered in a later chapter discussing interpretation and evaluation methods.

A more reasonable relationship than merely ρ_b vs. ash content, is

$$\rho_b = V_c \rho_c + V_v \rho_v + V_{sh} \rho_{sh} + V_w \rho_w \qquad (8.20)$$

where the subscripts, c, v, sh, and w, in this case, represent coal (carbon), volatile (hydrocarbons), ash (shale), and moisture (water solutions), respectively. This takes into account most of the factors commonly affecting the physical aspect of coal.

8.11   Scattered Gamma Ray Spectroscopy

The scattered gamma ray spectrographic system is an expanded and improved new generation of the compensated formation density systems. (The Schlumberger Well Services trade name for this system is the

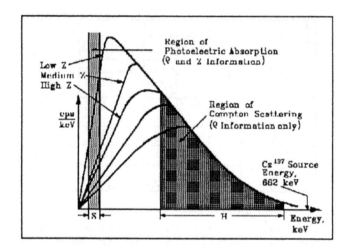

FIGURE 8.15
Hypothetical scattered gamma spectrum (after Schlumberger, 1987).

Litho-Density Log.) The scattered gamma ray spectrum from the detector is electronically divided into two windows. One is centered in the Compton scattered region (region H in Figure 8.15) and is used to determine density, ρ_b. The other is centered at the low end of the spectrum (region S in Figure 8.15), where the gamma rays are most affected by photoelectric absorption. The response in this latter region is used to calculate a photoelectric absorption index, Pe.

The Schlumberger version of the spectroscopic scattered gamma ray system, the Litho-Density, uses a shorter spacing and a stronger source than their standard density system, the FDC. Therefore, counting rates are higher, the probable error is smaller and repeatability is better for the density portion of the Litho-Density. Also, the elimination of the photoelectric absorption effects from the density measurement improves the accuracy of the density determination.

At a distance from the source (i.e., at the detector), the spectrum might be as shown in Figure 8.15. The number of gamma rays in the Compton scattering region (H) is related inversely to the logarithm of the electron density of the formation material, only. This signal is excellent for determining the bulk density value of the formation material.

The number of gamma rays in the lower energy region (S), the region of photoelectric absorption, is inversely related to the photoelectric absorption. The electron density also has an influence in this region but its effect is small. A significant part of the photoelectric absorption is due to the tool housing. This, however, is constant and can be accounted for. The signals in this region can be used to determine the photoelectric index, *Pe*.

Table 8.3 lists the photoelectric absorption cross sections, in barns, for several elements at the incident gamma ray energy level. The atomic number, Z,

TABLE 8.3

Photoelectric cross sections of a few elements

Element	Photoelectric Cross Section barns	Atomic Number Z
Hydrogen	0.00025	1
Carbon	0.15898	6
Oxygen	0.44784	8
Sodium	1.4093	11
Magnesium	1.9277	12
Aluminum	2.5715	13
Silicon	3.3579	14
Sulfur	5.4304	16
Chlorine	6.7549	17
Potassium	10.0810	19
Calcium	12.1260	20
Titanium	17.0890	22
Iron	31.1860	26
Copper	46.2000	29
Strontium	122.2400	38
Zirconium	147.0300	40
Barium	493.7200	56

After Schlumberger, 1989.

for each of these elements is also listed. The photoelectric cross section index, P_e, in barns per electron, commonly used is,

$$Pe \approx \frac{Z^{3.6}}{10} \qquad (8.21)$$

A photoelectric absorption cross section index, Pe, for a molecule made up of several atoms, may be determined from its atomic fractions:

$$Pe = \frac{\Sigma A_a Z_a P_a}{A_t Z_t} \qquad (8.22)$$

where A_a is the atomic number of each atom of each atom in the molecule. Z_a is the atomic weight and P_a is the fraction present of each atom. A_t and Z_t are the total atomic number and weight, respectively.

The value of P_e is relatively independent of the porosity of the formation material. Since P_e units are barns per electron and ρ_e units are electrons per cubic centimeter, U has units of barns per cubic centimeter. By definition,

$$U = P_e \rho_e \qquad (8.23)$$

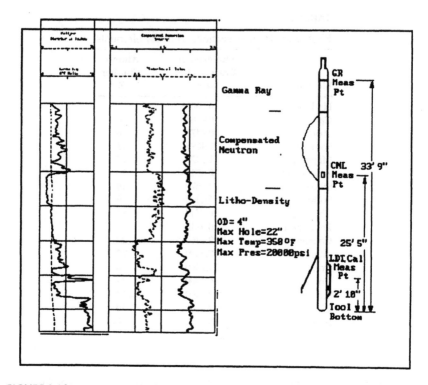

FIGURE 8.16
The Schlumberger Litho-Density tool and log (after Schlumberger).

This permits the cross sections of the volumetric components of a formation to be added in a simple volumetric manner:

$$U = \phi U_f + (1-\phi) U_{ma} \qquad (8.23a)$$

where U, U_f, and U_{ma}, are the photoelectron cross sections of the formation, the pore fluid, and the matrix material respectively. The units are barns per cubic centimeter. Table 8.2 lists the parameters for some of the formation materials.

Figure 8.16 shows the outline of the downhole Schlumberger spectrometric scattered gamma ray (Litho-Density) sonde and a sample log from Schlumberger Well Services.

Figure 8.17 shows the Matrix Identification Plot.* This is a variation of the MID Plot, which will be discussed in a later chapter. Explanations can also be found in the various manuals published by Schlumberger Well Services, Inc. This version uses the density and photoelectric index, with a

---

* Matrix Identification Plot and MID Plot are trade names and/or trademarks of Schlumberger Well Services, Inc.

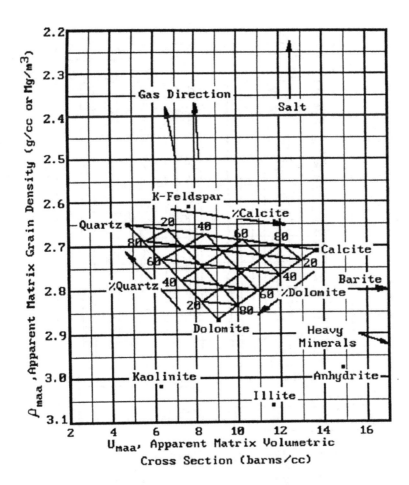

FIGURE 8.17
The Litho-Density Matrix Identification (MID) Plot (after Schlumberger, 1988).

porosity value. The result of the plot is chart positions which are functions of specific rock matrix types, almost independent of porosity effects. This MID Plot method uses the bulk density (ρ_b) and the acoustic-derived porosity (ϕ_t) to determine the apparent grain density (ρ_{maa}) with the aid of a special chart (SWC Chart CP14). This is the same parameter used in the older, conventional MID* Plot. It also used the photoelectric index (P_e), the bulk density (ρ_b), and the apparent porosity (ϕ_a) to determine the value of U_{maa}. These two parameters are then cross-plotted on a grid, such as Figure 8.14, to determine the rock type. There will be little or no error because of the formation porosity.

Figure 8.18 is a mineral-type density log. Both it and the log of Figure 8.13 were made with digital systems.

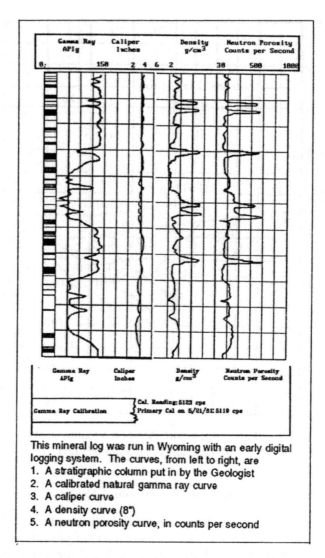

This mineral log was run in Wyoming with an early digital logging system. The curves, from left to right, are
1. A stratigraphic column put in by the Geologist
2. A calibrated natural gamma ray curve
3. A caliper curve
4. A density curve (8")
5. A neutron porosity curve, in counts per second

FIGURE 8.18
A mineral-type density log.

Since the photoelectric index (Pe) primarily uses the short spacing detector, it has a better resolution than the density measurement. Therefore, it is a good device for examining fracture systems and laminar structure.

9

Neutron Porosity Logging

9.1 Introduction

Neutron geophysical evaluation methods encompass a wide range of activities. Those we are chiefly interested in for formation evaluation purposes are mostly downhole methods. A few methods, which are not borehole methods, are important. For the most part, however, they are experimental or semi-experimental laboratory methods.

The borehole methods, employing neutrons, are common and very useful in both hydrocarbon and non-hydrocarbon logging. The most common neutron logging family is the neutron porosity group. See Table 9.1.

The neutron porosity systems usually each employ a continuous flux of neutrons, almost always from an isotopic source. A pulsed neutron porosity system is newly available. These systems are used to measure the porosity of the formation material through its hydrogen content. They are also used, especially in combination with other log types, to determine the probable rock matrix type.

9.2 Physical Description

The neutron (n) was discovered in 1930 by Bothe and Becker. It is a nuclear particle with a neutral charge. It has a mass of 1.6746×10^{-24} g or 1.008665 AMU. It may be considered to be a combination of a proton and an electron. Naturally occurring neutrons in the formation material are virtually all found in the nuclei of the formation atoms. There are almost no free neutrons normally present in the formations.

The energy of the neutron is manifested in its velocity. It is a function of the mass and the velocity of the neutron, as with any body:

$$E = \frac{1}{2}mv^2 \tag{9.1}$$

Energy of the neutron is lost through collisions with atomic nuclei of neighboring elements.

TABLE 9.1

Uses of neutron logging systems and methods

1.	Porosity determinations, ϕ_N,
2.	Hydrocarbon saturation determinations, S_o,
3.	Detection of chlorine presence and amounts,
4.	Elemental analysis,
5.	Rock type determinations when used in combination with the acoustic, t, and the density, ρ_b, measurements,
6.	Gas presence and saturation determinations when used with the density measurements, ρ_b,
7.	Porosity independent rock type determinations when used with the density, ρ, and acoustic, t, measurements,
8.	Rock type determinations when thermal neutron capture cross section, Σ, methods are used,
9.	Rock type determinations when used with the spectral density, LTD, and acoustic, t, measurements,
10.	Clay type determinations when used with the spectral gamma ray, KUT, measurements,
11.	Road and structure quality determinations,
13.	Water content for dewatering,
14.	Coal quality,
15.	Identification of shales with the thermal and epithermal neutron methods,
16.	Cased hole porosity, ϕ, and lithology.

A neutron, like other particles, has a wave length, λ, usually measured in Ångstrom units, Å, (see Chapter 8 of *Introduction to Geophysical Formation Evaluation Methods*) which is related to its energy, E, and velocity, v:

$$\lambda = \frac{h}{mv} = \frac{3.96}{v} = \frac{0.236}{E^{1/2}} \tag{9.2}$$

where h is Planck's Constant, m is the mass of the neutron, v is its velocity in kilometers per second, and E is the energy in electron volts (eV).

A neutron, $^1n_0$, will decay into a proton and an electron:

$$n_0 - {}^1p_1 + {}^0e_{-1} + n_0 + Q \tag{9.3}$$

where the particles, $^1p_1$, $^0e_{-1}$, and n_0 are a proton (hydrogen nucleus), a negative electron, and a neutrino, respectively. The value of Q is equal to the energy equivalent of the mass difference between the neutron and a hydrogen nucleus. This is 8.07×10^{-4} AMU or 0.75 MeV. Thus, the electron will have an energy of 0.75 MeV. The neutron has a halflife of about 12 minutes.

9.2.1 Operational Principles

All neutron logging methods put neutrons into the formation from an artificial source which is usually in the sonde. An absence of natural free neutrons

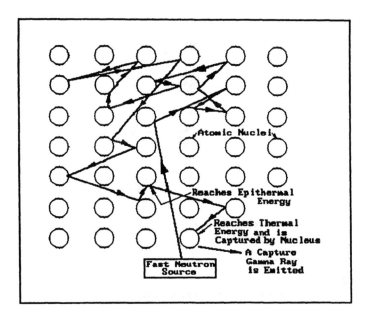

FIGURE 9.1
Neutron moderation in formation material.

is an advantage from a measurement standpoint, because it means that there is no natural background to interfere with or dilute the measurement, such as we find from gamma rays in density logging.

9.2.2 Moderation

Neutrons enter the formation, usually (but not with all tools) in 360° solid angle (omni-directionally) at high energy. Logging sources usually emit neutrons in the energy range between approximately 2 MeV and 14 MeV, depending upon the source type. They are moderated by collisions with the formation material (Figure 9.1), until they reach epithermal and/or thermal energy or are captured (with the emission of a capture gamma ray). Those that are not captured will eventually decay into a proton and an electron (Equation 9.3). Each collision between a neutron and an atomic nucleus will result in the transfer of some of the energy of the neutron to the nucleus. The most efficient collision process is head-on with a particle of the same mass (in the formation materials an atom of hydrogen has approximately the same mass as a neutron). If the collision is "head-on", (see Figure 9.2), the reduction of energy of the neutron, ΔE, will be

$$\Delta E = E\left[1 - \left(\frac{A-1}{A+1}\right)^2\right] \tag{9.4}$$

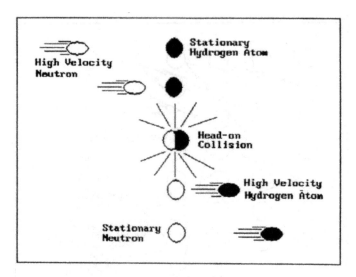

FIGURE 9.2
Direct collision between a neutron and a hydrogen atom.

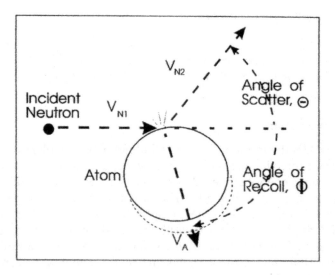

FIGURE 9.3
The angles of rebound of a neutron and an atom.

Of course, most collisions are not "head-on," and the impact will transfer a lesser amount of energy, depending upon the directness of impact. In this case, the neutron and the hydrogen atom will rebound at an angle, *I* for the neutron and an angle, ϕ for the nucleus. The total energy will be conserved. Figure 9.3 illustrates this case. The relationships are:

$$v_{n1} = v_{n2}\cos\theta + v_A\cos\phi \tag{9.5}$$

$$0 = v_{n2}\sin\theta - v_A\sin\phi \tag{9.6}$$

$$v_{n1}^2 = v_{n2}^2 + v_A^2 \tag{9.7}$$

$$\tan\theta = \cot\phi \tag{9.8a}$$

$$\therefore \theta + \phi = 90° \tag{9.8b}$$

where
θ = the angle of scatter of the neutron
ϕ = the angle of recoil of the atom
v_{n1} = the initial velocity of the incident neutron
v_{n2} = the neutron velocity after collision
v_A = the velocity of the atom after collision

On the average, 18 collisions with hydrogen (atomic mass ≈ 1.008) are needed to thermalize the high-energy neutron, compared to almost 50 collisions with the next geologically common atom, carbon (mass = 12.000 AMU). Therefore, the low energy or capture gamma population is mostly a function of the formation hydrogen content. Since almost all of the formation hydrogen is in the pore fluid, the measurement is a porosity measurement. But also, but to a lesser extent, these systems are sensitive to rock type. See Table 9.2. This is called the "moderation" (slowing down) of the neutron.

TABLE 9.2

The average number of collisions to thermalize

Element Atomic Number	Ave. Number of Collisions
1	18
2	25
4	40
9	50
12	110

9.2.3 Reactions

The high-energy neutron will moderate down through the epithermal range (<100 Ev). During this moderation process, there are several possible reactions which may occur between the neutron and the formation atomic nuclei. Refer to later in this chapter for more information on this.

The average minimum energy, $E_{n,min}$, in ergs, the neutron can have is that of the ambient thermal level:

$$E_{n,min} = \frac{1}{2}mv^2 = \frac{2}{3}kK \qquad (9.9)$$

where
　　m = the mass of the neutron
　　v = the velocity of the neutron
　　K = the absolute temperature of the surrounding material, in Kelvins
　　k = Boltzman's Constant (1.38×10^{-16} erg/K).

At 300K (80°F), these thermal neutrons would have an energy of 0.017eV ($1eV = 1.602 \times 10^{-12}$ erg).

Thermal neutrons are easily captured by nearby atomic nuclei. Table 9.3a shows the actual capture micro-cross sections (s, the apparent "target" size in barns) of the various elements, for thermal neutrons. (A barn is equal to 1×10^{-24} cm^2.) The capture process will release some of the mass as a capture gamma photon of a characteristic energy. Table 9.3b shows the macro-cross section, Σ, in barns. Σ is the cross section per cubic centimeter of material.

TABLE 9.3a

Thermal neutron micro-cross sections

Z	Element	A	Cross Section (Barns)
1.	Hydrogen, H	1.008	0.3326
2.	Helium, He	4.003	0.007
3.	Lithium, Li	6.939	70.45
4.	Beryllium, Be	9.013	0.0076
5.	Boron, B	10.811	767.4
6.	Carbon, C	12.011	0.0035
7.	Nitrogen, N	14.007	1.898
8.	Oxygen, O	15.999	0.0002
9.	Fluorine, F	18.998	0.0096
10	Neon, Ne	20.170	0.039
11	Sodium, Na	22.990	0.53
12	Magnesium, Mg	24.305	0.0635
13.	Aluminum, Al	26.981	0.231
14.	Silicon, Si	28.086	0.171
15.	Phosphorous, P	30.974	0.172
16.	Sulfur, S	32.065	0.513
17.	Chlorine, Cl	35.453	33.5
18.	Argon, Ar	39.948	0.66
19.	Potassium, K	39.098	2.06

TABLE 9.3a (continued)

Thermal neutron micro-cross sections

Z	Element	A	Cross Section (Barns)
20.	Calcium, Ca	40.080	0.43
21.	Scandium, Sc	44.956	27.2
22.	Titanium, Ti	47.90	6.09
23.	Vanadium, V	50.941	5.08
24.	Chromium, Cr	52.996	3.07
25.	Manganese, Mn	54.938	13.300
26.	Iron, Fe	55.847	2.56
27.	Cobalt, Co	58.933	37.18
28.	Nickel, Ni	58.71	4.49
29.	Copper, Cu	63.546	3.78
30.	Zinc, Zn	65.38	1.111
31.	Gallium, Ga	69.735	2.9
32.	Germanium, Ge	72.59	2.3
33.	Arsenic, As	74.922	4.5
34.	Selenium, Se	78.96	11.7
35.	Bromine, Br	79.904	6.9
36.	Krypton, Kr	83.80	25.0
37.	Rubidium, Rb	85.467	0.38
38.	Strontium, Sr	87.62	1.28
39.	Yttrium, Y	88.906	1.28
40.	Zirconium, Zr	91.22	0.185
41.	Niobium, Nb	92.906	1.15
42.	Molybdenum, Mo	95.94	2.55
43.	Technetium, Tc	98.906	*
44.	Rhuthenium, Ru	101.07	2.56
45.	Rhodium, Rh	102.905	14.5
46.	Paladium, Pd	106.4	63.3
47.	Silver, Ag	107.868	63.3
48.	Cadmium, Cd	112.41	2520.00
49.	Indium, In	114.82	193.8
50.	Tin, Sn	118.69	0.626
51.	Antimony, Sb	121.75	5.1
52.	Tellurium, Te	127.60	4.7
53.	Iodine, I	126.904	6.20
54.	Xenon, Xe	131.30	23.9
55.	Cesium, Cs	132.905	29.0
56.	Barium, Ba	137.33	1.20
57.	Lanthanum, La	138.905	8.97
58.	Cerium, Ce	140.12	0.63
59.	Praseodium, Pr	140.907	11.5
60.	Neodymium, Nd	144.24	50.5
61.	Prometheum, Pm	145	*
62.	Samarium, Sm	150.4	5911.
63.	Europium, Eu	151.96	4519.
64.	Gadolinium, Gd	157.25	49000.
65.	Terbium, Tb	158.924	23.
66.	Dysprosium, Dy	162.50	942.
67.	Holium, Ho	164.930	93.
68.	Erbium, Er	167.26	157.

TABLE 9.3a (continued)

Thermal neutron micro-cross sections

Z	Element	A	Cross Section (Barns)
69.	Thulium, Tm	168.934	106.
70.	Ytterbium, Yb	173.04	36.5
71.	Lutetium, Lu	174.971	85.5
72.	Hafnium, Hf	178.49	111.0
73.	Tantalum, Ta	180.948	21.1
74.	Tungsten , W	183.85	18.4
75.	Rhenium, Re	186.2	87.5
76.	Osmium, Os	190.2	7.6
77.	Iridium, Ir	192.2	420.
78.	Platinum, Pt	195.09	10.2
79.	Gold, Au	196.967	98.8
80.	Mercury, Hg	200.59	364.
81.	Thallium, Tl	204.37	3.02
82.	Lead, Pb	207.2	0.17
83.	Bismuth, Bi	208.980	0.034
84.	Polonium, Po	209	*
85.	Astatine, At	210	*
86.	Radon, Rn	222	*
87.	Francium, Fr	223	*
88.	Radium, Ra	226.025	*
89.	Actinium, Ac	227	*
90.	Thorium, Th	232.038	*
91.	Protactinium, Pa	231.036	*
92.	Uranium , U	238.029	7.59
93.	Neptunium, Np	237.048	*
94.	Plutonium, Pu	244	*
95.	Americium, Am	243	*
96.	Curium, Cu	247	*
97.	Berklinium, Bk	247	*
98.	Californium, Cf	251	*
99.	Einsteinium, Es	254	*
100.	Fermium, Fm	257	*
101.	Mendelium, Md	258	*
102.	Nobelium, No	259	*
103.	Lawrencium, Lr	260	*

TABLE 9.3b

Thermal neutron macro-cross sections

Material	Macro Capture Cross Section Σ, b/cm^3
Lead, Pb	5.61
Uraninite, UO$_2$	49.69
Cinnabar, HgS	7981.16
Iron, Fe	214.90
Galena, PbS	12.47

TABLE 9.3b (continued)

Thermal neutron macro-cross sections

Material	Macro Capture Cross Section Σ, b/cm³
Wulfenite, $PbMoO_4$	32.50
Arsenopyrite, FeAsS	165.22
Cobaltite, CoAsS	936.73
Chalcosite, Cu_2S	173.56
Hemitite, Fe_2O_3	100.47
Magnetite, Fe_3O_4	112.10
Bornite, Cu_5FeS_4	145.63
Pyrite, FeS_2	89.06
Illmanite, $FeTiO_3$	158.23
Zircon, $ZrSiO_4$	5.42
Stibnite, Sb_2S_3	17.82
Pyrrhotite, Fe_5S_6	90.52
Barite, $BaSO_4$	19.40
Chromite, $FeCr_2O_4$	102.20
Rutile, TiO_2	202.75
Chalcopyrite, $CuFeS_2$	80.95
Corundum, Al_2O_3	11.04
Carnotite, $K_2O.2UO_3.V_2O_5.2H_2O$	56.21
Rhodocrosite, $MnCO_3$	287.93
Sphalerite, ZnS	38.33
Siderite, Fe_2CO_3	68.81
Limonite, $2Fe_2O_3.3H_2O$	74.10
Dunite,	17.03
Olivine, $(Mg,Fe)_2SiO_4$	31.74
Magnesite, $MgCO_3$	1.48
Norite	12.88
Diabase	17.12
Gabbro	21.47
Anhydrite, $CaSO_4$	12.30
Aragonite, $CaCO_3$	8.12
Muscovite, $KAl_2(AlSi_3)O_{10}(OH)_2$	17.30
Biotite, $H_2K(Mg,Fe)_3Al(SiO_4)_3$	25.20
Dolomite, $CaMg(CO_3)_2$	4.78
Illite, $KAl_5Si_7O_{20}(OH)_4$	39.90
Diorite	14.33
Langbeinite, $K_2Mg_2(SO_4)_3$	78.87
Polyhalite, $2CaSO_4.MgSO_4.K_2SO_4.2HO$	21.00
Synite	16.43
Granodiorite	11.33
Chlorite, $(Mg,Al,Fe)_{12}(Si,Al)_8O_{20}(OH)_{16}$	17.56
Calcite, $CaCO_3$	7.48
Aluminum, Al	13.99
Plag. Feldspar, $xNaAlSi_2O_8,yCaAl_2Si_2O_8$	6.99
Limestone	8.72
Granite	11.62
Quartz, SiO_2	4.36
Sandstone	8.663
Kaolinite, $(OH)_8Al_4Si_4O_{10}$	13.06

TABLE 9.3b (continued)

Thermal neutron macro-cross sections

Material	Macro Capture Cross Section Σ, b/cm³
Albite, $NaAlSi_3O_8$	6.77
Orthoclase Feldspar, $KAlSi_3O_8$	16.00
Kieserite, $MgSO_4.H_2O$	12.77
Concrete**	
Montmorillonite, $(OH)_4Si_8Al_4O_{20}.nH_2O$	8.10
Gypsum, $CaSO_4.2H_2O$	19.40
Glauconite, $KMg(FeAl)(SiO_3)_6.3H_2O$	16.80
Graphite, C	0.38
Serpentine, $Mg_3Si_2)_5(OH)_4$	8.80
Halite, NaCl	752.36
Nahcolite, $NaHCO_3$	
Kainite, $MgSO_4.KCl.3H_2O$	196.13
Trona, $Na_2CO_3HNaCO_3.2H_2O$	16.21
Sulfur, orthorhombic, S	19.06
Potash, $KCO_3.2H_2O$	39.70
Sylvite, KCl	570.68
Cement**	~13
Sulfur, monoclinic, S	18.05
Kernite, $Na_2B_4O_7.4H_2O$	12,793.69
Magnesium, Mg	
Carnalite, $KMgCl_3.6H_2O$	370.92
Coal, anthracite**	1.08
Coal, bituminous**	1.54
Coal, lignite**	
Water, (3.0×10^5 ppm NaCl)	146.22
Water, (2.5×10^5 ppm NaCl)	122.55
Water, (2.0×10^5 ppm NaCl)	100.08
Water, (1.5×10^5 ppm NaCl)	78.75
Water, (1.0×10^5 ppm NaCl)	58.69
Water, (5.0×10^4 ppm NaCl)	39.02
Water, (3.0×10^4 ppm NaCl)	32.56
Water, pure, STP	22.08
Hydrocarbon oil, nCH_2, 10°API	28.02
Hydrocarbon oil, nCH_2, 30°API	25.89
Hydrocarbon oil, nCH_2, 40°API	24.22
Hydrocarbon oil, nCH_2, 50°API	22.23
N-Octane, C_8H_{18}, 70°API	22.12
N-Pentane, C_5H_{12}, STP	20.80
N-Pentane, C_5H_{12}, 200°F, 7k psi	22.02
N-Hexane, C_6H_{14}, STP	21.38
N-Hexane, C_6H_{14}, 200°F, 7k psi	20.37
N-Heptane, C_7H_{16}, STP	21.80
N-Heptane, C_7H_{16}, 200°F, 7k psi	20.84
N-Octane, C_8H_{18}, STP	22.12
N-Octane, C_8H_{18}, 200°F, 7k psi	21.12
N-Nonane, C_9H_{20}, STP	22.37

TABLE 9.3b (continued)

Thermal neutron macro-cross sections

Material	Macro Capture Cross Section Σ, b/cm³
N-Nonane, C_9H_{20}, 200°F, 7k psi	21.37
N-Decane, $C_{10}H_{22}$, STP	22.55
N-Decane, $C_{10}H_{22}$, 200°F, 7k psi	21.65
N-Undecane, $C_{11}H_{24}$, STP	22.71
N-Undecane, $C_{11}H_{24}$, 200°F, 7k psi	21.87
Methane, CH_4, STP	0.028
Methane, CH_4, 200°F, 7k psi	10.88
Ethane, C_2H_6, STP	0.051
Ethane, C_2H_6, 200°F, 7k psi	16.34
Propane, C_3H_8, STP	0.067
N-Butane, C_4H_{10}, STP	0.085
Helium, He, STP	0.0000
Carbon Dioxide, CO_2	0.0001
Nitrogen, N_2, STP	0.004
Oxygen, O_2, STP	0.00001
Hydrogen Sulfide, H_2S, STP	0.029

The "hydrogen index", H_W, is a comparative measure of the porosity. If the formation material is calcite and the pore and borehole fluids are fresh water, it is equal to the actual porosity. Table 9.4 shows the hydrogen index of several common materials.

$$H_w = 1 - 0.4P = \rho_w(1 - P) \qquad (9.10)$$

where

P = the NaCXl concentration in ppm $\times 10^{-6}$.

Table 9.5 evaluates the Hydrogen Index.

A detector placed at the proper distance from the source will detect the capture gamma rays or the thermal neutrons (depending upon the detector type). If it is closer and modified to ignore thermal neutrons, it will detect epithermal neutrons. See Figure 9.4 for a representation of this. Figure 9.5 shows the placement of the detector for optimum thermal neutron or capture gamma detection.

The epithermal neutron population will only depend upon the rate of moderation. The thermal neutron population will depend upon the rate of moderation and upon the rate of capture of the thermal neutrons (the rate of depletion). The capture gamma ray intensity will depend upon the population and the rate of capture of thermal neutrons. Of course, the population of all types will depend upon amount of the source flux, also.

TABLE 9.4

The hydrogen index, H_w, of some materials

Material	Hydrogen Atoms $\times 10^{23}$ per cc	Hydrogen Index
Pure Water		
60°F, 14.7psi	0.669	1
200°F, 7k psi	0.667	1
Salt Water, 200k ppm NaCl		
60°F, 14.7psi	0.614	0.92
200°F, 7k psi	0.602	0.90
Methane, CH_4		
60°F, 14.7psi275	0.0010	0.0015
141°F, 3.8 kpsi	0.275	0.41
200°F, 7k psi	0.329	0.49
Ethane, C_2H_6		
60°F, 14.7psi	0.0015	0.0023
200°F, 7k psi	0.493	0.74
Average Natural Gas		
60°F, 14.7psi	0.0011	0.0017
200°F, 7k psi	0.363	0.54
N-Pentane, C_5H_{12}		
68°F, 14.7psi	0.627	0.94
200°F, 7k psi	0.604	0.90
N-Hexane, C_6H_{14}		
68°F, 14.7psi	0.645	0.96
200°F, 7k psi	0.615	0.92
N-Heptane, C_7H_{16}		
68°F, 14.7psi	0.658	0.99
200°F, 7k psi	0.632	0.95
N-Octane, C_8H_{18}		
68°F, 14.7psi	0.667	1.00
200°F, 7k psi	0.639	0.96
N-Nonane, C_9H_{20}		
68°F, 14.7psi	0.675	1.01
200°F, 7k psi	0.645	0.97
N-Decane, $C_{10}H_{22}$		
68°F, 14.7psi	0.680	1.02
200°F, 7k psi	0.653	0.98
N-Undecane, $C_{11}H_{24}$		
68°F, 14.7psi	0.684	1.02
200°F, 7k psi	0.662	0.99
Bituminous Coal,		
84% C, 6% H	0.442	0.66
Carnalite	0.419	0.63
Limonite	0.369	0.55
Cement	~0.334	~0.50
Kernite	0.337	0.50
Gypsum	0.325	0.49
Kainite	0.309	0.46
Trona	0.284	0.42
Potash	0.282	0.42
Anthracite Coal	0.268	0.40

TABLE 9.4 (continued)

The hydrogen index, H_w, of some materials

Material	Hydrogen Atoms $\times 10^{23}$ per cc	Hydrogen Index
Kaolinite	0.250	0.37
Chlorite	0.213	0.32
Keiserite	0.210	0.31
Serpentine	0.192	0.29
Nahcolite	0.158	0.24
Glauconite	0.127	0.19
Montmorillinite	0.115	0.17
Polyhalite	0.111	0.17
Muscovite	0.089	0.13
Illite	0.059	0.09
Biotite	0.041	0.06

TABLE 9.5

Hydrogen Index

The number of hydrogen atoms per cubic centimeter (cc) of a substance may be determined:

1. Determine the molecular weight of the material,
2. Divide the molecular weight by the density of the material,
3. Divide Avogadro's Number by the result of step 2 ($A = 6.025 \times 10^{23}$)
4. Multiple the result of step 3 by the number of hydrogen atoms per molecule of the material.

Hydrogen Index is equal to the porosity, if the formation material is calcite and the formation pore fluid is pure water.

9.3 Sources

Neutron porosity systems make use of a neutron source in either an omni-directional tool or a sidewall tool. The source is usually isotopic (but increasingly, an electronic neutron generator). It is usually an isotope which emits alpha particles, such as radium, $_{88}Ra^{226}$ (1620y), polonium, $_{84}Po^{210}$ (138d), plutonium, $_{94}Pu^{239}$ (22,400y), or americium, $_{95}Am^{241}$ (458y). These alpha emitters are mixed with beryllium (or bismuth) and the reaction is

$$\left(^9Be_4\right) +^4 He_2 -^{12} C_6 +^1 n_0 + Q \tag{9.11}$$

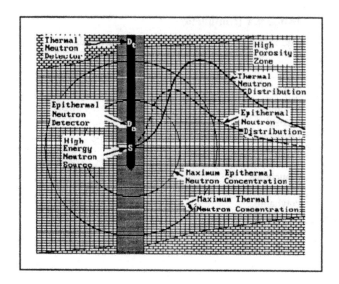

FIGURE 9.4
Relationship between thermal and epithermal neutron distribution.

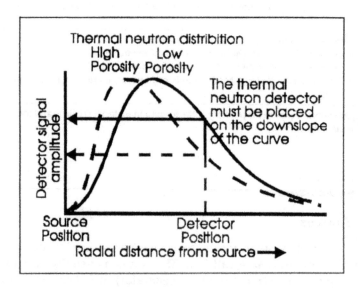

FIGURE 9.5
Placement of neutron porosity detectors.

Table 9.6 lists many of the commonly used source materials and some of their characteristics emissions.

The halflife of a source is the length of time for half of the radioactive material to decay. The time may be in microseconds or millions of years. The relationship is

TABLE 9.6

Isotopic neutron source materials

Radioactive Material	Symbol	Halflife	Neutrons Per Curie Per Second	Neutron Energy, MeV (Ave)	(Max)	Other Emissions Type MeV	Gammas Per Curie Per Second	Gamma Radiation (R/h/m/c)
Polonium-Beryllium	Po^{210}-Be	138 d	2.5×10^6	4.2	10.87	α 5.3 γ 0.80	1.5×10^6	10^{-4}
Californium	Cf^{252}	2.65 y	4.4×10^9	2.3	6	α 6.11 γ 0.2–1.8	0.15	
Americium-Beryllium	Am^{241}-Be	458.1 y	2.2×10^6	4.0	11	α 5.44(13.8%) 2.4×10^{10} 5.48(85%) γ 0.017(48.2%) 13.7×10^9 0.026(3.52%) 10.0×10^8 0.043(0.08%) 22.2×10^6 0.060(48.2%) 13.7×10^9 0.099(0.026%) 7.4×10^6 Total 2.8×10^{10}		
Radium-Beryllium	Ra^{226}-Be	1620 y	1.5×10^7	3.9	13.08	α 4.8(94%) 4.6(6%) γ 0.2–2.2 1.5×10^{11}	0.974	
Plutonium-Beryllium	Pu^{239}-Be	24,400 y	2.2×10^6	4.5	10.74	α 4.89(75%) 4.85(25%) Daughters α 0.017(99.36%) 10.7×10^8 0.039(0.07%) 7.4×10^5 0.053(0.24%) 25.9×10^5 0.060 Am daughter build-up 0.1 (0.18%) 20.2×10^5 0.124(0.09%) 9.2×10^5 0.384(0.05%) 5.6×10^5 Total 10.8×10^8	0.011	

$$\log(0.5) = Halflife \times \log\left(1 + \frac{rate}{unit\ time}\right) \qquad (9.12)$$

See also, the definitions in Chapter 8, Volume 1.

9.3.1 A Brief History of Neutron Logging Sources

The history of neutron sources for geophysical logging is rather interesting. The original source, used for many years, was $_{88}Ra^{226}$ mixed with powdered beryllium. This source was unsatisfactory because of its emission of large numbers of gamma rays. This meant that a large, dense material shield was more important for safety reasons than shielding from the neutrons. One type shield used a large ball of lead in a steel case (about 7 in, 18 cm) and a steel frame to prevent anyone getting too close. The frame was also a handle for 2 men to carry it. With the source in the shield, it weighed well over 150lbs (69kg).

When reactor by-products became available commercially, $_{84}Po^{210}$ was used as the alpha emitter. This was used, mixed with beryllium. It has no gamma emission. Unfortunately, it has a half life of only 138 days. Thus, its output decreases 5% every week. This required careful and constant bookkeeping by the operator and other field people and was an excellent opportunity for making errors. It only required, however, a light paraffin (high-hydrogen content) shield and was nearly a point source.

$_{94}Pu^{239}$ was next tried. It also, is an alpha emitter and is used mixed with beryllium for neutron production. It emits low-level gammas over a wide range of energies. It has a half life of 24,400 years, so strength decrease was no problem. This is a decrease of strength of only 0.004% per year. Its activity was so low, however, that it required a large volume of source material. The sources were large enough that they were no longer point sources. The large size meant, also, a large container, which, for logging uses, are always designed to withstand extremely high external pressures and are of double-sealed stainless steel. Eventually, the helium (from the alpha particles) gas pressure would distort the container enough that it no longer fit into its socket in the sonde. Also, $_{94}Pu^{239}$ is a fissionable material. Laws, regulations, and storage are serious problems. One major contractor had enough plutonium, at one time, to make a small bomb. The Mohole Project obtained an act of the U.S. Congress declaring that its ocean vessel was U.S. territory, so that fissionable material would still be in the United States when the vessel was at sea, and possession would comply with the law. These sources are now seldom used for logging.

$_{95}Am^{241}$ is still in use. It has a half life of 458 years, so bookkeeping is no problem, even with analog systems. Its decay rate is 0.15% per year. It emits mostly low energy gamma rays (virtually all at 59.6 keV or lower).

Thus, shielding is not a serious problem. It can be shielded with a light-weight paraffin shield. It is not fissionable. Since it is a prolific α emitter, it also is mixed with beryllium for neutron emission.

$_{94}Cf^{252}$ has been used often since it became commercially available about 1970. It is a prolific spontaneous neutron emitter. It has a neutron flux about 100 times higher per curie that any of the alpha/beryllium sources. Its activity is about 4.4×10^8 n/s/Ci. Its decay rate is 23% per year. Because of its high activity (half life is 2.65 years) it has a small physical size and is almost a point source. Its gamma emissions are all below 570 kev and of low intensity and can be controlled easily. Bookkeeping is not as serious a problem as with the polonium sources. The U.S. Geological Survey conducted extensive experiments with the first $_{94}Cf^{252}$ source that was available.

Each isotope has advantages and disadvantages. The energy of the output neutrons will vary type to type, depending upon the kinetic energy of its alpha particles. The neutrons from an alpha/beryllium source will usually have an energy of about 4 MeV (the $_{95}Am^{241}$/Be source is the most common one, at this time). The yield of an AmBe source is about 2.2×10^6 n/s/Ci.

Electrical sources (linear accelerators; also called "neutron generators") were developed as atomic bomb triggers. They were easily adapted to geophysical use. Their engineering problems are great, however, because of the confined space in a logging tool, the high ambient operating temperatures and the very high voltages (>123,000 volts) required. Linear accelerators or generators are being used more frequently in geophysics. In this use they have no competitor in the generation of short (a few microseconds long) pulses of neutrons. They have the advantage that the can be turned off when they are not being used. In the "off" mode their radioactivity is low. It is due to the $_1H^3$ in the target; a β emitter with a halflife of 12.26 years at 18.6 kev. They may also be operated in the pulse mode, with very high flux rates. They usually use one of two reactions:

$$\left(_1H^2\right)+\left(_1H^3\right)-\left(_2He^4\right)+\left(_0n^1\right)+Q \qquad (9.13)$$

which has a yield from about 10^7 n/s to about 10^{13} n/s, depending upon the accelerating voltage. Q has an approximate value of 14 MeV. And,

$$\left(_1H^2\right)+\left(_1H^2\right)-\left(_2He^3\right)+\left(_0n^1\right)+Q \qquad (9.14)$$

which has a yield of about 5×10^5 n/s to 10^7 n/s, also depending upon the accelerating voltage. Q, in this case, has an approximate value of 2.8 MeV. When the accelerators are operated in the continuous mode, their output is held to low levels.

9.3.2 Detectors

Capture gamma rays are detected with a standard gamma ray detector, such as NaI(Tl). They are also sensitive to natural gammas, which thus can interfere with the measurement. Chlorine has a large cross section (33 barns) compared to hydrogen (0.33 barns) and causes an important interference to the normal neutron porosity measurement. However, this may be used to tell the difference between the response of oil (hydrogen, but little chlorine) and formation water (lots of chlorine and hydrogen) with the chlorine log. The chlorine log is described later in this chapter.

Neutron detectors are usually $_2He^3$ proportional counter type. In the energy range of neutrons used for geophysical logging, $_2He^3$ has a cross section of 150 barns at 14 MeV to over 10,000 barns at 0.03 eV. The reaction used is

$$\left(_2He^3\right)+\left(_0n^1\right)-\left(_1H^3\right)+\left(_1p^1\right)+Q \tag{9.15}$$

where $_1p^1$ is a proton ($_1H^1$) and Q is the released energy (~0.8 MeV). Since the proton, $_1p^1$, is an energetic, charged particle, its passage through the gas of the chamber will ionize a path. This path conducts a current from the anode (the shell of the chamber) to the cathode (the central wire) of the counter tube. The reaction is that of a proportional counter. The resulting negative electrical pulse is recognized by the circuitry as an "event". Its amplitude is a function of the energy of the neutron, thus, the proton.

$_2He^3$ detectors are sensitive to thermal and epithermal neutrons, but not to gamma rays. Therefore, these are popular types detector. $_2He^3$ detectors are available with a wide range of sensitivities, depending upon the density of the $_2He^3$ gas filling. This filling may be as high as 10 or more atmospheres (1.7×10^{-4} to 1.7×10^{-3} g/cm³). These detectors may be made primarily sensitive to epithermal neutrons by shielding out the ambient thermal neutron flux with cadmium. Cadmium has a cross section to thermal neutrons on the order of 2,520b.

Some $_2He^3$ detectors are lined with $_{92}U^{235}$ to increase their sensitivity. $_{92}U^{235}$ has a large cross section (100b) to thermal neutrons and will undergo fission:

$$\left(_{92}U^{235}\right)+\left(_0n^1\right)-\left(\text{fission fragments}\right)+x\left(_0n^1\right)+Q \tag{9.16}$$

Q, in this case, ~150MeV. Since these fragments are ionized, the detector will respond to them with a G-M (Geiger-Muller) type reaction.

9.4 Systems Currently in Use

Neutron porosity systems which are in use at this time are single spacing (detecting, essentially, thermal neutrons), dual spacing (essentially detecting thermal neutrons; borehole compensated), dual spacing and function (thermal and epithermal system combination), and sidewall (epithermal detection) systems.

Several types of neutron porosity systems are in commercial use at this time. Most of them now employ the $_2He^3$ proportional counter tube. Lithium crystals and boron trifluoride crystals have occasionally also been used.

The neutron-gamma system (capture gamma ray detection) is the oldest and simplest type to be used. This system is highly sensitive to any neutron absorber in the formation material, including chlorine. A salinity (chlorine) correction must be applied for the formation and for the borehole fluid content. It is sensitive to natural gamma rays also. Most of the older systems are of this type.

The present-day neutron-thermal neutron system uses a $_2He^3$ detector which will primarily detect thermal neutrons. It uses a long source-to-detector spacing so that only thermal neutrons reach the detector. It is not affected by natural nor capture gamma rays. Thus, it may be used in high gamma ray environments, such as uriniferous zones. It is sensitive to thermal neutron absorbers, but not to the extent that the neutron-gamma system is. The compensated systems are sometimes of this type.

The third type is the neutron-thermal/epithermal neutron system. It is similar to the previous system, except that a somewhat shorter spacing is used to increase the counting rate. A $_2He^3$ detector is usually used. Thus, it is sensitive to both thermal and epithermal neutrons, but not to gamma rays. It has a slightly better linearity of response with respect to the logarithm of porosity than the previous two types. Its sensitivity (counting rate) is the highest of any of the systems. Thus, its probable error is smallest. Many compensated systems and mineral systems are of this type.

The fourth type is the neutron-epithermal neutron system. It is quite linear with respect to the log of porosity. It has, however, a much shallower lateral depth of investigation and much lower counting rates than the other types of systems. The depth of investigation is the result of the short spacing necessary rate and strategic placement of the detector where it will read the optimum number of epithermal neutrons. The detector is shielded with a material which has a high-capture cross section to thermal neutrons (i.e., cadmium). These systems are often designed as sidewall tools to minimize borehole effects.

A recently offered service employs a thermal neutron system combined with an epithermal system, both in compensating arrays. The combination of these two measurements results in much more useful information.

9.4.1 Single-Spacing Systems

A single-spacing system uses an array consisting of a single source and single detector. It is usually operated without any attempt to center or eccenter it. The detectors may be gamma ray or neutron types. If they are gamma ray detectors, the older logs were made with G-M detectors. All of the modern systems use He^3 neutron detectors. The standard downhole have diameters of 1in (2.5cm) to 4in (10cm) or more. Some specialty tools have larger diameters. Borehole size, drilling mud type, and salinity correction charts are published by the several contractors for use with their own systems.

Usually, the older logs displayed a simultaneous neutron porosity curve and natural gamma ray curve. These were often scaled in counts per second, API units, or arbitrary units. Many times these systems were used in cased holes as a "perforation" log. The gamma ray and neutron pulses were transmitted to the surface as opposite polarity pulses on a monocable. The same, single conductor also supplied the power to the downhole tool. A pulse polarity discriminator separated the pulses at the surface. This type system is subject to severe coincidence losses and should not be used quantitatively. These systems have the advantage of simplicity and low cost. They can also be quite rugged. Thousands of old "neutron" logs exist which were run with the "perforation" systems.

The great disadvantage to the single-spacing systems is their sensitivity to borehole size and variations. The borehole is filled with material which is rich in hydrogen (water and/or hydrocarbons). The borehole also occupies the most sensitive portion of the system's measurement volume. Borehole correction is mandatory for any quantitative porosity use.

Single-spacing systems may use gamma ray, BF^3, lithium crystal, or $_2He^3$ detectors. It is wise to ask the contractor the type of detector in use. Most of the them use a thermal-epithermal detection. No attempt is made to be selective. The counting rates are high and are primarily due to the thermal neutron population. Thus, counting statistics are good.

Epithermal detection is difficult to achieve with the single spacing tool because of the severe neutron moderation by the hydrogen-rich borehole fluid. Thus, except for the sidewall neutron system, epithermal detection is not a viable option with single spacing tools.

9.4.2 Sidewall–Neutron Porosity Systems

Sidewall–neutron porosity probes usually use epithermal detection because the counting rate of an epithermal system is nearly linear with the log of porosity. Because the counting rates are low and the borehole is such a dominating factor, they are run against the sidewall of the borehole and are shielded from the mud column. The responses of these systems must be corrected for the thickness and type of mudcake. Some modern equipment enters an automatic borehole caliper factor to correct for the mudcake thickness. Mudcake correction charts will be found in the contractor's chart books. Schlumberger identifies their's as the SNP. These systems are essentially free of interference from the chlorine content of the borehole and formation fluids.

9.4.3 Borehole Compensated Systems

Compensated neutron porosity are the most commonly used instrument for porosity measurements at this time. The compensated systems are substantially better than the single spacing systems. They have become the "standard" neutron porosity log.

The compensation instrumentation uses a single source and two detectors at different distances from the source. As with the BHC density systems, the borehole and mudcake occupy a relatively larger portion of the shorter spacing measurement volume than the longer spacing volume. Therefore, a factor is applied to the single of one of the detectors to make the borehole contributions of the two equal. One is subtracted from the other. The difference then, is only a function of the formation and its porosity.

Most of the "compensated" tools use thermal-epithermal detection. These systems are less sensitive to chlorine interference and insensitive to natural gammas. One contractor offers a combination thermal neutron log with an epithermal neutron log.

9.5 Neutron Porosity Measurements

Neutron porosity measurements make use of the much more efficient moderation of fast neutrons to epithermal or thermal energies by hydrogen, compared to the moderation by heavier elements. Table 9.2 shows the differences of moderation by several of the low atomic number elements. Therefore, the neutron response is a function of the amount of hydrogen in the formation material. Since almost all of the hydrogen in the formation occurs in the water and hydrocarbon of the pore space, the neutron response is a function of porosity. Table 9.4 shows the hydrogen content of a number of common materials found in earth sediments.

9.5.1 Shale Correction

Unfortunately, in addition to the pore space, the shales and clays also contain water. Corrections must be made to the porosity measurement for the influence of the bound water of the clay in the shale. Neutron systems are very sensitive to the presence of shale in the sand or carbonate. This is caused by the large quantities of water which are usually found in clays of shales. In addition, there may be substantial quantities of chlorine compounds (chlorine has a large-capture cross section to thermal neutrons) in the clay. Therefore, it is imperative that a shale correction be made whenever shaliness is suspected. The neutron derived porosity, if it is uncorrected for shale, will read very near total porosity. If the effect of the bound water of the clay is eliminated, the porosity value will be essentially the effective porosity, $\phi_{N,eff}$, after rock-type corrections.

The neutron shale correction may be made with the help of the several methods for determining the volume of shale. To do this, determine the volume of shale, V_{SH}, with, for instance, the gamma ray method. Determine the value of the apparent neutron porosity in the shaly sand, $\phi_{N,a}$, and the apparent neutron porosity of a nearby pure shale, $\phi_{N,SH,a}$. Then obtain a volume of shale from the gamma ray log:

$$I_\gamma = \frac{\gamma - \gamma_s}{\gamma_{sh} - \gamma_s} \tag{9.17}$$

$$V_{sh} = 0.083\left(2^{3.71 I_\gamma} - 1\right) \tag{9.18a}$$

for Tertiary or younger formations, and

$$V_{sh} = 0.33\left(2^{2 I_\gamma} - 1\right) \tag{9.18b}$$

for older formations. Then subtract the shale effect from the apparent, rock type corrected, neutron porosity:

$$\phi_N = \phi_{N,a} - V_{sh}\phi_{N,sh,a} \tag{9.19}$$

9.5.2   Rock Type Corrections

Also, the rock matrix contributes a significant portion of the total signal. This contribution varies from one type rock to another. Thus, rock type corrections must be made, also.

Since the neutrons react with the atom nuclei, neutron logs are sensitive to lithology. See Table 9.4 for effective cross sections. The log measurement reflects the macro cross section of the rock matrix, as well as that of the hydrogen content. Therefore, neutron log values must be corrected for rock type in order to determine porosity correctly. Figures 9.6 and 9.7 show sets for a single spacing, thermal/epithermal system. Figure 9.8 is a set of modern rock type correction curves for two of the Dresser Atlas systems. When logging in limestone formations, of course, a rock type correction need not be made because the system was calibrated in limestone model. Notice that each curve set is different. The correction curves are *not* interchangeable.

Table 9.3a and 9.3b shows the thermal neutron capture cross sections. This, with the porosity and instrumentation, is what determines is what determines the shape of each curve. The micro-cross section is the unit

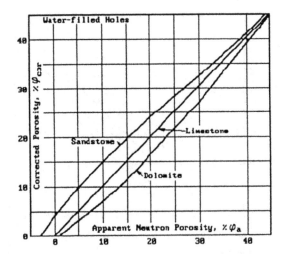

FIGURE 9.6
A set of neutron rock-type correction curves for a single spacing T/ET system. After Schlumberger Well Services, Inc.

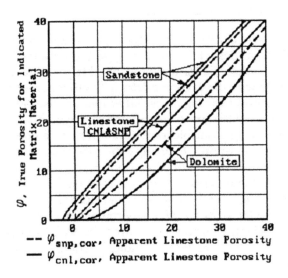

FIGURE 9.7
A modern set of rock-type correction curves. (after Schlumberger Well Services, Inc., 1988).

usually found in texts and reference tables. It is the apparent cross sectional area of the atomic nucleus which is presented to the thermal neutron. Its unit is the "barn", b (1b = 10^{-23} cm^2).

The macro-cross section is used extensively in geophysics. It is the total cross section area presented in a cubic centimeter of a material. Its unit is "barns per cm^3", b/cm^3.

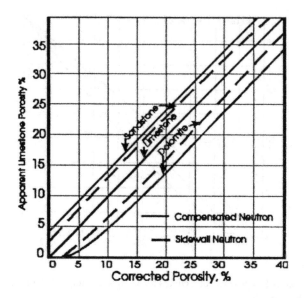

FIGURE 9.8
A modern set of neutron porosity rock-type correction curves. (Courtesy of Atlas Logging Services Division of Western Atlas International, Inc.)

Neutron porosity curves are frequently recorded in apparent limestone porosity unit, regardless of the real rock type. It is a scaling device and not a real porosity, unless the formation material happens to be water-filled limestone. It allows a direct comparison with other porosity curves, such as the density or acoustic log. These latter two use the same device. This feature may be used to start to identify the rock type, also. If an independent porosity value is available for the zone, it may be possible to use the error of the neutron reading to identify the rock material or mixture proportions. Thus, this neutron-derived porosity is only correct in fresh water-filled limestone. This uncorrected reading is called the "porosity index" or "hydrogen index". In fresh water-filled limestone, the porosity index is equal to the porosity (after environmental corrections have been made). In this case only, correct porosity may be read directly from the log. These correction charts are called "lithology" charts or "chemical correction" charts.

9.5.3 Borehole Corrections

All neutron measurements must be corrected for the presence of the borehole and contents. Since the borehole fluid contains large amounts of hydrogen, even the compensated neutron systems must have small corrections made for the maximum accuracy. The sidewall neutron systems must be corrected for mudcake thickness, since there is substantial amounts of hydrogen in the mudcake. Figures 9.9 and 9.10 show the corrections for

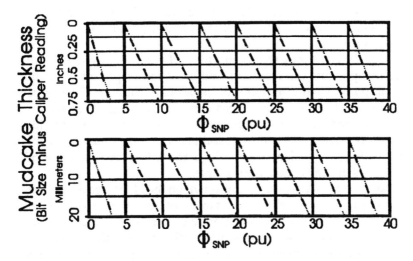

FIGURE 9.9
Schlumberger mudcake thickness charts for the SNP system. Standoff correction: 0 in ϕ_a = $\phi_{corrected}$; 0.5 in $\phi_{corrected}$ = $1.0041(\phi_a - 1.875)$; 1.0 in $\phi_{corrected}$ = $1.0390(\phi_a - 3.790)$; 1.5 in $\phi_{corrected}$ = $1.0588(\phi_a - 5.500)$. (after Schlumberger, 1988).

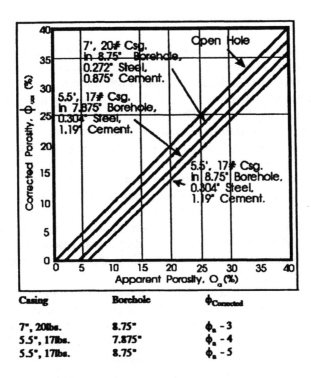

Casing	Borehole	$\phi_{Corrected}$
7", 20lbs.	8.75"	$\phi_a - 3$
5.5", 17lbs.	7.875"	$\phi_a - 4$
5.5", 17lbs.	8.75"	$\phi_a - 5$

FIGURE 9.10
Neutron standoff correction curves. (Courtesy of Western Atlas Logging Services Division of Western Atlas International, Inc.)

the Western Atlas compensated neutron for wall standoff and mud density. Other corrections may be for mud salinity, mud cake thickness, and temperature.

Just as different rock types can cause different log responses, different fluid components can also cause different log responses. Probably the most important in this respect is the presence of chlorine. Chlorine content is important because chlorine has a moderately high cross section, but also, it is a very common component in the form of sodium chloride. It is almost universally present in quantities which may vary from small to large. Salinity corrections are usually made with a separate set of calibration curves; for salty mud, for example.

Salinity corrections are most important on the older logs which were run with instruments which used gamma ray detectors. Thermal neutron detection is not as sensitive to chlorine content as gamma ray detection. Epithermal detection logs may require little or no salinity correction.

Trace amounts of some elements with large capture cross sections (i.e., gadolinium) can cause serious aberrations. This factor should be considered when repeatable, but unusual or unexplainable, responses show up in the log.

9.5.4 Cased Hole Use

Neutron porosity systems have a great advantage over many other systems. The casing material of cased holes is quite transparent to neutrons. The only exceptions are the plastics with high chlorine contents and fiberglass with borosilicate glass fibers. Even when logging in these, a simple field correction may be made. Thus, neutron logs are readily made in cased holes. There are uncertainties due to the unknown position of the sonde, with respect to the borehole wall. This simply degrades the confidence level in the log somewhat. Small corrections must be made for the presence of cement in the annular space outside the casing. Figure 9.11 shows the Western Atlas curves for making casing corrections to the logs made with their borehole compensated system.

Figure 9.12 is a log of neutron porosities made in a well that was open and then overlaid with another made after it was cased. Figure 9.13 is a similar situation, except the log was made in the open hole and then continued up into the casing.

One of the great advantages to the compensated neutron systems is its relative independence of most borehole conditions. We saw, in Figures 9.12 and 9.13, that steel casing has a minimal effect. The compensation feature makes the system relatively immune to borehole size and fluid type. Figure 9.14 is a log which was started in a water base drilling mud, below x065 ft. Above this, the hole was air-filled. There is a noticeable change in the density curve, but virtually none on the compensated neutron log.

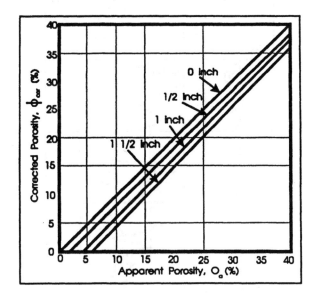

FIGURE 9.11
Neutron cased hole corrections. (Courtesy of Western Atlas Logging Services Division of Western Atlas International, Inc.)

Figure 9.15 is a log made with a properly calibrated and corrected compensated neutron porosity system. It is compared with the core laboratory report of the core porosities. Note that the correspondence is quite good. It is exceptionally good if the greater probable error of the core sample (because of the smaller core sample size) is taken into account. The measurement volume of this type neutron porosity system is probably a hollow, thick-walled sphere, with a major diameter of about 40 in (1 m), with an inner diameter of about 10 in (25 cm). Calibrations and corrections are necessary, as with *all* measurements of this type.

9.6 Calibration

The half-life of the AmBe sources is long enough (458y) that source strength corrections usually need not be made. This source will decrease in output about 1% per year.

As with all systems that are used quantitatively, the neutron porosity systems must be calibrated carefully. The usual calibration procedure starts with an initial calibration in a model designed to imitate well conditions adequately. Then, in the field, the systems should be calibrated in a

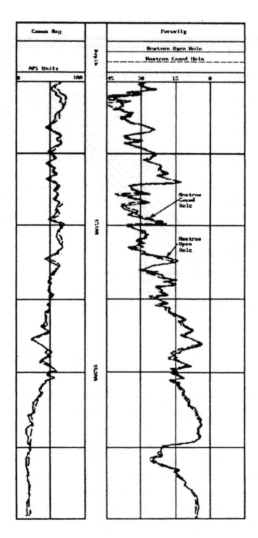

FIGURE 9.12
Two neutron logs in the same hole, one open hole and the other cased. (Courtesy of Atlas Logging Services Division of Western Atlas International, Inc.)

good secondary calibrator at regular, frequent intervals. For most types of equipment, a truck calibrator should be carried and used before and after the log to insure that no change of sensitivity has taken place since the secondary calibration and during the logging.

All neutron porosity systems are initially calibrated in limestone porosity models. The reason is that limestone is a very suitable material. It can easily be obtained in large, uniform blocks in a pure calcium carbonate form. It has a wide range of porosities. Impurities with large cross sections, such as boron compounds, do sometimes exist naturally in the limestone, but they are usually easily detected. The blocks are chemically assayed for

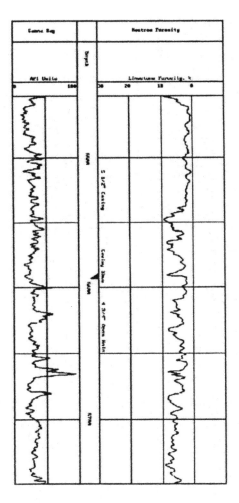

FIGURE 9.13
A neutron porosity log in the open and cased portions of the hole. (Courtesy of Atlas Logging Services Division of Western Atlas International, Inc.)

trace elements. The density is carefully measured. Then, the pore space is vacuum saturated with fresh or saline water. The density is again measured and used to calculate the porosity. Plugs are independently measured in the laboratory for dry and wet density and for porosity. The calibration standard for neutron porosity for petroleum gamma ray systems is the API Facility at the University of Houston. Figure 9.16 shows a neutron calibration model which is similar to the API Facility, but incorporates two borehole of different diameters. Note that, in a multi-borehole model, the boreholes must be separated by enough rock that they will not interfere with each other.

The API neutron models contain limestone blocks with three different water-saturated porosities with a 7 7/8 in (20 cm) diameter borehole in

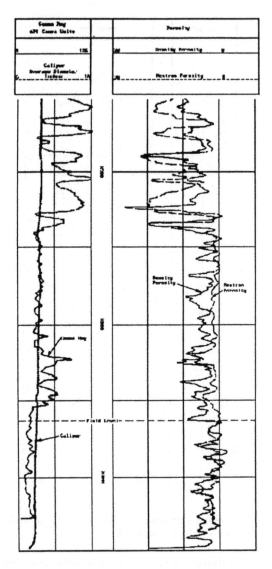

FIGURE 9.14
A neutron log started in mud and logged up into air (after Schlumberger).

the center. There is also a water zone (100% porosity) and a crushed marble zone. A zone of Austin chalk has a porosity of about 28%. Indiana Limestone, at 19% porosity is defined as 1000 APIn units. The third layer is Carthage marble, at 2% porosity. The borehole is filled with fresh water.

Figure 9.17 shows a set of actual response curves for a single spacing, 1 11/16 in, thermal/epithermal, mineral type neutron system in the Uni-

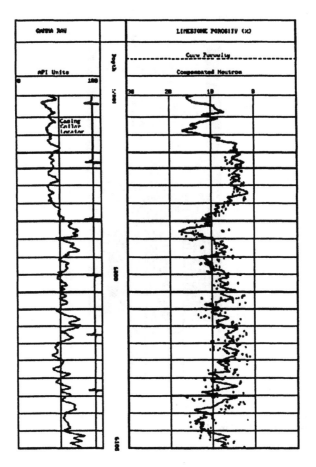

FIGURE 9.15
A properly calibrated and corrected neutron porosity log and core comparison. After Schlumberger.

versity of Houston API Neutron Calibration model. This system responded to both thermal and epithermal neutron flux. It was, however, primarily sensitive to thermal neutrons. There simply was no effort made to eliminate the epithermal response, which was very low. API Neutron units are linear units which are a function of the actual counting rate. The 19% porosity Indiana Limestone zone is defined as indicating 1000 API Neutron Units.

Note the shape of the curves. The response of most thermal neutron porosity systems is assumed to be

$$N = C + D \log \phi \qquad (9.20)$$

where

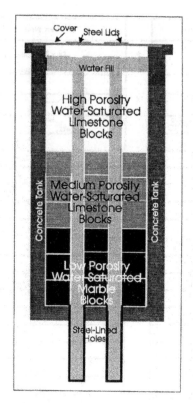

FIGURE 9.16
A Neutron Porosity Calibration Model.

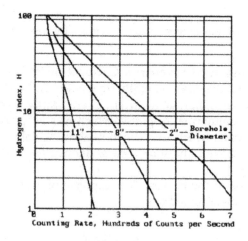

FIGURE 9.17
Calibration curves for a single S-D, thermal neutron porosity system.

N = the counting rate
C = a source strength constant
D = a factor which depends upon the borehole size and fluid type
ϕ = the total porosity

Figure 9.17 shows a measured set of thermal/epithermal neutron porosity curves, as a function of Hydrogen Index, H_w. You can see from the curves of Figure 9.18 that Equation 9.20 is only approximately true. The curves in Figure 9.17 were determined for a mineral logging system, for environments where the actual and apparent porosities can be quite high. The log-linear response assumption (a straight line on a semi-logarithmic grid) is not a valid one for this type application.

Figure 9.19 has been redrawn from an old set of Schlumberger curves for their neutron-gamma system. It has been simplified for clarity by removing most of the hole size curves. The gamma sensitive systems were the earliest commercial neutron porosity systems. These systems were quite sensitive to the presence of chlorine (i.e., in NaCl) and to the natural gamma radiation. Note that the response curves are nearly straight lines on a semi-logarithmic grid. This validates Equation 9.20. It is a valid approximation for deeper measurements, such as most petroleum logging.

9.7 Data Reduction

On modern logs, you generally will not see the raw counting rate curve nor will you have to convert to hydrogen index. This is usually done by the circuitry. The modern logs are usually scaled in apparent limestone porosity units.

The idealized neutron response equation is given in Equation 9.20. It is evident from Figure 9.19, that the Equation 9.20 does not describe the neutron porosity response exactly. It does, however, describe the curves of Figure 9.18. Over the range from about 2% to about 45% porosity, Equation 9.20 is a very good approximation. And, this is what has been done in Figure 9.18. If you use the system outside of the range of 2% to 50% porosity, as you often will in non-hydrocarbon applications, you must get the exact response charts from the contractor. These more exact charts may not always be available.

On older neutron porosity logs, the log may be scaled in counts per second or in API_n units. Conversion to porosity index must be made using a chart which was specifically designed for that particular tool and system, *in effect at the time the log was run*. Charts and instruments are upgraded frequently and the proper chart must be used. Contact the contractor for information.

A useable porosity scale can be generated for the older logs which are not scaled in apparent porosity units and for which conversion curves are

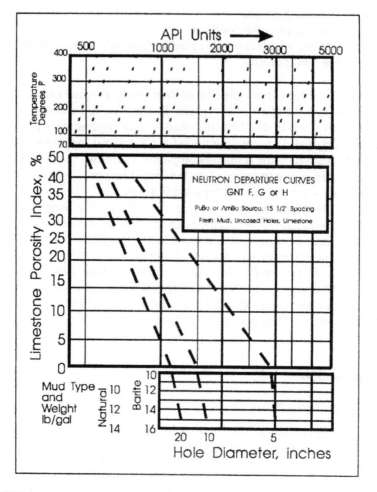

FIGURE 9.18

Calibration curves for the GNT system with PuBe or AmBe sources (after Schlumberger, 1959).

not available. This approximation is patterned on the situation shown in Figure 9.20. To use this method, select the neutron readings in the nearest solid shale and in the tightest limestone available. These readings, on older logs, will be in counts per second, API_n units, or some other non-porosity unit. Occasionally, on old logs, the scale will be in some other, often arbitrary, unit. Roentgen units, counts per second (uncalibrated), neutron scaling units, radiation units, and various others were used. Assign a probable porosity value of 40% to the shale reading, if the zone is deep. If it is shallow, assign 45% to 50% porosity to it. Plot this on a two-cycle, semi-logarithmic grid as in Figure 9.20. If it is very tight, as in the Hunton or Viola limestones in the Anadarko Basin in Oklahoma, assign a porosity value of 1% to 2% to the zone. Plot this point and connect the two. Scale the log axis

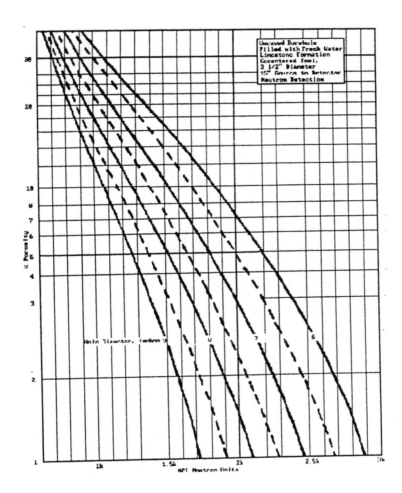

FIGURE 9.19
A curve set for a small diameter, single spacing, T/ET equipment.

in apparent porosity from 1% to 50%. If a porosity value is known for the lower end, use it.

A transparent logarithmic scale is often available in one of the chart books. It is intended for the purpose explained in the previous paragraph. If it is done correctly, a reasonable approximation of the apparent neutron porosity may be read from the log.

This same approximation method may be used to solve the neutron porosity response Equation (9.20) for the Constants C and D:

$$N = C + D \log \phi \qquad (9.20)$$

Merely put the log counting rates (corrected for rock type) N into two simultaneous equations and solve for C and D.

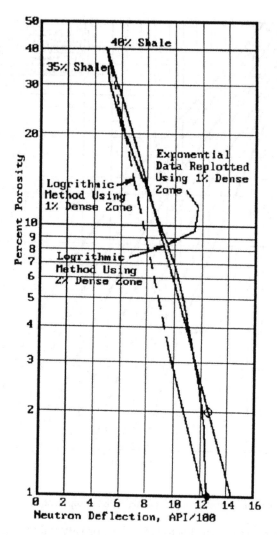

FIGURE 9.20
A logarithmic approximation (after Schlumberger).

The porosity value of the shale may tolerate a wide range of value. Thus, a ±5% porosity unit error at 50% to 45%, or 55% porosity has less effect upon the method than does a ±1% porosity-unit error at the low porosity end.

The modern neutron log is usually plotted in limestone porosity units, $\phi_{LS,a}$ (Hydrogen Index). The reduction, correction, and interpretation procedures in the chart books, usually assume that $\phi_{LS,a}$ is available. In general, if an equation is available to do the job, charts are not the most accurate way to determine $\phi_{LS,a}$ or to make a correction to a log value. An equation is usually also easier to program into a computer or calculator process. Sometimes, however, only empirical charts are available with older logs

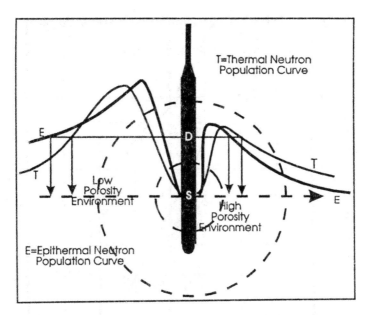

FIGURE 9.21
Distributions of neutron populations.

and *some* non-petroleum logs. The above process can easily be reversed, using chart values, to determine a valid approximation equation which can be put into a program. Be sure to use the correct chart. *The correction charts of the various systems and the different contractors are not interchangeable.*

9.7.1 Depth of Investigation

Neutron penetration depths are governed by the porosity of the formation material and the hydrogen content of the pore space. In zones with low hydrogen content (low porosity, clean zones) thermalization of the neutrons may occur several feet or more than a meter from the source. In high porosity and/or shaly zones, thermalization may occur a few inches or centimeters from the source. Therefore, the radius of investigation of neutron devices will depend upon the porosity and water content of the formation material. See Figures 9.5 and 9.21.

Neutron systems, in water-filled sediments, are all moderately shallow depth of investigation methods. In a 20% water-filled sand, the maximum concentration of thermal neutrons occurs at a distance of about 6 in (15 cm) from the source. Figures 9.22, 9.23, and 9.24 show the effect upon the spacial distribution of neutrons with three different variables. Figure 9.22 shows the influence of the formation material upon thermal and epithermal neutrons. The environment is a 39% porosity sand with an unknown sodium chloride content (except that the formation water is "fresh"). Notice that there is already a high degree of thermalization as close as 4 in

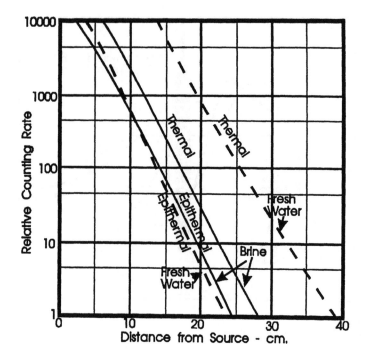

FIGURE 9.22
Thermal and epithermal neutron distribution, the effect of NaCl in the borehole.

(10 cm) from the source. Figure 9.23 shows the effect of the porosity upon the slowing down length of neutrons of three different source energies. Figure 9.24 shows the effect of the source-to-detector spacing upon neutron from the three most common energy sources, approximately 2.4 MeV, 4 MeV, and 14 MeV.

9.7.2 Source-Detector Spacing

The distance between the source of neutrons and the detector (S-D spacing) has a profound effect upon the signal the detector produces. This is because of the different and nonlinear distributions of the thermal and epithermal neutrons, as noted in the previous section of this chapter. This factor is as important, for example, as the type of detector used. Since these systems observe the low energy neutrons (epithermal neutrons, at and below 100 ev, and thermal neutrons, below about .05 ev), the spacing effects are relatively independent of source energy. The signal amplitude, of course, is directly proportional to the source strength (neutron flux). In addition, the spacing affects the signal and tends to favor measurements in a particular porosity range. In addition, to some extent, the shorter S-D

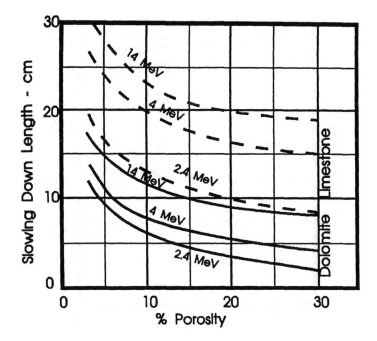

FIGURE 9.23
The effect of porosity upon thermal neutrons of various energies.

is, the higher the counting rate will be. Within limits, this approximately follows the inverse square law.

Figure 9.24 illustrates the effect on the count rate of the detector and the response linearity, as a function of the spacing for a wide range of porosities for a dual spacing thermal neutron system. Notice that the device with the longer spacings has a greater response contrast to porosity. This, however, is at the expense of counting rate and linearity. The shorter spacings, on the other hand, tend to have a slightly better linearity and a much better counting rate. This results in a smaller statistical uncertainty. In tool design, the spacing selections are based on,

1. Total count rate, favoring a shorter S-D
2. Porosity resolution, longer S-D for low ϕ, shorter for high ϕ
3. Borehole effects, favoring a longer spacing and borehole compensation

The count rate is important because the statistical probable error is a function of the reciprocal of the square root of the counting rate. Thus, higher count rates tend to reduce the ambiguity of the signal. The count rates generally become higher as the spacing is decreased.

The porosity resolution is important because the porosity parameter is the purpose of the measurement. A tool designed for low porosity carbon-

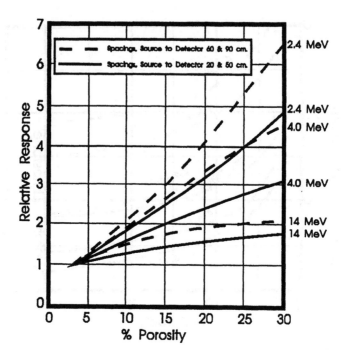

FIGURE 9.24

The effect of source-to-detector spacing upon thermal neutron distribution.

ates might favor a longer S-D and a stronger source. One designed for soft, porous sediments might use a longer S-D.

The borehole effects are interfering components and must be removed from the total signal to obtain a porosity related signal. The lower the interfering components are, proportionally, the less important they are. The borehole effects usually become less important at the longer spacings. In addition, the diameter of the tool is a constant, non-hydrogenous factor. Larger holes should use the largest diameter downhole tools possible.

The optimum spacing is determined during the engineering of each system. Thus, the optimum spacing is likely to be different for each device. The spacing will be affected by the source output, detector sensitivity, tool diameter, the type and diameter of the hole, and the type of measurement desired (thermal, thermal/epithermal, epithermal, or capture gamma ray). Spacing generally are from about 11 in (30 cm) to 18 in (46 cm) for thermal neutron measurements.

9.7.3　Interpretation Methods

Apparent neutron porosity is *approximately* total porosity. This will include, of course, any water in the clays. Thus neutron porosity values are quite sensitive to any clay or shale. The effect of the clay or shale must be

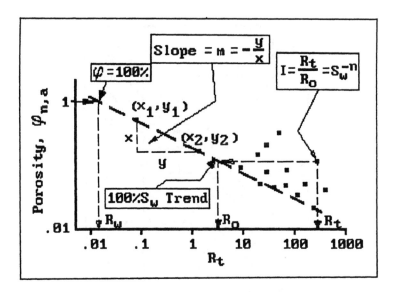

FIGURE 9.25
The R_t vs. ϕ crossplot, using ϕ_n (after Schlumberger, 1986).

removed to obtain effective porosity. This procedure is shown in an earlier section of this chapter.

Neutron porosity values, ϕ_n, may be substituted for the core porosity, ϕ_c, in the R_t vs. ϕ (Pickett) crossplot. The apparent limestone neutron porosities, $\phi_{n,LS,a}$, must be corrected for environmental factors before use. It is probably not wise to try to use this method in shaly sands. This is shown in Figure 9.25.

Neutron porosity values, $\phi_{n,LS,a}$, may be cross-plotted with the log values from the other porosity systems to obtain a corrected porosity value and an indication of lithology. An example of the neutron-density crossplot is shown as Figure 9.26.

Since the neutron-derived porosity tends to be the value of the total porosity and the density-derived porosity is nearly that of the effective porosity, then the difference is approximately that of the volume of shale, V_{sh}:

$$V_{sh} \approx \phi_{N,a} - \phi_{p,a} \tag{9.21}$$

9.8 Chlorine Logs

Thermal neutron reactions are very sensitive to the presence of chlorine because of chlorine's relatively large thermal neutron capture cross section. When gamma detection porosities are compared with those measured by a method unaffected by the presence of chlorine, such as the epithermal

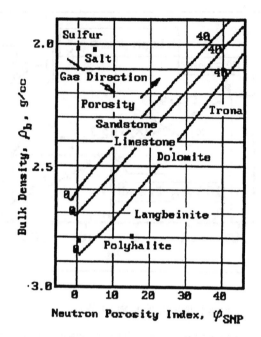

FIGURE 9.26
The ϕ_n vs. ρ_b crossplot. After Schlumberger.

detection system, the difference is a function of the amount of chlorine present. Since there is virtually no chlorine in natural hydrocarbons, the existence of chlorine can be equated with the presence of water, instead of hydrocarbon. Various devices are used on the systems to enhance the sensitivity to chlorine and reduce the non-chlorine signal.

Neutron porosity logs are quite effective in gassy zones. Since the hydrogen density in a gas is low, the neutron porosity system has a tendency to interpret this as low porosity. The bulk density of the total formation material is reduced by the extremely low density of the gas. Therefore, the density porosity tends to be high in a gas-filled zone. This signature of the two curves is indicative of the presence of gas. The presence of shale has the opposite effect. The neutron porosity reading responds to the high water content of the clay of the shale. It will read a high apparent porosity. The density system is only slightly affected by the shale because the shale density is so near that of a sand. Thus, its density will be nearly correct. This separation of the curves is diagnostic of the presence of a shale. This combination is called a "gas log". Figure 9.27 is a copy of a neutron and a density log (gas log) made in a gas well.

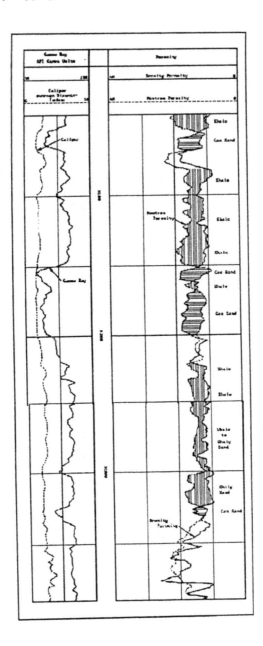

FIGURE 9.27
A gas log.

10

Neutron Activation Methods

10.1 Introduction

Neutron activation methods have been interesting techniques since the production of neutrons in 1930 by Bothe and Becker. They produced the neutrons by bombarding beryllium with alpha particles. This, of course, is exactly the reaction which is most commonly used today in isotopic sources for neutron porosity logging. Bothe and Becker, however, thought that the neutrons were high-energy gamma photons because of their penetrating power. Cork (1957) reports that Curie–Joliot determined that the radiation was non-ionizing. The neutron, as a distinct particle, was identified by Chadwick, (1932) who named the particle the "neutron" because of its neutral charge. This neutral charge accounts for its penetrating ability. Gamma ray production by interaction of the neutron with an atomic nucleus was recognized early. Curie and Joliot verified the neutron as a distinct particle by reacting it with nitrogen and measuring the energy of the resulting gamma rays. This, by the way, is the operating principle behind the recently developed explosive detector for use in airport security.

Recognition in the 1930s, of the role of thermal neutron capture by nuclei resulting in atomic fission, led, in the next decade, to the development of practical, controlled, atomic fission, with the release of large amounts of energy. Elemental analysis detecting neutron-produced gamma rays became a valuable chemical analytical tool, including some very accurate and sensitive forensic applications during the 1950s.

With the introduction of the neutron porosity geophysical systems early in the 1950s, it was obvious that there was enough activation of the formation materials by the logging source passage to interfere seriously with the simultaneous natural gamma ray measurement. Since all logging is done coming up hole, this problem was solved by putting the gamma ray detector above the neutron source so the gamma ray would be measured before the source reached the zone. Possible uses were suggested by the activation, however.

During the 1960s, the Soviets experimented in Saudi Arabia with activation logging for sulfur. They used a standard neutron source (of about 5×10^6 n/s) and a gamma ray detector. One sulfur isotope, with a half-life of several minutes, was produced. By placing the gamma ray detector about 3 m (10 ft) to follow behind the source, as it was trolled very slowly, the detector could examine the irradiated zone about one half-life after irradiation.

In the course of field testing the PGT high resolution gamma ray spec-trographic logging system, Goldman successfully logged for gold in Colorado, using an isotopic source of 2×10^6 n/s (personal communication). He achieved a sensitivity of 40 ppm of gold, *in situ*, with an accuracy of ±20%. With a higher neutron flux, this could become a viable commercial service.

Further experiments were done when the $_{94}Cf^{252}$ source, with its much higher neutron output, was introduced. One of the interesting investigations by Sentfle (U.S.G.S.) was the analysis for heavy metal content of the mud of the Potomac River. This also involved one of the early uses of simultaneous equation analysis technique for multiple log analysis.

All of the sources used in these trials were isotopic sources. Because they cannot be effectively shut off, it is necessary to either make the measurement during the high neutron flux interval or to mechanically interrupt the flux or remove the source. This latter, of course, took enough time that the prompt reactions could not be studied. There were successful systems using mechanical source removal, however (see the description of the IRT DFN system later in this chapter).

When the neutron generator was released for public use late in the decade of 1950, it became possible to make measurements with the source turned off within a few microseconds after the initiation of the neutron flux. Since this was done electronically, it was also simple to turn many of the monitoring and control functions over to the circuitry and the computer. Thus, it is possible to put an intense pulse (10^8 to 10^{14} n/s) of neutrons into the formation for 5 to 15 microseconds, turn the generator off and turn the detector on, accurately, at any time desired. Figure 10.1 is a schematic diagram of a linear accelerator type neutron generator. Continuous monitoring and correction of the operational parameters of the system are usually assigned to the computer.

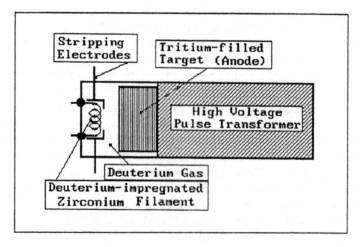

FIGURE 10.1
A diagram of a neutron generator.

Most of the neutron logging systems, including the neutron porosity systems in present commercial and experimental use, put a short burst of high energy neutrons (2 MeV to 14 MeV) into the formation materials. The reactions and/or the neutron populations are then examined as the neutrons moderate to thermal energy and decay or are subsequently captured. Any of the inelastic reactions produce gamma rays whose energies are characteristic of the reaction and the element involved. Therefore, most of these systems employ gamma ray detection. Sometimes the detection is spectrometric. In other systems it is a gross count detection. A few systems use thermal and/or epithermal neutron detection.

10.2 Types of Systems

The types of systems in use can be subdivided into several categories, frequently by source character:

1. Isotopic source systems are not used commercially for activation methods to any great extent, except for neutron porosity systems. They have some potential as special or "one-time" measurements. The source is usually alpha-beryllium (αBe) or $_{94}Cf^{252}$. They usually use unique source-to-detector spacings and carefully controlled logging velocities in order to enhance detection of the desired reaction.

2. Continuous or dc generator systems are the equivalent of the isotopic systems except that the source can be turned off when not in use. Commercial use of the dc mode is mostly restricted to neutron porosity systems.

3. Pulsed neutron systems:

   a. Saturation systems measure the probable capture cross sections of the hydrocarbon and water of the pore space to determine the saturations.

   b. Elemental analysis systems use some form of energy-sensitive gamma ray detection.

   c. Neutron population systems measure the numbers of neutrons present at a particular time to determine the amount of the reacting material present.

Isotopic sources used in activation systems are generally the same type of sources which are used in neutron porosity logging. They were discussed in more detail in Chapter 9. Usually, when an isotopic source is designed specifically for activation use (as contrasted with simply using a porosity tool source), it is considerably stronger than those normally used in porosity logging. This is especially true of those sources used in the laboratory.

Porosity equipment uses sources of 2×10^5 to 2×10^6 n/s output. Activation sources may be almost any output, depending upon the source type, the particular use to which it will be put, the handling and storage facilities, and the training and competency of the operators.

10.3 Neutron Generators

The neutron generator used in geophysical applications, is usually a small, linear accelerator which uses the reaction between deuterium, $_1H^2$, and tritium, $_1H^3$ (DT). Other reactions may be used, but the DT reaction is used most commonly, because of its high-neutron output. Neutron generators were also discussed in Chapter 9.

Neutron generators have several disadvantages, but for activation applications they have many advantages:

1. They may be turned off when not in use. This minimizes transport, storage, health, safety, and licensing problems.

2. The emitted neutrons are very energetic. The DT reaction results emits neutrons of about 14 MeV energy. This allows examining and/or using many more reactions (i.e., a reaction with $_8O^{17}$ at about 8 MeV) than with the usual isotopic neutron energies of about 2 to 4 MeV.

3. Generators can generally be operated in a dc (continuous) mode or they may be operated in a pulsed mode. Thus, they may be used for operations, such as neutron porosity, in the dc mode. Or, they may be operated in a pulsed mode, with a strong pulse of neutrons being emitted, the generator turned off and a detector gated on to detect a whole range of elastic and inelastic, short term, delayed, and prompt reactions.

4. The neutron flux density depends upon the accelerating voltage. The flux of one design of generator, operating at 125,000 volts is about 5×10^8 n/s. Some similar units are operated with outputs as high as 10^{13} n/s in the pulsed mode.

5. The cost of such generators is probably competitive with common isotopic sources, when costs of education, storage, transport, and licensing are considered. The average life of a typical generator is greater than about 10^7 pulses (almost 300 hours at 10 pps).

The most serious problem in the use of these generators in logging tools is an engineering one. It is the need to confine the high voltages (120,000 volts and higher) in the small, steel, pressure housings of the sonde. These are 1.7 to 4 in (4.3 to 10 cm) outside diameter. A 5 μs pulse in a system whose average power is 100 watts, has an instantaneous pulse power of about 20 megawatts. Its pulse current, at 125,000 volts, is 160 amperes. When this type of system fails, it is often catastrophic to the transformer assembly.

Commercial systems, in addition to the previously mentioned neutron porosity system, which use these generators are,

1. *The Pulsed-Neutron Petroleum Saturation Systems.* These have trade names such as TDT (SWC), Thermal Decay Time Log (SWC), Neutron Lifetime Log (WA), LLD (WA), TMD (Welex), and Thermal Multigate Decay Log (Welex). These systems measure the die-away rates or lifetimes of the thermal neutron population around the sonde by means of gross count gamma ray readings of the capture gamma rays over predetermined time intervals. It is also possible to use neutron detection to determine the population of thermal neutrons. Gamma ray detection is usually used because it has a statistical advantage over the neutron detection method. A major use of this system is determine the water and hydrocarbon saturation in cased holes where a zone may have been bypassed initially or where a zone has been partially depleted.

2. *Neutron-Induced Gamma Ray Logs.* These have names such as the GST (SWC), Gamma Spectrometry Log (SWC), and Spectrolog (WA). These usually use a gamma ray spectrographic detector and are used for elemental analysis. Sometimes the elemental analysis logs are run in conjunction with the logs of item 1. These systems measure numbers of capture gamma photons at specific gamma ray energies.

3. *The Direct Uranium Systems.* These are the PFN (prompt fission neutron) system and the delayed fission neutron (DFN) system. They were developed by Mobil Research, the Sandia National Laboratories, and the IRT Corporation. They measure the thermal neutron population at a predetermined time after the initial high energy neutron pulse. These systems are in limited use at this time, because of the slow pace of uranium exploration.

4. *The Thermal Neutron Formation Temperature Determination.* This system was developed by the IRT Corporation for the U.S. Department of Energy and measures the average energy of the thermal neutron population for determining the formation tem-

perature (as opposed to the normally measured borehole temperature). This system has limited availability, at this time.

10.4 Detector Types

Detection systems are conventional gamma ray, gamma ray spectrograph, and/or neutron systems. These have been described in previous chapters in this text and will not be repeated here. A unique feature of many detection systems is gating the detector on and off at specific times; off, to avoid the initial high-energy neutron pulse or on, after the pulse, at a specific time and for a given interval of time.

The type of detector in use will depend upon the system and the type of measurement desired. For example, the saturation systems use gamma ray detector or thermal neutron detection because either will be a function of the change in the thermal neutron population. The neutron induced gamma ray systems use gamma ray detectors and narrow energy band windows or multichannel analyzers because the goal is to identify the gamma emission of a particular energy. The direct uranium systems detect thermal neutrons because their population is a function of the amount of fission which has taken place. The formation temperature system of IRT Corporation uses a ^{3}He detector in the proportional (energy sensitive) mode.

10.5 Action Within the Formation Material

After the high-energy neutron leaves the source, it is moderated by collisions with the nuclei of the atoms of the formation materials. If the collision between the incident neutron and an atomic nucleus is an elastic one, the neutron bounces off the nucleus with an energy loss which is a function of the angles of rebound and the relative masses. There is no gamma ray emission in this type collision. This is the primary moderation type considered in neutron porosity logging. (See also Chatper 9.)

Inelastic scattering occurs only at higher neutron energies. In this case, the nucleus rebound does not take all of the energy lost by the neutron. The reaction appears as if the neutron were momentarily absorbed by the atomic nucleus, then released. Part of the incident neutron energy causes an excitation of the nucleus. This is quickly dissipated by the emission of at least one gamma photon by the nucleus. The number of photons and their energies are characteristic of the nucleus element and reaction involved. Thus, the photon energy analysis can identify the originating element.

With these two processes contributing to the moderating of the fast neutrons, about ten to several hundred milliseconds are required for the neutron to reach a thermal equilibrium with its environment. Fast neutrons can also be absorbed in some reactions, resulting in the emission of other particles and/or gamma photons. An example of this is the reaction of $_8O^{17}$ with a (~8 MeV) neutron, resulting in the emission of a proton and the generation of $_7N^{17}$. The nitrogen decays with the emission of a neutron. There are many other possible reactions also.

Thermal neutrons can scatter elastically with no net loss of energy because of the heat energy available from the environment. The energy of the neutron at this stage is

$$E_n = \frac{1}{2}m_n v^2 = \frac{3}{2}kK \tag{10.1}$$

where

m_n = the mass of the neutron
v = the velocity of the neutron
k = the Boltzman's Constant (1.38032×10^{-16} erg/K)
K = the absolute ambient temperature in Kelvins

Eventually the neutrons will be captured by a nucleus (or will otherwise decay into a proton and a negative electron with an average half-life of about 12 min). When a thermal neutron is captured, part of its mass is converted to energy. This mass conversion is described by Einstein's mass-energy equation:

$$E = m_n c^2 \tag{10.2}$$

where c = the velocity of an electromagnetic wave (light).

This capture excites the nucleus, which promptly emits one or more gamma photons of characteristic number and energy. This takes place, typically, 100 to 1000 microseconds after the emission of the neutrons from the source.

10.6 Saturation Systems

The most extensive use of pulsed neutron systems, at this time, is in the family of saturation determinations, (often combined with measurements of the emission ratios of carbon to oxygen (C/O), and silicon to calcium (Si/Ca), which are discussed later in this chapter).

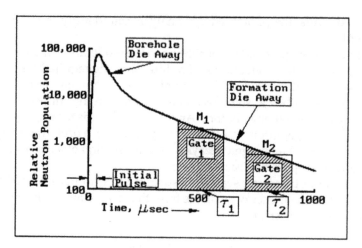

FIGURE 10.2
The pulsed neutron die-away curve. Original copyrighted by S.P.E.

In Figure 10.2, the thermal neutron population increases during the pulse from the generator. After the generator is turned off, the thermal neutron population will peak about 10 to 15 microseconds later. It then begins to decrease because the rate of moderation of the high energy neutron population to thermal energy decreases because of subsequent capture by atoms of material surrounding the probe. The initial slope of the die-away curve (out to about 250 microseconds) is principally a function of reactions within the material immediately surrounding the probe; the borehole. After about 350 μs, the slope is a function of the formation material farther out, around the borehole. The detector is gated at two or more times after about 400 μs. The difference of the average values, ΔM, of two of these measurements, divided by the difference in their times, Δt, is the slope of the curve:

$$Slope = \frac{\Delta M}{\Delta t} = \frac{M_2 - M_1}{t_2 - t_1} \tag{10.3}$$

where M = the average measurement value and t is the median time of that measurement, after the initiation of the pulse. The slope is a function of the total capture cross section of the formation material, Σ_t.

The thermal neutron lifetime logging term (or decay time; the reciprocal) is the rate at which thermal neutrons are captured or absorbed by neighboring material atoms or decay. This is controlled by the average thermal neutron capture macro cross section, Σ_t, of the material. Table 8.3b (in Chapter 8 of this text) list macro cross sections, Σ, of a number of materials. The lifetime is an inverse function of Σ and is a "half-life". It is the time, in microseconds, for an existing neutron population to be reduced one half.

A sigma unit, Σ (sometimes abbreviated as "su"), is equal to 10^{21} barns per cubic centimeter. The value of Σ_t is a function of the formation capture cross section, s, and the formation density:

$$\Sigma_t = \sigma A \rho \qquad (10.4)$$

where

σ = the thermal neutron nuclear capture cross section
A = Avagadro's Number
ρ = the material density in g/cm³

Chlorine, usually as NaCl, has the largest sigma value of the elements *commonly* found in sediments (Σ_{halite} = 752 Σ). Since the formation waters almost invariably contain some sodium chloride, Σ_w of formation water is usually high, due to its chloride content. The value will vary with the chloride concentration. The value of pure water is about 22 Σ.

The saturation systems measure the average capture macro cross sections of the mixture of the formation atoms. The value of Σ_t will reflect how rapidly thermal neutrons disappear from the neighborhood of the neutron or capture gamma ray detector. If the cross sections and amounts of most of the components can be identified, then it will be possible to interpret the total cross section, Σ_t, in terms of formation water saturation.

The early part of the neutron die-away is complex. Refer to Figure 10.2. It is a combination of the near environment of the logging instrument (sonde housing, borehole fluid, casing, and cement) and the formation material. Since the borehole and housing are so close to the detection system, their contribution forms and dies away quickly (usually within about 300 microseconds). The remaining reaction is only a function of the formation material cross section.

Fresh water and dead oil have almost the same capture cross section for thermal neutrons. Fresh water has a value of about 22 Σ and dead oil has slightly above 22 Σ. Light fractions and dissolved gas will lower the cross section value below 22 Σ. See Figures 10.3 to 10.7.

As the salinity of the formation water increases, the thermal neutron capture cross section of the water will rise rapidly to much higher values, up to about 250 Σ at saturation, because of the NaCl cross section. Thus, the contrast of Σ_w, with respect to Σ_{hc}, will increase rapidly. See Figure 10.7. In some cases, boron compounds may be added to the borehole fluid to increase the cross section. The additional cross section may be as high as 300 Σ for invasion detection and injection use.

Common, permeable, sedimentary rock matrix materials have cross sections less than 4 Σ to about 9 Σ. Shales and clays have cross sections of 5 to 40 Σ. Refer to Table 9.3b. Σ values may also be calculated from the values in Table 9.3a.

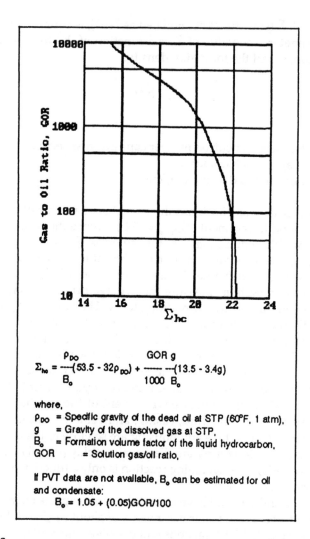

FIGURE 10.3
Σ curve, G/O ratio (Courtesy of Atlas Logging Services Division of Western Atlas International, Inc.).

The relationships which describe the total cross section, Σ_t, are,

$$\Sigma_T = \Sigma_{ma}(1-\phi) + \Sigma_w \phi S_w + \Sigma_{hc}\phi(1-S_w)$$ (10.5)

$$\Sigma_w = V_w \Sigma_{w,pure} + V_{NaCl}\Sigma_{NaCl}$$ (10.5a)

$$\Sigma_{ma} = V_s \Sigma_s + V_{sh}\Sigma_{sh} + \ldots + V_{LS}\Sigma_{LS}$$ (10.5b)

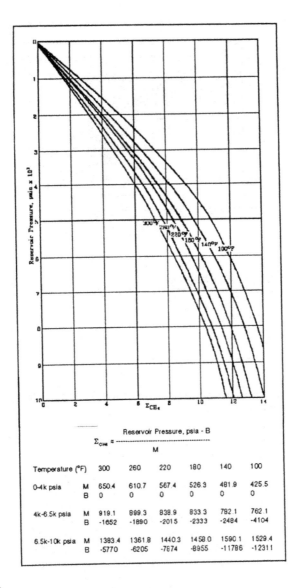

Reservoir Pressure, psia - B

$$\Sigma_{OH4} = \frac{\text{Reservoir Pressure, psia - B}}{M}$$

Temperature (°F)		300	260	220	180	140	100
0–4k psia	M	650.4	610.7	567.4	526.3	481.9	425.5
	B	0	0	0	0	0	0
4k–6.5k psia	M	919.1	899.3	838.9	833.3	782.1	762.1
	B	-1652	-1890	-2015	-2333	-2484	-4104
6.5k–10k psia	M	1383.4	1361.8	1440.3	1458.0	1590.1	1529.4
	B	-5770	-6205	-7874	-8955	-11786	-12311

FIGURE 10.4
Σ curve, methane, CH_4. (Courtesy of Western Atlas Logging Services Division of Western Atlas International, Inc.)

For dry clay shale this becomes

$$\Sigma_T = \left(V_s\Sigma_s + V_{sh}\Sigma_{sh}\right)(1-\phi) + \Sigma_w\left(\phi S_w + 1 - V_{sh}\right) + \Sigma_{hc}\phi(1-S_w) \qquad (10.5c)$$

For wet clay shale,

$$\Sigma_t = \left(1-\phi-V_{sh}\right) + V_{sh}\Sigma_{sh} + \phi\Sigma_w S_w + \phi\Sigma_{hc}(1-S_w) \qquad (10.5d)$$

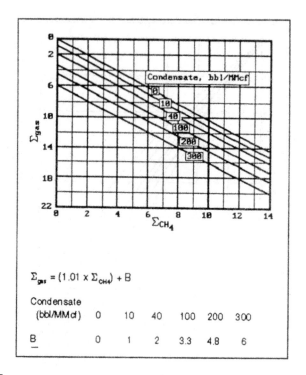

FIGURE 10.5
Σ curve, gas and condensate (Courtesy of Western Atlas Logging Services Division of Western Atlas International, Inc.).

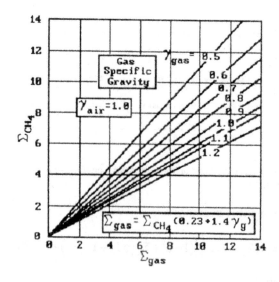

FIGURE 10.6
Σ curve, wet gas (Courtesy of Western Atlas Logging Services Division of Western Atlas International, Inc.)

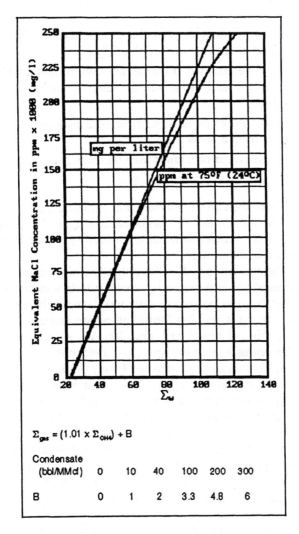

FIGURE 10.7
Σ curve, water and NaCl (Courtesy of Western Atlas Logging Services Division of Western Atlas International, Inc.)

where

ϕ = the fractional effective porosity
S_w = the fractional water saturation
V_s = the fractional volume of sand

Subscripts:

ma denotes matrix
w   denotes water
hc denotes hydrocarbon
s   denotes sand
sh denotes shale

The other values in Equations 10.6 to 10.7 may be found:

1. Σ_{ma} may be estimated from the core analysis, lithology cross plotting, or area experience (Table 9.3).

2. Σ_w may be determined from salinity determinations of the formation water and from salinity charts, such as shown in Figure 10.7.

3. Porosity, ϕ, may be determined from core analysis, from a porosity log, or from a cross plot.

4. Σ_{hc} may estimated or a chart may be used, such as Figures 10.3 to 10.6.

Equation 10.5 may then be solved for water saturation, S_w.

Because of the large cross section of NaCl ($\Sigma_{Nacl} = 752 \ \Sigma$) the normal amount of salt in the formation water will usually give good contrast with the hydrocarbon in the pore space.

In spite of the shallow depth of investigation of neutron systems, this method (and most neutron methods) is very effective in cased holes. The reasons are that the invaded zone of a permeable region begins to dissipate as soon as the casing is set. The formation region just outside the casing returns to near original or ambient condition very soon after casing. And, the steel of the casing is quite transparent to neutrons.

In actual use, Σ_t vs. ϕ can be plotted as a family of straight lines on a linear grid, with Σ_t as the y-axis (vertical) and ϕ as the x-axis (horizontal). The highest values of Σ_t will often show a trend of $S_w = 100\%$, especially if values are picked to deliberately include some which are suspected to be 100% water saturated. See Figure 10.8. Extrapolating the trend of $S_w = 100\%$ to zero porosity will show the value of Σ_{ma}, the y intercept. If these values are substituted into the Equation 10.7 with $S_w = 100\%$ and an assumed porosity, the probable value of Σ_{hc} may be calculated. Assumed values of ϕ and $S_w = 25\%$ and 50% will then define the trends of these saturation values.

Figure 10.9 shows some typical logs. These logged values must, of course, be corrected for various environmental factors, such as hole size, etc. Charts will be found for this operation in the chart books of the contractor who supplied the service.

If the pulsed neutron saturation systems are run in an open hole, the measurement will be in the invaded zone. The value of S_{xo} will be determined and the value of the residual oil saturation, ($S_{or} = 1 - S_{xo}$), may be calculated. The formation water will be, of course, the mud filtrate. Accurate measurements are quite possible here, because the salinity and salt type of the mud filtrate can be measured independently on the surface. In cases where contrast is low, borax compounds can be added to the mud in a known amount to increase Σ_w.

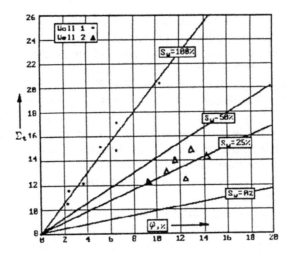

FIGURE 10.8
A crossplot of Σ_t vs. ϕ.

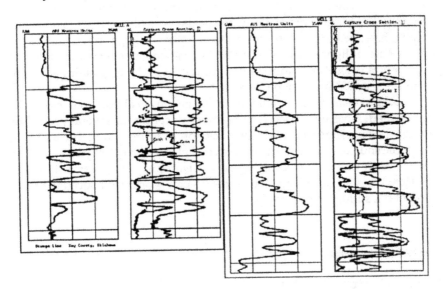

FIGURE 10.9
Neutron porosity and pulsed neutron logs (Courtesy of Western Atlas Logging Services Division of Western Atlas International, Inc.)

10.7 Neutron-Induced Gamma Ray Logs

The combination of the pulsed neutron generator and a gamma ray spectrograph allows examination of the characteristic emissions of the various reactions. It is obvious that this is a logical extension of the measurements of the previous section; the capture cross section.

Reactions between neutrons and atoms fall into several categories. The reaction may be elastic. That is, the neutron and the atom will still exist after the reaction. Each will rebound elastically, at angles predicted by elastic collision calculation, with no emission of gamma energy.

The reaction may be inelastic. Part of the energy of the incident neutron may excite the nucleus of the target atom. In this case, the neutron and nucleus will rebound after the collision, but not to as great a degree as in the elastic collision case. The energy of excitation will be immediately dissipated by the emission of one or more gamma photons of characteristic energy and number.

Thermal neutrons may scatter elastically, with no loss of energy, because they can take energy from the ambient heat. Or, they may be "captured" by a neighboring nucleus. In this case, the target atom may be made unstable and will change into another atom. Particles may also be emitted. One or more gamma ray photons of an energy characteristic of the reaction and the atom will be emitted, due to the energy from the reaction the mass difference. The particle may be any one of the possible particles, such as an alpha, beta, neutron, etc. It may, in some cases, be larger fragments of the atom, such as lighter elements. When the emitted gamma rays are detected by a good resolution gamma ray spectrograph, the originating element and the reaction type can be identified.

This is a relatively new field, as far as geophysical borehole logging is concerned. It has been used for laboratory purposes for several years. The Russians experimented with the detection of formation sulfur using an isotopic source, in the Middle Eastern oil fields a number of years ago. The half-life of one of the resulting sulfur isotopes was several minutes. Selected logging speed and source-to-detector spacing enhanced the reading of the sulfur isotope decay.

10.8 Capture Spectra

The rate, f_γ^t, of production of capture gamma rays per unit volume of material by the capture of thermal neutrons at point (r,z), in Figure 10.10, is, according to Tittman (1986).

$$f_\gamma^t = f_n^t \left(r^2 + z^2\right)^{\frac{1}{2}} \Sigma N_i \sigma_{ai} \gamma_i \left(E_\gamma\right) \tag{10.6}$$

where

f_n^t = the thermal neutron flux
N_i = the number of atoms of the *ith* kind per unit volume

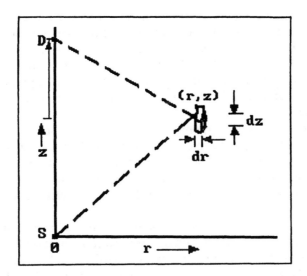

FIGURE 10.10
Incremental diagram for capture spectra.

σ_{ai} = the thermal neutron absorption cross section
$\gamma_i(E_\gamma)$ = the produced number of gamma rays produced of energy E_γ

The gamma rays are attenuated, E_γ, almost exclusively, by Compton scattering on their way to the detector by the factor

$$E_\gamma = \frac{\exp\left[-\mu_m(E_\gamma)\rho_b\left\{r^2+(I-z)^2\right\}^{\frac{1}{2}}\right]}{4\pi\left\{r^2+(I-z)\right\}} \tag{10.7}$$

Combining Equations 10.6 and 10.7 and integrating through all space will give the unscattered flux at the detector, produced by the thermal neutron capture.

10.9 Ratio Logs

If the ratio of the counting rates from the inelastic reactions of two different elements is taken, the formation parameters are equal in the numerator and denominator and drop out. This is because the natural environmental factors affect each peak equally. Thus, they are canceled out. For example, the measurement is less sensitive to natural gamma radiation. Then the ratio becomes merely the ratio of the number of atoms of each element.

The saturation systems described in an earlier section of this chapter often have energy detection windows and circuitry set up to determine the ratios of carbon to oxygen (C/O) and silicon to calcium (Si/Ca). The ratio of chlorine to hydrogen (Cl/H), and others, is also sometimes offered. These may be presented on the saturation log or may be presented separately. These ratios are of the inelastic scattering reactions and occur at several gamma energy levels.

The process that is undergone for creating inelastic spectra is much the same as for elastic spectra, except that the inelastic cross section must be used and the integration must be performed over the neutron energy, as well. This is because the process is energy dependent. The ratio of the emissions of two elements is also specific for the neutron parameters, ρ_b, and E_γ.

The C/O ratio log will show the relative amounts of carbon and oxygen in the formation material. Carbon occurs in large amounts in the hydrocarbons, in the pore space, and in the carbonates (X_nCO_3) (i.e., limestone and dolomite). Oxygen occurs in the pore water and in the carbonates. Thus, a high ratio reading is indicative of possible hydrocarbon. Some degree of possible saturation would be indicated. A medium level reading could be a carbonate. A low reading would suggest water.

The Si/Ca log is used as a companion to the C/O log. A high reading of the Si/Ca log could indicate a silica sandstone (usually some amount of $Si)_2$). Since silica also contains a large amount of oxygen, it would depress the C/O ratio value. A low Si/Ca reading would indicate a limestone. This would be accompanied by a medium reading on the C/O log. Figure 10.11 shows the response curves for a Si/Ca log.

Because of the oxygen content of water, the C/O ratio is also sensitive to porosity. And, because of the carbon content of hydrocarbons, it can indicate the degree of saturation. It is, of course, necessary, in this case, to have an independent measure of porosity. Figure 10.12 shows the SWC plot of C/O vs. porosity for determining the water saturation value, S_w. The effect of the rock type is also shown. A saturation log and a porosity log must be examined with these ratio logs. These measurements are sensitive to natural gamma radiation and should not be used in a radioactive sand.

The chlorine-hydrogen ratio, Cl/H is also available for determining fluid salinity. This can confirm the mud filtrate value in an open hole situation or the formation water salinity in a cased hole situation. Figure 10.13 shows the SWC response curves of this method for one sample hole size.

Figure 10.14 shows a Schlumberger GST log of absolute values of the gamma radiation iron, chlorine, calcium, silicon, sulfur, and hydrogen. Figure 10.15 is also a *Dresser Atlas* log showing ratios of Si/Ca for rock type identification in sediments. Also shown are C/O ratios, to differentiate between water and hydrocarbons, and a Ca/Si ratio curve. Schlumberger offers similar curves for ratios of Cl/H for salinity assessment, H/(Si+Ca) for porosity determinations, Fe/(Si+Ca) for relative iron indications, and Si/(Si+Ca) for easier rock type identification in sediments.

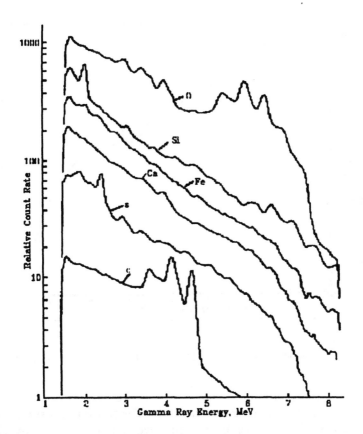

FIGURE 10.11
A spectrum with the silicon and calcium capture peaks identified. (Original copyrighted by S.P.E.)

Princeton Gamma Tech has experimented with their high-resolution (ultrapure or intrinsic germanium detector) gamma ray spectrograph logging system and a small (2.4×10^6 n/s) isotopic source in the *in situ* assaying for gold. The experiment was quite successful, but the source needs to be stronger, such as a generator source, to be viable commercially. Figure 10.16 shows their recorded spectrum.

10.10 Uranium Systems

Sandia National Laboratories, Mobil Research Corporation, and Century Geophysical Corporation successfully operated neutron-induced $_{98}U^{235}$ fission logging systems for the direct quantitative detection of *in situ* uranium. A short pulse (5 to 10 µs width) of high energy neutrons in the formation will moderate to thermal energy in about one millisecond. The

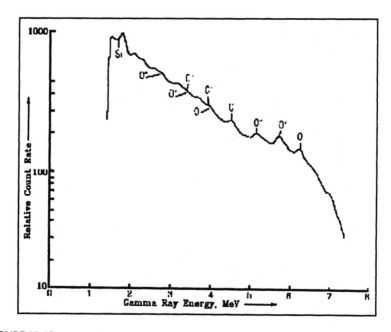

FIGURE 10.12

A spectrum with the carbon and oxygen capture peaks identified. (Original copyrighted by S.P.E.)

$_{98}U^{235}$ in natural uranium deposits can undergo a non-sustained fission reaction when exposed to thermal neutrons. The excess population of neutrons in the neighborhood of the detector, over that due to the source, indicated the presence of $_{98}U^{235}$. Since the ratio of $_{98}U^{235}$ to $_{92}U^{238}$ is usually constant, the total amount of uranium present in a formation can be calculated.

The reaction used was

$$U^{235}+n_t \rightarrow F_p+Q+n_p \rightarrow F_d+Q+n_d \tag{10.8}$$

where

n_t = a thermal neutron
F_p = the prompt fission fragments
n_p = the prompt fission neutrons
F_d = the delayed fission fragments
n_d = the delayed neutron

The fission fragments appear with the initial reaction.

If the neutron detector is gated to come on within a very few milliseconds (1 to 3 ms after the initial pulse), the prompt neutrons, which accompany the prompt fragments, can be detected. Their population is a function of the amount of $_{98}U^{235}$ present. This technique was suggested by

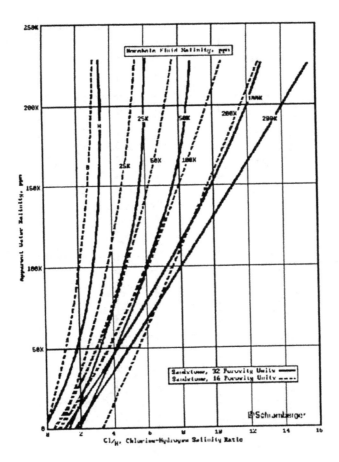

FIGURE 10.13
A spectrum with the carbon and oxygen capture peaks identified. Original copyrighted by
S.P.E.

Czubek (personal communication). If the waiting period is extended to 10
to 13 ms, the prompt reactions will have died away. In the meantime, the
prompt fission fragments will disintegrate, with the release of additional
delayed neutrons. A detector can be gated on for 300 to 350 ms and this
delayed fission neutrons can be counted. This, also, is a function of the
amount of $_{98}U^{235}$ present. This system was developed and published by
Mobil in about 1976 by Givens, Mills, Dennis, and Caldwell (1976).

Another system to expose the formation with fast neutrons from cali-
fornium ($_{94}Cf^{252}$) was developed by Eberline and Shreve at Kerr-McGee in
1971. This was further developed by Steinman, (1977), at IRT Corporation.
This system exposed the formation material to fast neutrons from a $_{94}Cf^{252}$
source on a rapidly moving belt. After exposure, the belt quickly removed
the source. The detector was then turned on, detecting the delayed fission
neutrons.

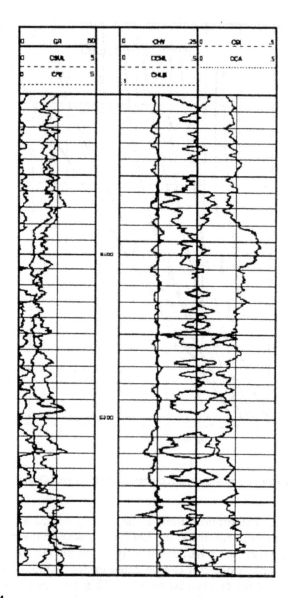

FIGURE 10.14
A neutron-gamma ray log for sulfur, iron, hydrogen, silicon, calcium, plus a gross count gamma ray (Courtesy of Western Atlas Logging Services Division of Western Atlas International, Inc.)

Interfering reactions are the fission of $_8O^{17}$ and $_{90}Th^{234}$. Both of these will also produce neutrons. $_8O^{17}$ will react with 8 MeV or higher energy neutrons, so this is not a factor when a $_{94}Cf^{252}$ source is used.

Sandia National Laboratories developed the prompt fission system (PFN) and Mobil and Century Geophysical Corporation developed the

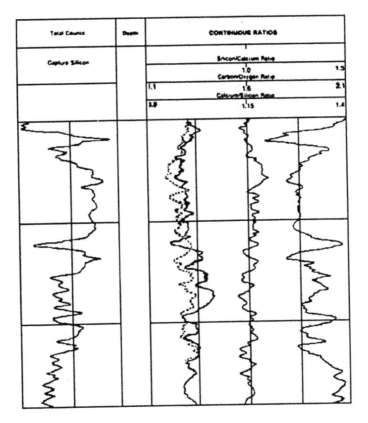

FIGURE 10.15
A ratio log. (Courtesy of Atlas Logging Services Division of Western Atlas International, Inc.)

delayed fission (DFN) system. Both systems were viable commercially, but were dropped when the uranium exploration slump started.

Figure 10.11 shows the time spectrum of neutrons.

10.11 Thermal Neutron Formation Temperature Log

Since the thermal neutrons possess energy by virtue of the heat content of the ambient material, their average velocity is a function of the temperature of the material. Fast formation neutrons moderate to thermal state in about 6 to 12 in (15 cm to 30 cm) from the source, depending upon the porosity (hydrogen density) of the formation material. The IRT Corporation formation temperature system spectrometrically measured the energy of the formation thermal neutrons. Their average velocity was interpreted as a function of the formation temperature. Even at the shal-

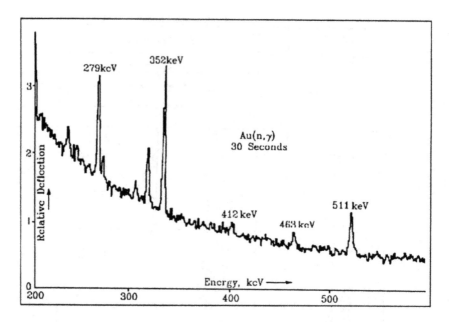

FIGURE 10.16
A Princeton Gamma Tech n-γ spectrum for gold (after Princeton Gamma Tech).

low penetration into the formation material of the neutrons, this was a more realistic value of the formation temperature than the usual borehole temperature measurement value.

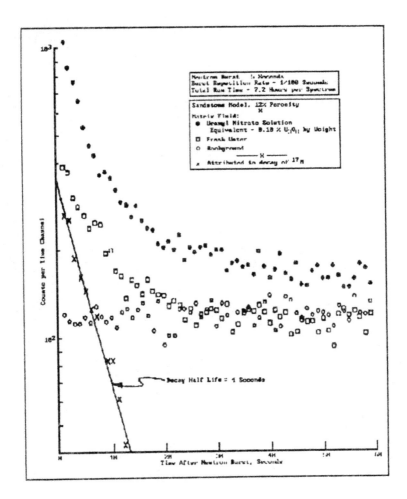

FIGURE 10.17
A DFN time spectrum. Courtesy Society of Exploration Geophysicists, 1976.

11

Acoustical Methods

11.1 Introduction

If any material is subject to a varying energy input, such as a blow or a continuous vibration, a mechanical (acoustic) wave will be set up. If the input is a blow (or a pulse, as shown in Figure 11.1) it will still be a wave, but one which dies out; which is damped. It will be a "ringing" wave. We will consider these phenomena in formation and borehole materials.

Acoustic energy is mechanical energy in wave (or pulse) form in the "acoustic" frequency range (usually presumed to be about 10 to 10^5 Hertz). It is transmitted through a material by mechanical oscillatory movement of the molecules of the material. Several types of acoustic waves can exist in materials. The velocity of a wave type through a single material is characteristic of that material and its physical environment. The velocity of a particular wave type in a multi-mode material is a function of the sum of the partial velocities of the material. Analysis of the velocities can identify some of the physical properties of the mixed material; in this case, the formation material.

11.2 Principles

In usual borehole geophysical logging practice, a mechanical pulse is put into the borehole fluid from a transducer. In some cases the pulse source may be on the surface or even in another hole, as in the case of cross-hole logging. The pulse travels radially as a pressure wave (Pressure-wave) through the borehole fluid to the wall of the hole and up and down the hole (see Figure 11.2). When the pulse contacts the wall of the hole, its velocity changes and the pulse continues in the solid formation material as several different types of waves, each at a different velocity. This is shown in Figure 11.3. The velocity will depend upon the average physical parameters of the formation material and upon the type of wave being considered.

At a position or positions some distance from the transmitter, the wave is detected by one or more receivers. The velocity, travel time, amplitude,

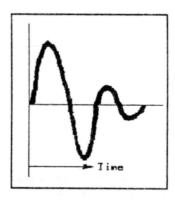

FIGURE 11.1
Acoustic pulse form.

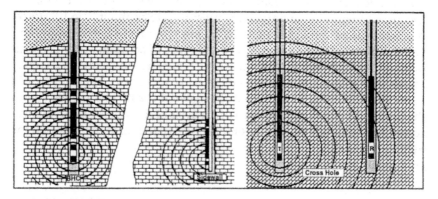

FIGURE 11.2
Types of acoustic porosity logging.

attenuation, and/or shape of the wave are measured. These parameters are then interpreted to indicate the conditions resulting in them (data inversion).

The acoustic waves obey all of the physical laws describing wave motion. The principles of geometric and physical optics describe the acoustic phenomena well. One principle, Huygens' principle governing reflection states,

$$\theta_r = \theta_i \qquad (11.1)$$

which says that the angle of reflection (θ_r) of the wave is equal to its angle of incidence (θ_i); Snell's Law of refraction which states that the ratio of the sine of the incident wave angle, θ_i, to its velocity, v_i, is equal to the same ratio of the reflected wave, r, and the transmitted wave, t:

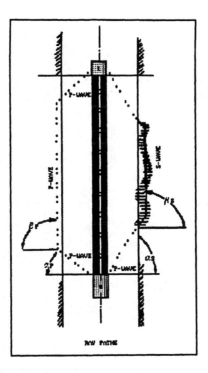

FIGURE 11.3
Wave paths of the P-wave and S-wave.

$$\frac{\sin\theta_i}{v_i} = \frac{\sin\theta_r}{v_r} = \frac{\sin\theta_t}{v_t} \tag{11.2}$$

This is illustrated in Figure 11.4. The wavelength, λ, of the pulse wave is a function of its velocity, v:

$$\frac{\lambda_1}{\lambda_2} = \frac{v_2}{v_1} \tag{11.3}$$

The amount of energy, W, transmitted across an interface plus the amount of energy reflected equals the amount of incident energy:

$$W_i = W_r + W_t \tag{11.4}$$

The direction of the oscillatory movement or vibration of the molecules within a solid, rigid material can be parallel to the direction of travel of the wave front (called "pressure waves", "compressional waves", or "P-waves")(Figure 11.5). These are compressional in nature. Or it can be normal to the direction of travel (called "shear waves" or "Shear-waves" or

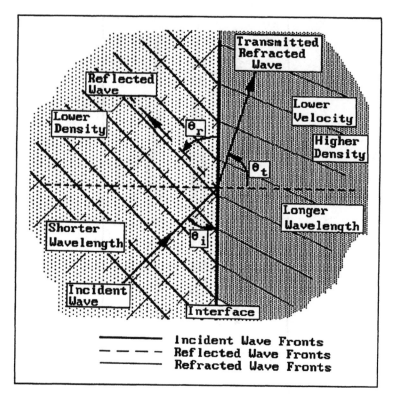

FIGURE 11.4
Reflection and transmission of a wave.

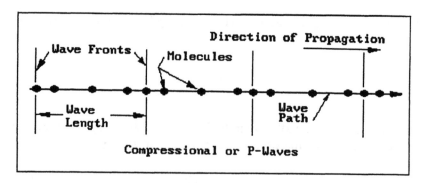

FIGURE 11.5
A schematic of a P-wave.

S-waves")(Figure 11.6). The Shear-waves are generally slower than the Pressure-waves. In a rigid material, such as a rock, both can exist simultaneously. In fact, more than one Shear-wave, in directions normal to the direction of propagation, can be supported.

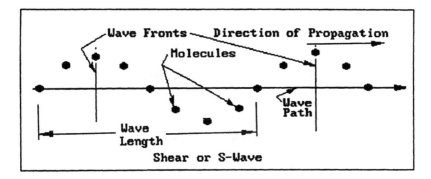

FIGURE 11.6
A representation of the molecular movement in a S-wave.

TABLE 11.1

Guided wave properties

Pseudo-Rayleigh Waves
  Dispersive (normal)
  Cutoff frequency $v_F < v_R \leq v_S$
  No geometrical spreading
  Multiple modes
  Do not exist if $v_S < v_F$
  Amplitude decreases into formation
  Amplitude oscillatory in borehole fluid
Stoneley Waves
  Slightly dispersive (normal and reverse)
  No cutoff frequency
  $v_{ST} < v_F$
  No geometrical spreading
  Tube wave at low frequency
  Highly attenuated if $v_{ST} < v_S$
  Amplitude decreases exponentially from borehole wall into borehole fluid and formation

v_{ST} = velocity of Stoneley waves, v_F = velocity in borehole fluid, v_R = velocity of Rayleigh waves, v_S = velocity of shear waves.

The interface between the hole and formation, the wall of the hole, will support a circular wave. This is the "Stoneley wave" or tube wave". The Stoneley waves may travel slower than the direct waves and their velocity is a function of their frequency. In addition, in a borehole, a direct compressional wave will travel up and down the borehole, but be confined to it by total reflection. This is "direct wave". These travel at the velocity of the wave in the borehole fluid. A wave will travel through the body of the instrument. This is usually deliberately attenuated and/or slowed by the design of the instrument, so as to not interfere with the detection of the other waves. See Table 11.1.

One can picture each molecule being attached to each neighbor with a spring (Figure 11.7). The springs represent the intermolecular force fields.

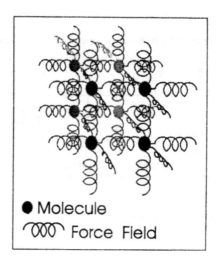

FIGURE 11.7
A representation of the intermolecular forces.

If one molecule is struck, the pulse will travel, via the springs, from the first molecule all through the system. It will radiate in all forward directions. It will reflect from any discontinuity, such as an edge or a change of density. Any change of spring character, such as rigidity, will change the velocity, amplitude, and direction of propagation of the wave front and result in a reflection. The oscillation may be either as a P-wave or as S-waves. All motions can exist simultaneously.

In a solid or rigid material, the intermolecular coupling is stiff and close. Any pulse is transmitted at a high velocity as both a pressure-wave and shear-waves. In a fluid, the coupling is loose. The pulse is transmitted more slowly. An shear-wave does not exist in a fluid because of the nature of the coupling in a fluid. In gels, such as clays, the material acts as a fluid when the water content is high. At very low water contents, the material begins to act more like a rigid solid.

The distance between successive, like peaks, is the wave length, λ. The length of time for one full oscillation (peak to peak) to pass is the period, T. The rate at which the wave front passes is its velocity, v. These factors are also related:

$$v = \nu\lambda \qquad (11.5)$$

$$T = \frac{1}{\nu} \qquad (11.6)$$

The rate of vibration, the frequency, ν, is the number of vibrations or cycles per second which pass a fixed spot. Frequency is given in Hertz (Hz); wave peaks or cycles per second. The number of wave peaks per centimeter is called the wave number, σ:

$$\sigma = \frac{3.28 \times 10^{-2}}{\lambda} feet \qquad (11.7a)$$

if λ is in feet, or

$$\sigma = \frac{10^{-2}}{\lambda} meters \qquad (11.7b)$$

if λ is in meters.

In downhole geophysical methods, the "travel time", t, is used. It measures the length of time, in microseconds, μs, for a wave front to travel 1 unit length. Thus it is a function of the reciprocal of the velocity:

$$t = \frac{10^6}{v} \qquad (11.8)$$

The inverse travel time is called "slowness". The depth of penetration of acoustic waves is the distance into the formation material where their amplitude has been attenuated to $1/e$ ($e = 2.718...$, $1/e = 0.368...$) will depend upon the velocity and the frequency. This is the skin effect and the depth is called the "skin depth", δ:

$$\delta = f\left(\lambda, \frac{1}{v}\right)^{1/2} \qquad (11.9)$$

It follows, then, in a typical logging situation, if the velocity in limestone is 19,000 ft/sec (5792 m/sec), and the frequency, v, is 20 KHz,

1. The wave length, λ, will be $v/v = 19000$ ft sec^{-1}/20000 sec^{-1} = 0.95 ft = 5792 m sec^{-1}/20 × 10^3 m = 29 cm
2. The period, T, will be $1/v = 1/2 \times 10^4$ sec^{-1} = 5 × 10^{-5} sec
3. The wave number, σ, will be 3.28 × 10^{-2}/λ = 0.034 cm^{-1}
4. The travel time, t, will be 10^6/v = 52.6 μs/ft = 173 μs/m.

The frequencies of the waves used in geophysical methods are in and near the acoustic range. This is the reason for the name, "acoustic". Surface seismic methods use low frequencies, in the range of <0.1 Hz to <100 Hz. Those used in downhole methods, such as wireline logging, are usually much higher; they range from about 5 KHz to around 100 KHz.

In a rigid medium, such as the formation rock, the acoustic waves will exist as pressure-waves and as shear-waves, simultaneously. These will have different velocities, the pressure-waves being faster. When a borehole is introduced, there will also be a fluid or direct wave. This wave propagates as a pressure-wave only (a shear wave cannot exist in a fluid), in the borehole fluid. Because of the relatively small diameter of the borehole, compared to the wave length of the tube wave, the wave is confined to the borehole by total reflection. That is, it always hits the wall of the hole at less than its critical angle, and thus is totally reflected back into the borehole. Total reflection is achieved when the angle of refraction equals or exceeds 90 degrees. thus, it depends upon the wavelength, λ, and the relation between the indices of refraction of the borehole fluid and the formation material. This is the same principle that speaking tubes use. It is also part of the operating principle of optical fibers.

Associated with the borehole also, are Stoneley-waves or tube waves. These waves travel along the borehole wall at the formation-borehole interface. They have the same rolling motion that an ocean wave has in shallow water. They can be considered to be a combination of one Pressure-wave and two Shear-waves (see Figure 11.8). These two types of waves travel at velocities slower than the formation P-wave.

Also present is a deep wave which is a formation P-wave which has apparently traveled farther than the first P-wave. Its path goes deeper into the formation, away from the borehole, than the normal first arrival of the Pressure-wave (which is assumed to travel parallel to the borehole wall). It frequently has a higher velocity than the one closer to the borehole, in the invaded zone, but travels farther. Thus, it arrives later than the one in the invaded zone.

The factors which affect acoustic waves in a formation are listed in Table 11.2. Of these, there are comments and qualifications:

1. The petrologic composition has less effect upon the elasticity than does the rock texture and the geologic history of that particular zone.

2. An increase of the porosity of a sedimentary zone element will usually result in decreases of the elastic moduli and wave velocity.

3. The relationships involving the water content are complex. In consolidated sediments, an increase of water content will frequently result in an increase of the wave velocity. In unconsolidated sediments, it usually results in a decrease of velocity.

4. Diastrophism generally tends to increase wave velocity and elastic constants. Thus, we see higher velocities in older rocks because of the greater diastrophism.

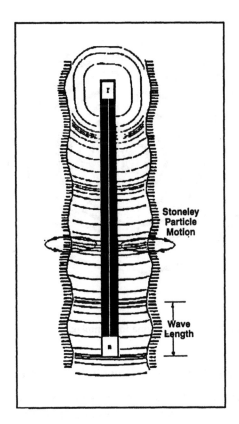

FIGURE 11.8
A diagram of Stoneley waves.

TABLE 11-2

Factors affecting acoustic waves

Six factors affect acoustic waves in rock:
1. Petrologic composition (mineral type)
2. Formation porosity
3. Water content
4. Diastrophism (deformation processes)
5. Elastic anisotropy
6. Depth of burial (overburden pressure)

5. Elastic anisotropy appears to be important only in metamorphic and igneous rocks. Sediments can be treated as isotropic because of the random grain orientation,

6. Dehydration and granulation tend to increase with depth of burial.

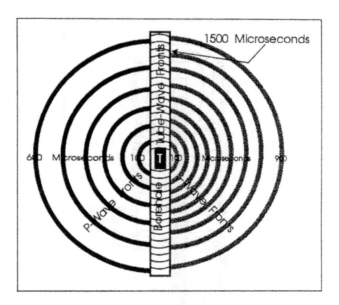

FIGURE 11.9
Wave types within the borehole and formation complex.

11.2.1   Operation Principles of Downhole Acoustic Systems

The acoustic or mechanical wave travels from the transmitter, T, through the mud to the wall of the hole as a P-wave. It enters the formation and propagates as a P-wave and as a slower S-wave. See Figure 11.9. The portion of each wave front (both P-wave and S-wave) which impinges on the borehole wall at a critical angle is refracted parallel to the axis of the hole, within the wall material. This component of each wave re-radiates some of its energy back into the borehole mud all along its path as a representative P-wave.

The receiver, R, is at a fixed distance from the transmitter and senses the portions of the parallel waves radiating into the borehole. This may be a short distance (3 to 7 ft or 0.8 to 2.1 m) in the standard systems or a greater length (8 to 12 ft or 2.4 to 3.7 m) in the long spacing systems.

The shorter spacing acoustic logging systems are very shallow investigation devices, because of the refraction of the waves parallel to the borehole and because of the skin effect. The usual acoustic device probably does not investigate more than 6 in (15 cm) into the formation from the borehole. Even, the "deep investigation sonic" systems do not penetrate horizontally more than about a meter. The waves which penetrate more deeply into the formation will also eventually re-radiate into the borehole as "slow waves". These last are detected with the long spacing systems.

The velocities, v_p, of the pressure-wave and v_s, of the shear-waves, are reciprocals of their travel times, t, and are determined by the sum of the proportional mechanical properties of the various materials through which they are passing:

$$t_{ave} = \frac{10^6}{\Sigma v_n V_n}$$

(11.10)

The velocity of the shear-wave, v_s, is

$$v_s = \sqrt{\left(\frac{\mu}{\rho}\right)}$$

(11.11)

where
   μ = the shear modulus of the path material
   ρ = bulk density of the material

The shear modulus, μ, describes the torsional stress relationship to the strain. For logging data, the value of μ is usually taken as,

$$\mu = 1.34 \times 10^{10}\left(\frac{\rho}{t_s^2}\right)$$

(11.12)

The relationship for the velocity of the pressure-wave, v_p, is

$$B = \rho_b v_p^2 - \frac{4}{3}\mu$$

(11.13a)

or

$$v_p = \sqrt{\frac{B + \dfrac{4\mu}{3}}{\rho}}$$

(11.13b)

where
   ρ_b = the bulk density
   B = the bulk modulus of the formation material

The bulk modulus, B, is the change of volume of a sample with applied pressure. It is volumetric compression factor.
   A shear wave will not exist in a liquid. Thus, in the drilling mud, the term, $4\mu/3$, of Equation 11.13b becomes equal to zero. In this medium,

$$v_p = \left(\frac{B}{\rho}\right)^{\frac{1}{2}}$$

(11.14)

The third mechanical modulus is Young's modulus, E. Young's modulus is the ratio of the longitudinal stress to the longitudinal strain. Thus, it is the extension or elastic modulus:

$$E = 2\rho v_s^2 (1 + P)$$ (11.15)

where P is Poisson's ratio:

$$P = \frac{0.5 \left(\dfrac{v_p}{v_s} \right)^2 - 1}{\left(\dfrac{v_p}{v_s} \right)^2 - 1}$$ (11.16)

Poisson's ratio is the relationship of the diameter change of a sample with compression or extension. For logging purposes, the value of P is usually

$$P = \frac{1}{2} \left(\frac{t_s^2 - 2t_p^2}{t_s^2 - t_p^2} \right)$$ (11.17)

where
    t_s = the shearwave travel time
    t_p = the pressure-wave travel time

    Normally, only the formation pressure-wave was used in the older porosity logging systems and many of the more modern ones. There are, however, many new systems that make use of the shear-wave. In these, the waves are separated by timing and with some dependance upon amplitude. Direct-waves, tube-waves, and tool-waves exist also. See Figure 11.10a,b,c.

11.3　Tool Configurations

11.3.1　Single-Transmitter, Single-Receiver Systems

Although tools with a single transmitter and a single receiver are in use, it is difficult to use them quantitatively because of the difficulty separating the mud path the signal from the formation signal. The total time from the initiation of the pulse at the transmitter until the first arrival of the P-wave at the receiver is measured. This includes two occasions of substantial and unknown time through the borehole mud. The use of these

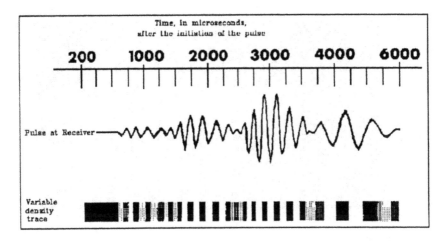

FIGURE 11.10a
The wave train at the receiver. (Courtesy of Western Atlas Logging Services Division of Western Atlas International, Inc.)

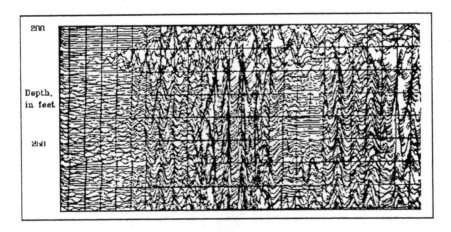

FIGURE 11.10b
A variable amplitude presentation. (Courtesy of Western Atlas Logging Services Division of Western Atlas International, Inc.)

systems is confined mostly to cement bond determinations and to qualitative investigations in some mineral work.

11.3.2 Multiple-Receiver Systems

The most common of the petroleum acoustic logging tools use two or more receivers and one or more transmitters. See Figure 11.11.

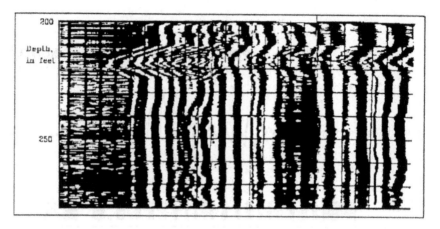

FIGURE 11.10C
A variable-density presentation. (Courtesy of Western Atlas Logging Services Division of Western Atlas International, Inc.)

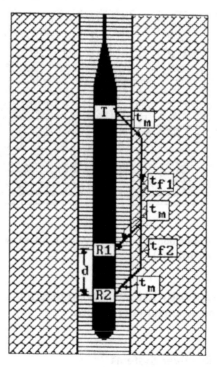

FIGURE 11.11
A single-transmitter, 2-receiver acoustic sonde.

The 2-receiver systems measure the pulse transit time from the transmitter to each receiver. The time for the wave front to reach the first receiver, R_1, is

$$T_1 = T_{m1} + T_{f1} + T_{m2} \qquad (11.18a)$$

where
T_m = the passage time through the mud to the formation
T_f = the passage time through the formation

Time to reach the second receiver, R_2, is

$$T_2 = T_m + T_{f1} + T_{f2} + T_{m3} \qquad (11.18b)$$

Then the total time to the first receiver is subtracted from that to the second receiver.

$$\therefore T_t = T_2 - T_1 = T_{f2} \qquad (11.18c)$$

This eliminates the time through the mud. The remainder is the time through the formation material. Then, the total time, T_t, is divided by the distance, d, between the receivers. The result is the average travel time for the formation in microseconds per unit distance:

$$t = \frac{T_t}{d} \qquad (11.18d)$$

The assumption is made that the logging tool is coaxial with the borehole and that the borehole wall is uniform and not caved. In large holes, stand-off devices might be used.

A typical system will have the first receiver about 3 ft or 1 m from the transmitter. There will be a second receiver about 4 ft from the transmitter. Often, there will be additional receivers at greater distances. Multiple receivers allow a choice of spacings for selection of the best signal for the hole conditions.

The resolution of the system will be equal to the distance between the receivers. The response shape will be a rounded trapezoid. The bed boundaries will be at the midpoints or inflection points of the trapezoidal slopes.

If the sonde is tilted in the borehole (not coaxial), the average travel time value will be in error, if the receivers are closest to the wall (compared to the transmitter) or visa versa. This is very difficult to detect on the log.

The long spacing systems are designed primarily to take advantage of the longer paths of the P-wave, the S-wave and St-wave. This will result in greater mode separation and better resolution than using those which remain at the wall of the hole. Once the travel time of the P-wave, has been reliably measured, Equations 11.12 to 11.17 can be solved (using also, the density measurements) for the mechanical relationships given earlier in this chapter. These mechanical relationships furnish valuable information for mining and engineering work. The data can be used to determine, non-destructively, such things as slope stability, column strength, and concrete quality.

In order for Equations 11.18 to be correct, when using a single transmitter, dual receiver system, the sonde axis must be coaxial with and centered in the borehole. There also must not be any changes of borehole diameter within the array length. Standoffs or bumpers on the tool help some to control these problems, but are not always completely satisfactory.

11.3.3   Borehole-Compensated Systems

To overcome the effects of tool position and (somewhat) of caves in the borehole wall, a borehole -compensated (BHC) system is used. This system basically uses two transmitters and a pair of receivers for each transmitter. It is effectively two systems, back-to-back. The average of the two signals tends to compensate for the tool position. The resolution of this type system is equal to the distance from the midpoint of R_1 to R_2 to the midpoint of R_3 to R_4. See Figure 11.12. The results of this are used in the same way as the results of the non-compensated tool.

Some BHC systems may have only two receivers with the two transmitters. These receivers serve both transmitters by switching, as shown also in Figure 11.12. The resolution of this type system is equal to the distance between the two receivers.

Another system uses one transmitter and two receivers with a computer program to properly invert and average the signals. Another uses two transmitters close together and, at a distance, two receivers close together. These are operated sequentially with a computer program to combine and average. These systems all appear to be valid. This last technique is also called "depth-derived borehole compensation" (DDBC) by Schlumberger Well Services, Inc.

The DDBC system minimizes tool position errors. It uses two transmitters and two receivers (see Figure 11.13). Readings of the formation are taken at two different positions of the sonde; once when the two receivers straddle the measure point and again when the two transmitters straddle the same point at the formation. The first reading, t_1, is from T_1 to R_1 minus T_1 to R_2. The second reading, t_2, is T_1 to R_2 minus T_2 to R_2. t_1 is memorized until t_2 is taken. The two are then averaged to get the equivalent of a BHC measurement value of the interval travel time:

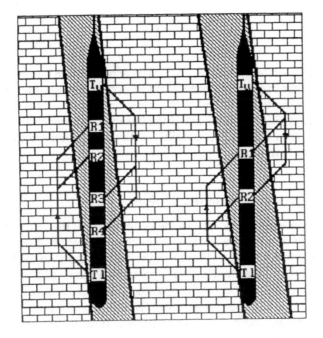

FIGURE 11.12
Borehole-compensated acoustic system sondes.

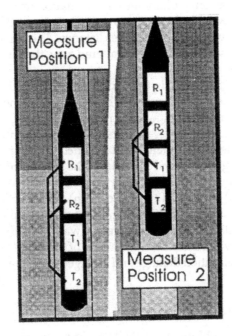

FIGURE 11.13
The DDBC system (after Schlumberger, 1989).

$$t = \frac{t_1 + t_2}{2d} \tag{11.19}$$

where d is the distance from R_1 to R_2. The actual device has several arrays available. If the upper transmitters and receivers are used, the spacings are 8 ft and 10 ft (2.4 m and 3.0 m). If the lower ones are used, the spacings are 10 ft and 12 ft (3.0 m and 3.7 m). If the borehole is large and the two-way mud transit time is large, it is possible, with these systems, for two consecutive signals to overlap.

11.3.4   Long-Spacing Acoustic Systems

A great advantage of the long-spacing systems is that they allow the slower Stonely-wave and the S-wave to gain greater separations from each other and from the faster P-wave. Therefore, the determination of the travel times for the S-wave and the Stoneley-wave are more accurate and reliable.

Drilling and invasion affect the acoustic wave in several ways. Mechanical damage due to the drilling process is often present near the borehole. The types of fluid components and their amounts are obviously different from those in the deeper zones. The zone immediately adjacent to the borehole wall can be altered by both the mechanical process and the chemical process. This is the zone in which the acoustic waves usually travel. The depth of invasion, radially into the formation from the borehole wall, must be >3M (at 30 kHz this is approximately 0.6 m or 21 in) to contain the Pressure-wave over any appreciable distance. The velocities of all of the wave types are usually lower in the altered zone than in the unaltered formation material.

There is a large difference in travel distance between the long spacing signal and that of the conventional (shallow) system. It should be mentioned too, that the apparent porosities determined by the long spacing system are frequently significantly lower than those of the conventional system. This last probably reflects the difference in the effects of the mechanical damage due to the drilling process and the relief of the stress (stress relaxation).

The long spacing systems of one service company have spacings of 8 and 10 ft (2.4 to 3.0 m) or 10 and 12 ft (3.0 to 3.7 m). This gives much greater penetration than the standard BHC system, whose signal is probably confined to the region immediately next to the borehole wall. Thus, the long spacing system is more likely to be free from the effects of the borehole and drilling process alteration, relaxation damage and enlarged borehole effects.

11.3.5   The Array Sonic System

The Array Sonic (Schlumberger Well Services, Inc.) is a multipurpose system. It will operate in several modes. It has two transmitters which are capable of transmitting pulses whose frequencies can be varied from about 5 KHz to 18 kHz. Two receivers are 3 ft and 5 ft (0.9 m and 1.5 m) from the upper transmitter. This combination allows standard 3 ft and 5 ft spacings in open hole logging. In a cased hole, they can be used for cement bond logging and 5 ft variable density logs. Figure 11.14 is a diagram of the downhole portion of this system. It also has an array of 8 closely spaced receivers, 3.5 ft (1.07 m) long, 8 ft (2.43 m) from the nearest of two transmitters, each 2 ft (0.61 m) apart. There is another transmitter-receiver array at the top of the tool whose function is to measure the travel time of the borehole fluid. The waveforms are collected from the receivers and digitized. Figure 11.15 is an illustration of the waveforms which are processed. The processing software finds and analyses all propagating waves in the composite waveform. It uses a digital semblance method algorithm to identify and align the various arrivals across the array and to determine the velocities of all coherent components.

A fixed length time window in the processing algorithm scans the wave forms in small overlapping steps. For each position on the first receiver waveform, the window position is moved out linearly in time across the array of receiver waveforms, beginning with the fastest wave expected and stepping to the slowest wave expected. For each of these movements, a coherence function is computed that measures the similarity in the waves within the windows. When time and movement outward correspond to the arrival time and slowness of a particular component, the waveforms within the windows will be almost identical and will yield a high value of coherence. In this way, the set of waveforms form the array is examined over a wide range of possible arrival times and slowness components. Figure 11.16 shows a plot of the processing coherence function contour plot and a log derived from it.

A "coherence" diagram plots the results of the "coherence algorithm". It shows the arrivals of the Pressure-wave, the Shear-wave, and the St-wave. A typical log of this mode of this system is also shown in Figure 11.16.

In fast formations (S-wave velocity > mud wave velocity), the Array Sonic system will determine the velocity values of the P-wave, the S-wave, and the St-wave. In slow formations, the P-wave, the St-wave, and the mud wave velocities will be determined and the S-wave velocity derived from them. The data which can be supplied by this system are shown in Table 11.3.

11.3.6 Log Presentation

The acoustic porosity log and the gamma ray log are usually run simultaneously. The acoustic log may be presented as travel time (μs per unit distance) or as apparent porosity ($\phi_{l,a}$). Very occasionally, the log will be presented in velocity, in distance per second. The usual scales are presented

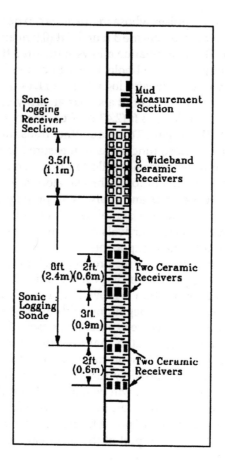

FIGURE 11.14
The array sonic sonde (after Schlumberger, 1989).

right to left in tracks 2 and 3 of the API standard log format. If $\phi_{t,a}$ is presented, it is done in the same way as the density-derived or the neutron-derived porosity; that is, scaled right to left and from –15% to +45% (or sometimes –15% to 60%), reading right to left. The negative apparent porosity value is indicated because the probability is high that the assumed matrix travel time will not be correct for the whole log. Track 1, in this presentation usually has the gamma ray curve and often has the caliper and bit diameter curves. If an *SP* is available, it will be recorded in track 1, also.

Track 2 often has an integrated travel time curve, in milliseconds. It takes the form of a straight line with pips spaced by depth. The vertical formation distance indicated between pips indicates 10 milliseconds of vertical formation travel time.

Many presentation formats are available. One contractor records a single, amplitude modulated wave trace every 6 ft of depth, on the left side

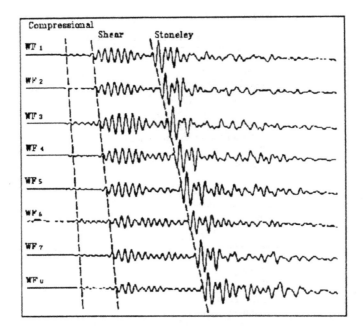

FIGURE 11.15
Array-sonic waveforms (after Schlumberger, 1989).

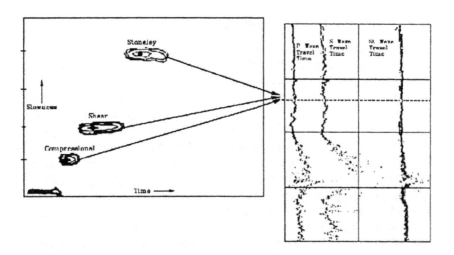

FIGURE 11.16
Contour plot and log (after Schlumberger, 1989).

of their full wave, variable density acoustic recordings. The P-wave, S-wave, and St-wave arrivals are identified and are easier to see than on the variable density recording. This presentation is an aid in locating these features on the variable density recording.

TABLE 11-3

Functions of the Array Sonic System

P-wave Transit Time
  6 inch spacing
  3 feet to 5 feet spacing
  5 feet to 7 feet spacing
  8 feet to 10 feet spacing
  10 feet to 12 feet spacing
Wave Train Derived Transit Time
  P-wave
  S-wave
  St-wave
Borehole Fluid Transit Time
Amplitude Values
Energy Values
Frequency Analysis
Cased Hole
  P-wave
  S-wave
  St-wave
Cement Bond
Variable Density

After Schlumberger, 1989.

11.3.7 Effect of Gas and Road Noise

The presence of gas in the formation and, especially, in the borehole can cause a measurement malfunction, called "cycle skipping". Cycle skipping is a sudden and erroneous large increase in the recorded apparent travel time. The increase will be an integral multiple of the wave period, T. It is usually quite evident on the log and it is caused by attenuation of the wave amplitude because of the compressibility of the gas. The amplitude of the first arrival can be attenuated enough that it has not enough amplitude to trigger the timing circuit. Thus, the second or third arrival will be detected instead of the first (Figure 11.17). Cycle skipping can also be caused by the improper setting of the amplitude discriminator in the acoustic instrument. This latter, however, is less likely than the gas effect.

The usual type of acoustic tool cannot be used in gas-filled holes and sometimes not in gas-filled formations because of the energy absorption of the compressible gas.

"Road noise" is the noise induced by the irregular accelerations from being dragged along the wall of the hole. It will usually take the form of spikes. The receiver, R_2, is more sensitive to this problem than is R_1, because of the lower signal amplitude at R_2. The spikes will usually be toward lower values of t_a, the average travel time.

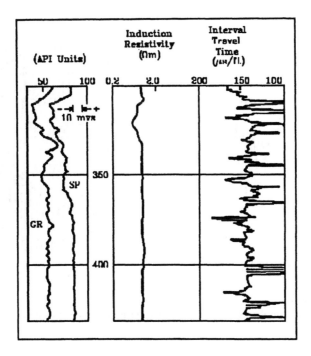

(API Units)	Induction Resistivity (Ωm)	Interval Travel Time (μs/ft.)

FIGURE 11.17
An example of cycle skipping (after Schlumberger).

11.4 Full-Wave Systems

The arriving (at the receiver) representative P-wave train is a complex mixture of the formation P-wave, the Pressure-wave representation of the formation S-wave, the direct wave (a Pressure-wave coming through the mud column and through the tool, itself), and the Pressure-wave representing the St-wave. There also may be some other waves and reflected waves present, usually at lower amplitude and later time.

The most successful of the original methods of recording the full wave train is the intensity modulation method shown in Figure 11.18. The recording was made by drawing a photographic film before the face of an oscilloscope at a rate which was proportional to the logging speed. The oscilloscope trace was intensity modulated by the signal. The intensity modulation came to be preferred because it is easier to read. Amplitude modulation was also used, but is seldom seen anymore.

Newer methods involve digital selection and comparison techniques to separate the signal components in time. In this case, each measured and separated component is recorded on the log as a single trace. Much effort

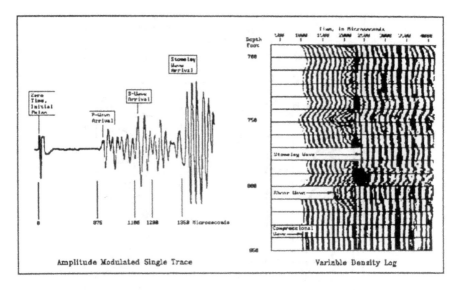

FIGURE 11.18
An early variable amplitude trace and variable density log in an open hole.

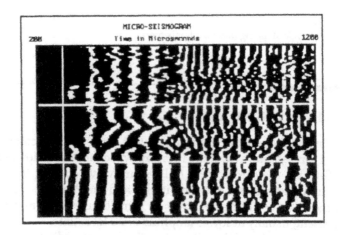

FIGURE 11.19
An early variable density log made in a cased hole.

has been and is being expended to make this method reliable, since it often promises to be much easier to use, more accurate, and faster than the older full-wave train method, such as the analog variable density or analog variable amplitude recordings. Also, it lends itself better to computer processing.

It should be noted that these systems are often very effective in cased holes. This is illustrated in Figure 11.19. Usually, the formation may be "seen" behind the casing.

11.5 Evaluation

11.5.1 Porosity Determinations

The original intent of acoustic borehole logging was to measure formation porosity. The most commonly used relationship for relating travel time to porosity is the Wyllie Time Average Relationship. It was proposed by M.R.J. Wyllie in 1956:

$$t = \phi t_f + (1-\phi)t_{ma} \qquad (11.20)$$

where
t = the logged travel time
t_f = the average fluid travel time
t_{ma} = the average rock matrix travel time

The fluid travel time, t_f, is the sum of the products of the component travel time, t_{fn}, of each component fluid times the partial volume, V_n, of that fluid:

$$t_f = V_1 t_1 + V_2 t_2 + \ldots + V_n t_n \qquad (11.21)$$

Figure 11.20 shows the effects of pressure, temperature, and water salinity upon the water in the pore spaces of the formation. Note that even the low compressibility of water still has a significant effect upon the acoustic velocity.

The matrix travel time, t_{ma}, is the sum of the products of each component matrix travel time, t_{man}, times the partial volume of that component, V_{man}:

$$t_{ma} = V_{ma1} t_{ma1} + V_{ma2} t_{ma2} + \ldots + V_{ma,n} t_{ma,n} \qquad (11.22)$$

The Wyllie "time average" relationship has a number of shortcomings. It responds well to the average porosity only in uniform, granular sands. Charts published by the major logging contractors show both the Wyllie response curves (straight lines) and the empirical "field experience" lines (curved lines). Figure 11.21 was extracted from the SWC chart, Por-3.

Compaction by the overburden pressure also has an influence on the measurement of acoustic travel time of a core sample. The overburden stress is relieved when the core is brought to the surface. This may tend to make the laboratory values of the core porosities higher than *in situ* porosities.

It is obvious from Figure 11.20 that some form of correction for pressure, compaction, and perhaps temperature, is frequently needed when reducing

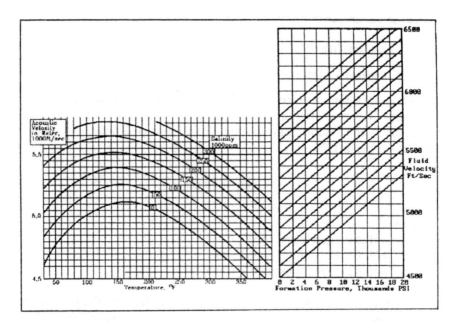

FIGURE 11.20
Pressure and temperature corrections for acoustic wave velocities.

the acoustic data. The usual form of the compaction correction, C_p, to obtain a corrected acoustic porosity, $\phi_{t,cor}$, is

$$\phi_{t,cor} = \frac{\phi_{t,a}}{C_p} \qquad (11.23)$$

where
  $\phi_{t,a}$ = the porosity from the acoustic log,
  C_p = the "compaction" correction factor.

C_p will always be greater than 1.0. It should be applied when the travel time in a shale, t_{SH}, is greater than 100 μs/ft (328 μs/m).
  There are several empirical methods for determining the value of C_p. The most common method is to determine the travel time in a nearby shale, t_{SH}, when it is greater than 100 μs/ft

$$C_p = \frac{t_{SH}}{100 \mu s / ft} \qquad (11.24a)$$

The value of C_p may be determined on the newer charts from the major contractors. Schlumberger publishes both the Wyllie type response curves

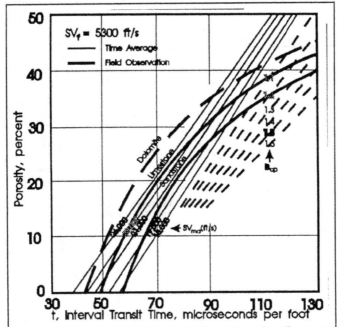

FIGURE 11.21
Chart Por-3 for acoustic-derived porosity.

(straight lines) and the empirical response curves determined from field experience (curved lines), as well as a family of curves for determining C_p (Schlumberger labels it B_p).

Brock lists several methods for determining C_p. Most of these involve comparing the acoustic-derived porosity with that found by some other method, both in the same shale-free sand. A known value of porosity from some other source (resistivity, neutron, density, or cores) must be used to find the compaction factor:

$$C_p = \frac{\phi_t}{\phi_{other}}$$

(11.24b)

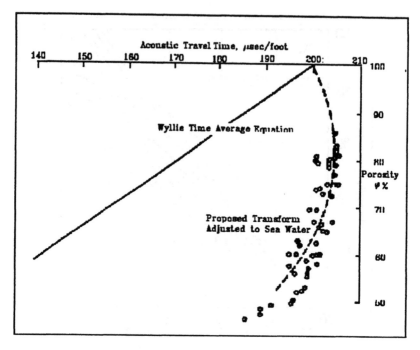

FIGURE 11.22
Comparison of acoustic transit times in ocean floor sediments. (After Schlumberger, 1988.)

Brock also suggests

$$C_p = 6.996 - 1.414 \log_{10} D \qquad (11.24c)$$

where D is the depth to the formation zone in feet. If D is in meters, use $3.28D$.

Raymer et al. (1970), demonstrated the poor fit of the Wyllie Time Average relationship to porosity in many cases and suggested an empirical transform to correct the problems. They suggested two alternate equations for wave velocity for the range of 0% to 37% porosity (see Figures 11.22 to 11.24),

$$V_t = \left(\frac{\rho_{ma}}{\rho_b} \right)^{1/2} \qquad (11.25a)$$

or

$$V_t = (1 - \phi)^2 V_{ma} + \phi V_t \qquad (11.25b)$$

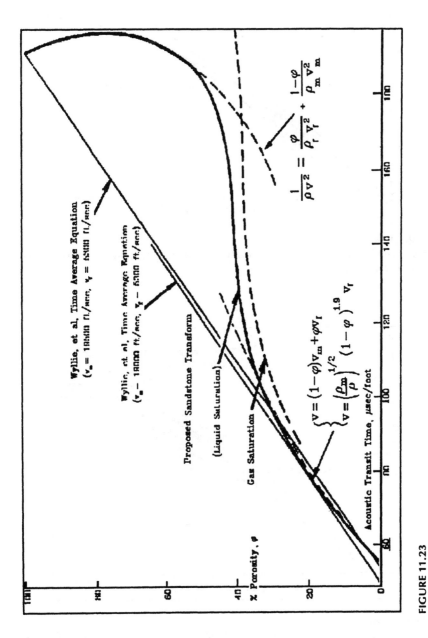

FIGURE 11.23
A proposed transform and suggested algorithms. (From Raymer, et al., 1988)

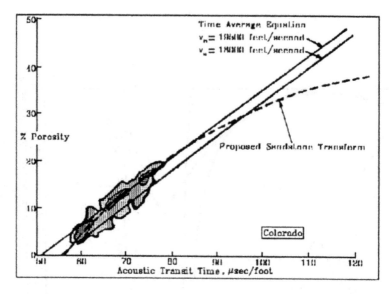

FIGURE 11.24a
Comparison of acoustic transit time to porosity in low porosity Colorado sandstones. (After Raymer, et al., 1988.)

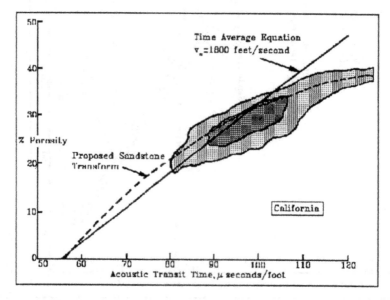

FIGURE 11.24b
Comparison of acoustic transit time to porosity in medium to high porosity sandstones. (After Raymer, et al., 1988.)

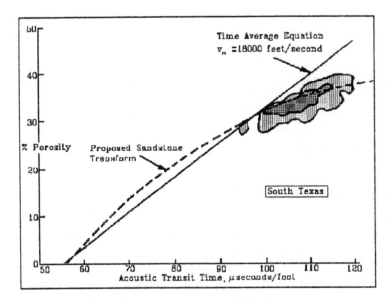

FIGURE 11.24c
Comparison of acoustic transit times to high porosities from Texas sandstones. (After Raymer, et al., 1980.)

These two equations give identical results over the usual range of porosities or up to about 100 μs per foot (328 μs per meter). They also suggested algorithms for higher porosity ranges:

$$\frac{1}{\rho V^2} = \frac{\phi}{\rho_f V_f^2} + \frac{1-\phi}{\rho_{ma} V_{ma}^2} \tag{11.26}$$

Raiga-Clemenceau et al. (1976) have suggested the use of a "formation factor" type exponent, x (Equation 11.27:

$$\frac{t}{t_{ma}} = (1-\phi)^{-x} \tag{11.27}$$

where the values of "x" are variable and depend upon the rock type. This is similar in form to Archie's resistivity-porosity relationship.

The acoustic waves tend to "ignore" secondary porosity, such as some forms of vugular and fractured porosity found in carbonates and cemented sandstones. This is probably due to small fracture widths, compared to the typical wavelengths used (~0.5 ft or 15 cm in a 20% porosity sandstone). Thus, it responds more to primary porosity.

Since the acoustic methods respond more to primary porosity, acoustic derived porosities frequently disagree with the core derived porosities. This may be due, in part, to the presence of micro-fracture damage incurred in the core sample during coring, retrieval, and handling. If the acoustic derived porosities are lower than the core derived porosities, this possibility should be investigated.

Schlumberger has adopted many of the Raymer suggestions in their data reduction procedures advanced in their late information manuals. (See Raymer, 1980.)

11.5.2 Mechanical Properties

The use of the acoustic information, when combined with the bulk density values to determine the mechanical properties of the formation zones, has already been discussed. The equations were presented as Equations 11.10 through Equation 11.17.

The determination of the formation mechanical parameters for formation evaluation has great significance and the prospect for considerable cost saving in mining, pre-mining, and engineering projects. Savings in drilling time and indications of choice of equipment are also indicated. The wireline methods should always be mixed with the conventional sampling and laboratory methods. The experience on a project will dictate what the proportional use of each of the methods should be. In the petroleum industry, too, these parameters are important because of fracturing, possibility of fracturing, hole caving and distortion, stress relief, and general hole and formation quality.

In a pre-mining situation, for example, where 100% coring is common, coring of 200 test holes could be replaced by coring 20 holes and logging all 200, at about 20% of the total cost. For use on subsequent projects, one method should always be checked against the other. Both should be compared with the final results at the end of the project. This same philosophy should apply to petroleum projects, as well. Figure 11.25 shows a Schlumberger Array Sonic log, which records the basic Pressure-wave, Shear-wave, and Stonely-wave information for computation of the mechanical parameters. A "rock strength" log can be generated from this.

11.5.3 Effects of Clay or Shale

Shale greatly affects the acoustic travel time because it has a long travel time (t_{SH} = 60 to 120 $\mu s/ft$). Moreover, the travel time in shale can vary greatly from one shale to another and from one depth to another. The effective porosity in a laminar shaly sand can be corrected in a manner similar to that used for the neutron porosity systems. That is, to determine the partial

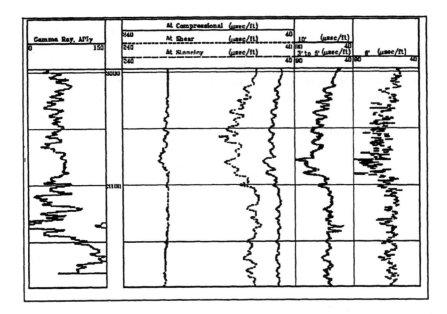

FIGURE 11.25
An Array Sonic log. (After Schlumberger, 1988.)

apparent porosity due to the shale and subtract it from the apparent acoustic derived porosity:

$$\phi_{t,cor} = \phi_{t,a} + V_{SH}\phi_{t,SH} \qquad (11.28)$$

where

$\phi_{t,cor}$ = the corrected acoustic derived porosity
$\phi_{t,a}$ = the apparent acoustic derived porosity after other corrections have been made
V_{SH} = the fractional volume of shale determined in the zone under consideration and derived from the gamma ray method or a method similar
$\phi_{t,SH}$ = the apparent acoustic porosity in a good, nearby shale

As in other determinations of this kind, the assumption is made that the shale in the zone of interest is identical to the nearby clean shale. This is usually a reasonable assumption in sediments.

If the shale is dispersed, the effective porosity is merely the difference of the apparent acoustic porosity, $\phi_{t,a}$ and the volume of shale, V_{SH},

$$\phi_{t,cor} = \phi_{t,s} - V_{sh} \qquad (11.29)$$

This will result in the effective porosity being eventually determined. It is also quite feasible to use the method of the sum of the partial travel times to determine the matrix travel time (Equation 11.28) before using any method.

The acoustic systems tend to measure the total porosity, because shale travel times are close to those of the formation water. This is not as close to the total porosity as the neutron-derived porosity. If t_{ma} is chosen properly, the acoustic log may be compared with the density log to determine the approximate amount of shale/clay (clay fraction) in the pore space:

$$\text{Clay fraction} \approx \frac{\phi_t - \phi_p}{\phi_t} \qquad (11.30)$$

or the total volume of shale, V_{SH}:

$$V_{SH} = \phi_t - \phi_p \qquad (11.31)$$

11.5.4 Primary Porosity

Acoustic measurement tend to favor primary porosity and to ignore secondary porosity. Therefore, in a clean formation, the secondary porosity, ϕ_{sec}, is approximately,

$$\phi_{sec} = \phi_p - \phi_t \qquad (11.32)$$

11.5.5 Fractured Media

Fractured media can cause problems with acoustic logs. For the same reason, use can be made of this to detect the possible presence of fractures.

The main cause of the problems in a fractured medium is the wave energy loss as it passes across the fracture interface. Some reflection will take place, subtracting from the transmitted wave energy. This can cause weak signals and possible cycle skipping. It can also cause multiple reflections.

The phenomena across a fracture interface can be described with Snell's law (Equation 11.2) because the fracture represents a change of impedance (a contrast of refractive index) at least two reflecting surfaces. In the case of a closed fracture, the incident Pressure-wave will be reflected as a Pressure-wave and a new Shear-wave. It will transmit to the other side of the fracture as a Pressure-wave. On the other side of the fracture it will become a new Pressure-wave and a new S-wave. This situation assumes that the two faces of the fracture are close enough together that the reflections from the two faces coincide with each other. That is, that the separation is much smaller than a wavelength. This is illustrated in Figure 11.26. If the fracture

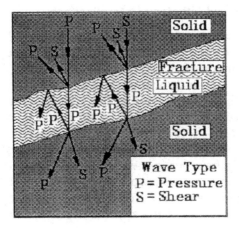

FIGURE 11.26
A simplified diagram of acoustic transmission across a fracture.

is wide open (separation greater than a wave length), the transmission coefficient, K_T, at the interface is

$$K_T = \frac{2}{\left(\dfrac{\rho_1 v_1}{\rho_2 v_2}\right)^{1/2} + \left(\dfrac{\rho_2 v_2}{\rho_1 v_1}\right)^{1/2}} \tag{11.33}$$

The acoustic impedance, Z_t, across an interface, is

$$z_t = \rho_b v_t \tag{11.34}$$

where
ρ_b = the bulk density of the formation
v_t = the wave interval velocity

In this case, there are two separate interfaces. Therefore, the total energy transmitted, E_t, will be

$$E_t = E_i \alpha_1 \alpha_2 \tag{11.35}$$

where α = the attenuation factor.
If $\alpha_1 = \alpha_2$, then $E_t = E_i \alpha^2$, since each interface transmits a fraction of the energy. The effect obeys the principle of reciprocity.
The reflection across the interface is indicated by the reflection coefficient, R:

$$R_{1,2} = \frac{Z_2 - Z_1}{Z_2 + Z_1} \tag{11.36}$$

11.5.6   Porosity and Lithology Determinations

The determination and variations of the porosity, using the acoustic information have already been covered. Equations 11.20 through 11.27 discuss
this subject. The important thing to remember is that the acoustic measurements determine a porosity value which is closer to the total, primary
porosity than to the effective porosity. This is because the acoustic P-wave
travel time in shales (60 to 170 μs/ft. or 117 to 558 μs/m) is near that of
water (180.5 to 207 μs/ft. or 592 to 679 μs/m) and oils (214 to 238 μs/ft. or
702 to 781 μs/m). It is difficult to differentiate partially shaly zones from
clean ones, if only the acoustic curve information is used. It is very difficult, if not impossible, to differentiate between water and liquid hydrocarbons. Also, it should be remembered that the water involved is probably
the mud filtrate, because of the shallow depth of investigation of the
acoustic systems.

One should use several of the evaluation and interpretation methods in
combination. This tends to use one method to correct the values of the
other. The result is usually a much more reliable end value. We have earlier
examined the density systems and the neutron systems. In the neutron
porosity chapter we started to combine methods for greater clarity, accuracy, and more information. We can carry that further, as we add more
types of logs to the analysis.

11.5.6.1 t_a vs. ϕ Cross-Plot

A simple cross-plot of the recorded interval travel time, t_a, vs. the porosity,
ϕ, can be effectively used in non-hydrocarbon bearing zones to fill in missing information and to separate zones by rock type. The rock type separation is because different types of rock have different travel times. This is
shown in Figure 11.27. Different matrix travel times, t_{ma}, will result in different slopes of the trends. The total travel time, t_a, is given by rearranging
Equation 11.20:

$$t_T = t_{ma} + \left(t_f - t_{ma}\right)\phi_t \qquad (11.37)$$

where the relationship is linear and the total apparent travel time, t_a, is the
y-axis value, t_{ma} is the y-intercept and will indicate the correct matrix travel
time when the trend is extrapolated to $\phi = 0\%$. The value of $(t_f - t_{ma})$ is the
slope. For any particular zone in a particular hole, $(t_f - t_{ma})$ is constant.
Either t_f or t_{ma} may be independently evaluated and the other determined.
This is illustrated in Figure 11.28.

11.5.6.2 ϕ vs. R_t Cross-Plot

The acoustic-derived porosity, ϕ_t, may be substituted for the porosity, ϕ, or
the core-derived porosity, ϕ_c, in the expression for the R_t vs. ϕ cross-plot:

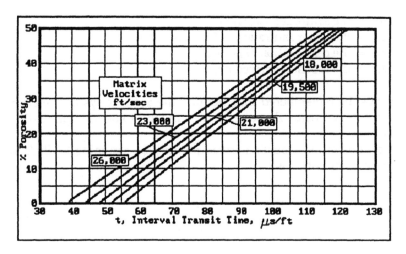

FIGURE 11.27
A plot of porosity, ϕ, vs. interval transit time (t),

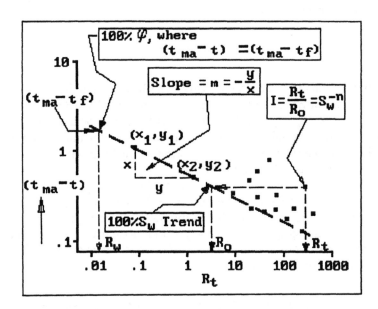

FIGURE 11.28
A Pickett plot using acoustic travel time.

$$\log R_t = \log R_w + n \log S_w - m \log \phi_t \qquad (11.38)$$

where

R_t = the true resistivity of the zone
R_w = the formation water resistivity

n = the saturation exponent
S_w = the water saturation of the pore space
m = the cementation exponent
ϕ_t = the acoustic-derived porosity

It must be remembered that, because the acoustic system determines nearly total porosity, Equation 11.37 will tend to be too optimistic in a shaly zone.

The value of the acoustic-derived porosity, ϕ_t, assuming Wyllie's Time Average relationship, is,

$$\phi_t = \frac{t_f - t_a}{t_f - t_{ma}} \tag{11.37}$$

The acoustic-derived porosity value, ϕ_t, may be substituted in the R_t vs. ϕ plot.

In any given zone, $(t_f - t_{ma})$ is constant. Therefore, the value of $(t_f - t_a)$ is a function of the porosity:

$$\left(t_f - t_a\right) = f(\phi) \tag{11.39}$$

and may be substituted as the porosity axis (the x-axis) in the R_t vs. ϕ cross-plot. This is shown in Figure 11.28. When the porosity is extrapolated to 100% ($\phi = 1.00$), $t_a = t_f$ and the fluid travel time may be evaluated. If the porosity is made equal to zero, $t_a = t_{ma}$ and the rock type may be identified, or at least, its character determined. *Remember*, the fluid is probably not the formation water but the mud filtrate.

In 1963, Pickett suggested cross plotting the travel times of the P-wave vs. the S-wave for lithology determination. This has an application, especially with the present methods of separately recording these two travel times in log (line-graph) form. This type plot is especially sensitive to the presence of gas. Thus, it makes a good method to locate or verify the presence of gas in the formation.

11.5.6.3   Travel Time vs. Bulk Density and Neutron Porosity

The value of the total travel time, t_a, may be plotted against the bulk density, ρ_b, as shown in Figure 11.30. If the density and acoustic velocity equations are solved for porosity and then set equal to each other, then,

$$\rho_b = \rho_{ma} - t_{ma}\left(\frac{\rho_f - \rho_{ma}}{t_f - t_{ma}}\right) + t_a\left(\frac{\rho_f - \rho_{ma}}{t_f - t_{ma}}\right) \tag{11.40}$$

where the equation is linear and of the slope-intercept form, ρ_b is the value of the y-axis, $t_a =$ the value of the x-axis, the value of the slope is

$$\frac{\left(\rho_f - \rho_{ma}\right)}{\left(t_f - t_{ma}\right)}$$

and the y intercept is

$$\left(\rho_{ma} - t_{ma}\right)\frac{\left(\rho_f - \rho_{ma}\right)}{\left(t_f - t_{ma}\right)}$$

which can be used to identify the rock type. The density and travel time of the fluid are assumed to be constant for any given zone. Therefore, the slope is a function of the matrix rock type for any given hole and zone. Figure 11.30 is a similar plot for t_a vs. ϕ_N.

In the plot of Figure 11.28, the separation of the trends for the usual three different rock types is not great. It is better in some of the other plots, such as in Figure 11.29. This illustrates the need to pick the measurement type to enhance the analysis method.

The sets of trends in Figures 11.27 to 11.29 indicate the Wyllie Time Average relationship as the straight lines and (in some of the charts) the field observations as the curved trend lines. The latter conform more to the Raymer transform (Equations 11.25a and b to 11.26) or the Raiga-Clemanceau algorithm (Equation 11.27) response of the acoustic information.

The plot of Figure 11.30 shows the trends of travel time, t_a, vs. the apparent neutron porosity, ϕ_N, from the SNP. Note the clear, widely separated rock type trends, as well as the porosity correction. The actual mixture of the rock components can easily be estimated. In this case, since both the neutron porosity system and the acoustic system tend to indicate total porosity, the corrected, cross-plotted porosity will be total porosity, unless the values from the neutron and the acoustic systems have been corrected for shale content before entering.

The equation for Figure 11.30 is similar to Equation 11.38a in form:

$$t_T = \left(t_{ma} + t_f - t_{ma}\right)\phi_N \tag{11.41}$$

where the equation, again, is linear and of the slope-intercept form. The value of t_{ma} is the y-intercept and $(t_f - t_{ma})$ is the slope of the trend. The value of t_{ma} is a function of the rock type, as is $(t_f - t_{ma})$ because each is assumed to be constant for any one zone.

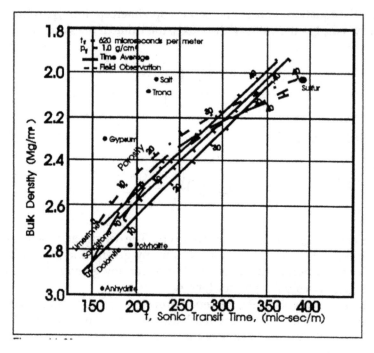

FIGURE 11.29
A crossplot of t_a vs. ρ_b. (After Schlumberger, 1988.)

Figure 11.31 shows an old, but representative, P-wave acoustic log. The curves are the gamma ray and the *SP* in track *I*, the 16-in normal resistivity and the 6FF40 induction resistivity (the reciprocal of the recorded conductivity in track III) in track II, the 6FF40 conductivity in track III, and the acoustic travel time in tracks IV and V. The log and an analysis of the zone from 5330 ft to 5336 ft is shown on the following several pages.

11.5.7   Analysis of the Log of Figure 11.31

Acoustic values:

$$t_{ma} = 55.5 \; \mu s/ft \; (182 \; \mu/m)$$
$$t_f = 189 \; \mu s/ft \; (620 \; \mu/m)$$
$$t_a = 82 \; \mu s/ft \; (269 \; \mu/m)$$

Mud and temperature values:

$$T_{form} = 115°F \; (46°C)$$
$$R_m = 1.3 \; ohms \; at \; 115°F \; (46°C)$$
$$R_{mf} = 0.591 \; ohms \; at \; 115°F \; (46°C)$$

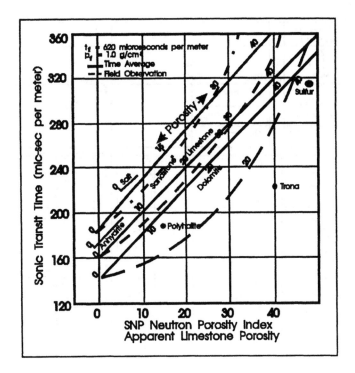

FIGURE 11.30
A porosity crossplot of t_a vs. ϕ_{SNP} (After Schlumberger, 1988.)

Value of the formation factor, F_R:

$$\therefore \phi_t = \frac{\left(t_f - t_a\right)}{\left(t_f - t_{ma}\right)} = \frac{(82 - 55.5)}{(189 - 55.5)} = 0.197$$

$$F_R = \frac{\cdot}{\phi^{1/2}} = \frac{\cdot}{\left(0.197\right)^{1/}}$$

Spontaneous potential and R_w:

SP	= –78 mvs
R_i	= 26 Ωm
SP correction	= 1.01
$\therefore$ SSP	= –78(1.01) = –79 mvs
K	= 60 + 0.133(115) = 75.3 mvs
$\log(R_{mf}/R_{we})$	= 1.05
R_{mf}/R_{we}	= 11.2
R_{we}	= 0.591/11.2 = 0.053 Ωm at 115°F (46°C)
$\therefore R_w$	= 0.068 Ωm at 115°F (46°C)

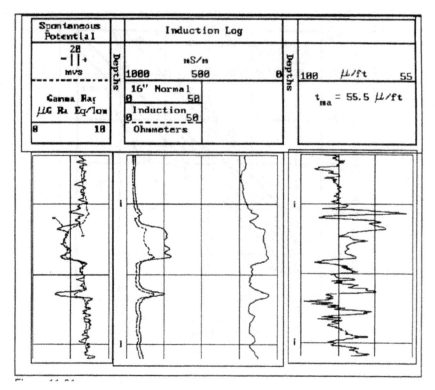

FIGURE 11.31
An electric, acoustic porosity log combination.

Resistivity data:
R_{IL} = 14.0 Ωm. Correction is negligible for this borehole size and bed thickness

Invasion correction,
Assume d_i = 40 in (1m)
G_i = 0.28
G_t = $(1 - G_i)$ = 0.72

Value of R_t :

$$\frac{1}{R_{IL}} = \frac{G_i}{R_i} + \frac{G_t}{R_t}$$

$$\frac{1}{14} = \frac{0.28}{26} + \frac{0.72}{R_t}$$

∴ R_t = 11.87 Ωm.

Saturation calculation:

$$S_w = \left(\frac{F_R R_w}{R_t}\right)^{1/2} = \left(\frac{(25.8)(0.068)}{11.87}\right)^{1/2} = 0.384 = 38.4\%$$

11.6 Determination of Permeability from Acoustic Logs

Attempts have been made for many years to measure formation permeability with acoustic logs. There exists a correlation between the acoustic amplitude and the formation permeability, lithology, pore fluid character, and matrix continuity. Fons found a correlation between formation permeability and the first period of the P-wave and with the total energy of the acoustic signal. These relationships have been occasionally successfully used empirically. In separate studies, the V-shaped patterns of the full-wave, amplitude-modulated acoustic log appear to be caused by open and closed fractures in the formation material. Zemanek suggested that the ratio of the amplitudes of the s-wave to the p-wave was an index of permeability. It does not, however, discriminate between the clean permeable intervals and shaly, impermeable intervals.

In 1978, Leberton, et al., proposed the permeability index, I_c,

$$I_c = \frac{A_2 + A_3}{A_1} \tag{11.42}$$

where A_1, A_2, and A_3, are the amplitudes of the first three half cycles of the arriving p-wave. They found a correlation between I_c and the formation permeability. They further proposed

$$I_c = \alpha \log\left(\frac{K_v}{\mu}\right) + \beta \tag{11.42a}$$

where

K_v	= the vertical permeability
μ	= the viscosity of the rock-wetting fluid
α and β	= constants for a given system and a given borehole

Mobil's long-spaced acoustic logging system, LSAL used two wide-band receivers spaced 15 ft (4.6 m) and 20 ft (6.1 m) from the transmitter. The transmitter was a magnetostriction device operating at 15 KHz.

The transmitter section and the receiver section were separated by a cable. Signals were digitized at the surface. They found that the Stoneley wave (St-wave) formation may have been facilitated by the absence of a large sonde body in the borehole. They also found that the increased horizontal depth of investigation (the result of the long spacing and the lower frequency) tended to give better permeability results than the conventional shorter spacing systems. This was, the decided, because the wave path was deeper than damage to the formation material by drilling and the invasion by hole fluids. They obtained good separation of the P-wave and S-wave arrivals because of the longer spacing. The Stoneley-wave amplitudes were measured at each receiver as representative of the signal energy. The ratio of the amplitudes was taken as a measure of the energy lost in that interval of formation. This was related to the permeability. The results were compared with the core permeabilities.

11.7 Geological Uses of Acoustical Logs

Figure 11.32 shows the log information in a Paleozoic carbonate sequence. The zone is a naturally fractured limestone-dolomite. Permeability was 0.1 to 20 millidarcies. Water saturation was 30% to 50%. The permeability and porosity were much higher in the dolomite. Porosity was 0 to 20%.

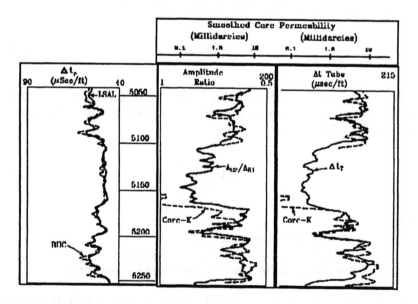

FIGURE 11.32

A Paleozoic, fractured, carbonate sequence (Courtesy of Western Atlas Logging Services Division of Western Atlas International, Inc.).

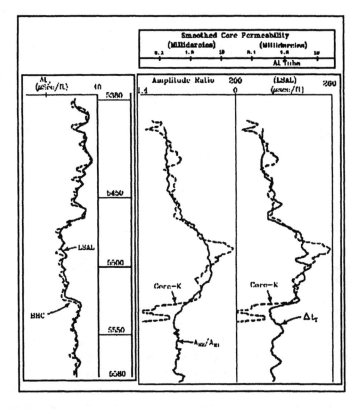

FIGURE 11.33
A Cretaceous, interbedded, oolitic carbonate (Courtesy of Western Atlas Logging Services Division of Western Atlas International, Inc.)

Figure 11.33 is the data from a Cretaceous carbonate. The top 71 ft (21.6 m) is interbedded oolitic limestone and sandstone. Porosity and permeabilities were poor in this interval. The next 51 ft (15.5 m) was more sandy and had better porosity and permeability. The zone had a thick mudcake (3/4 in; 1.9 cm) which did not seem to affect the log values, but appears to have smoothed out some of the detail. There was no oil in these zones.

Figure 11.34 shows the information from a shallow sandstone. The hole was for a uranium leaching project. The cores were completely unconsolidated to well consolidated. Permeabilities were up to 3 darcies. The authors observed that the wave amplitude measurement gave somewhat better permeabilities than the travel time in the high-permeability zones.

11.8 Cement Bond Logging

A modification of the acoustic logging system, which measures the amplitude of the received signal, rather than the travel time, is regularly used to

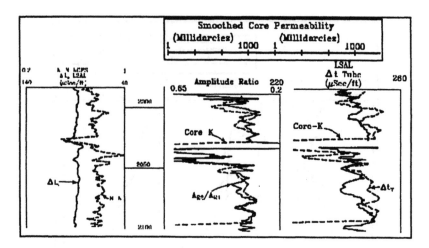

FIGURE 11.34
A shallow sandstone. (Courtesy of Western Atlas Logging Services Division of Western Atlas International, Inc.)

determine the quality of the cementing behind the casing (in the annular space) in a well. This is the cement bond log.

The acoustic wave will tend to remain within the body of the metal casing because of the high contrast between refraction indices of the well fluid and annular fluid on one hand and the casing steel on the other. Total reflection will tend to keep most of the wave energy in that channel. If, however, there is a good bond between the annular cement and the casing, the high refractive index contrast no longer exists. Therefore, a large amount of the energy enters the cement and the formation. Thus, the wave amplitude is lower at the receiver. If the amplitude of the acoustic wave at the receiver is measured, it is an index of the amount of energy transmitted through the steel casing. A high amplitude indicates a possible poor bond or no bond, or no cement in the annulus. A low amplitude indicates the possibility of a good bond. In this case, a full wave recording of the signal at the receiver will give information about the bond between the formation and cement and some of the characteristics of the formation. The principle of the downhole tool is shown in Figure 11.35.

Recent system introductions have displayed variations of tool design to overcome some of the shortcomings of the earlier systems. Multiple readings around the circumference of the borehole help to detect spotting bonds and cavities in the cement. This has been named the "Cement Quality Log".

One of the serious problems encountered with the cement bond logs is the presence of a micro-annulus. This is a situation where there is not a good bond between the casing and the cement, but the resulting annulus is extremely thin. Possibly, the lower relative internal pipe pressure during logging will cause the higher formation pressure to compress the casing. This will separate the casing and cement and induce an indication of poor bonding and possible communication vertically through the annulus.

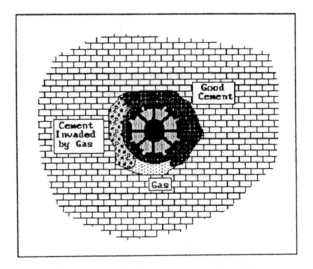

FIGURE 11.35
A plan diagram of the cement quality tool.

Then, because of the higher internal casing pressure during production, the casing will be expanded. Then, no serious problem really exists, because the micro-annulus is closed.

11.9 Determination of the Shear Wave Velocity using the Stoneley Mode

The velocity of the Stoneley-wave, v_{ST}, at zero frequency, is

$$\rho_b v_{ST} = \left(\frac{1}{\mu} + \frac{1}{B}\right)^{-1} \tag{11.43}$$

where
ρ_b = the bulk density of the formation material
μ = is the shear modulus of the formation zone, B is the bulk modulus of the borehole fluid

The shear wave velocity as presented in Equation 11.11:

$$v_s = \sqrt{\left(\frac{\mu}{\rho}\right)}$$

Combining Equations 11.47 and 11.11 and solving for the shear wave slowness.

$$v_s = \left[\frac{\rho_b}{\rho_{ma}} \left(\frac{1}{v_{ST}^2} - \frac{1}{v_{ma}^2} \right) \right]^{-1/2} \qquad (11.44)$$

In the frequency range of 5 to 20 kHz, six independent measurements are needed to determine the shear-wave slowness (or velocity) from the log of the Stoneley mode:

1. Stonely-wave slowness at a particular frequency
2. Borehole fluid density
3. Formation bulk density
4. Borehole size
5. Borehole fluid slowness
6. The pressure-wave slowness of the formation material

Accurate measurement of the Stoneley-wave slowness is required, because the Stoneley mode is quite dispersive, especially in the slow, unconsolidated zones. Stoneley amplitude is a function of the shear modulus, μ, and the frequency, v. The Stonely-wave amplitude will be low in slow, unconsolidated zones in the 10 to 20 kHz frequency range. A calculated S-wave is compared with a recorded one in Figure 11.36.

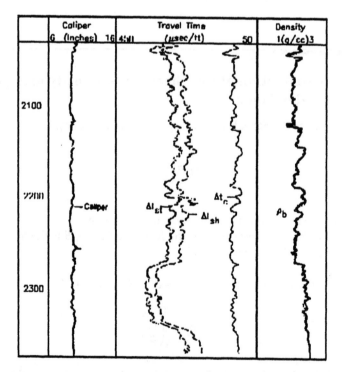

FIGURE 11.36
A calculated shear wave compared with a recorded one (After Schlumberger, 1989).

12

Formation Evaluation — Standard Methods

12.1 Introduction

The application of geophysics, geology, economics, and other disciplines to determine the identity, character, value, and dimensions of the subsurface features is called "formation evaluation". At this time, the major tools of formation evaluation are borehole geophysics, petrology, rock mechanics, subsurface geology, and core analysis. Many other methods, such as surface geology and surface geophysics, also enter into formation evaluation. Of course, if the goal is commercial, economics plays a major role. In this chapter we will look at methods for petroleum, mineral exploration, engineering applications, environmental assessments, and several other lesser applications.

Other texts have examined the background of some of the common, standard formation evaluation tools (especially borehole geophysical logging methods), some of the data reduction methods, necessary corrections, and some of the uses of them. In earlier chapters we have briefly looked at the methods of analysis and evaluation of the individual methods and a few combining methods. The most effective use of any of these methods, however, is to combine the data from several of them; to use all of the pertinent data available. In the past, with manual methods of data processing, this could be a difficult task, even with a calculator.

In this chapter we will look at using some of the data available from various sources. We will still restrict our examination to the standard methods, because those are the only ones we have examined to this point. We will examine some of the newer methods and more exotic methods in *Nonhydrocarbon Methods of Geophysical Formation Evaluation*, a companion volume to this text. This chapter will examine the data from core and cuttings analyses, surface geology and geophysics, electric logging, radioactivity logging, and the porosity log methods. Some of the mapping and normalizing methods will be briefly examined, and the necessary economic factors will be touched upon.

Table 12.1 lists a few of the measurements which can be made with standard methods of geophysics and formation evaluation.

TABLE 12.1

Measurements available with standard methods

1.	Thin sample
2.	Tables of properties
3.	Literature, records and experience
4.	Laboratory-derived relationships
5.	Borehole effects
6.	Mud measurements
7.	Core analyses
8.	Cuttings analyses
9.	Sampling
10.	Electrical resistivity
11.	Spontaneous potentials
12.	Natural radioactivity
13.	Density determinations
14.	Neutron porosity methods
15.	Activation analysis
16.	Acoustic porosity methods

12.2 Assumptions

We will assume that the reader is familiar with the data gathering process. We will also assume that all of the necessary corrections have been made to the raw values such as hole size, thin bed, coincidence losses, and any others required to obtain finished data. Which tool or method is best will be discussed, and why, for a specific purpose, Reference is made to some of the previous chapters to emphasize any necessary corrections to obtain finished data.

12.3 Methods Available

These "standard methods" have been around a long time. They are readily available. Most of the problems have been worked out of them. Existing files contain largely these methods. Supporting information is available and most of them have been continuously updated and refined.

12.4 Multiple Parameters

Whenever possible (time and money permitting) measurements should be repeated. Repeats which should be considered should be by the same system and by different types of systems which measure the same parameter in different ways.

The question arises, "Why duplicate measurements simply because the duplicate method is by a different type tool?" There are several good reasons:

1. The field and laboratory equipment in use today is reliable, for the most part, but, even the best equipment sometimes fail and the best operators occasionally make mistakes. A close repeat is assurance that the measurement is valid.

2. Different scales or presentations may be better on the second run. This is not important with much of the modern digital equipment now in use; with analog equipment it is vital.

3. All of our measurements are indirect. We do not measure porosity. We measure hydrogen content or gas volume or electrical resistivity and interpret that in terms of the probable porosity.

4. Repeats of the same measurement with the same system increases our confidence in the measurement. If a repeat closely duplicates the original, our level of confidence in the measurement is about 140%. A second good repeat increases our confidence to about 170%.

In mathematics there is a well-known principle which can be adapted to our work. That is, there must be one equation for each unknown. We usually work with many unknowns. *One source of information severely restricts our analysis. Multiple sources extent our horizons and confidence in the result, tremendously.*

We also must remember that, so far, we have only examined the older, standard methods. Some of the older methods, especially calibration methods on analog systems, were not as good as are modern techniques. Drift is always a problem with analog systems. In the next volume when we get to newer, more specialized, and more exotic methods, we will find virtually no end to the determinations which can be made.

12.5 Scope

Each source of information at our disposal has limitations. Each type has advantages. Each type has specific areas where it is better suited than in other areas. These factors should be examined carefully when a method is specified or data from a particular method is available. This is an area which has been somewhat neglected by most authors. It is also an area where you cannot always rely on the judgment and advice of some of the contractor's field people. You must know what a given measurement can do and why it will benefit you.

Ideally, we would know what parameters are needed. We would specify (or have in our possession) only the information best suited to determine those parameters. Each method would be used under ideal conditions and the data would be unambiguous. Unfortunately these ideal conditions are seldom realized. It is the nature of our work that we will often work with existing material, stored cores, old thin samples, existing logs and reports, and someone else's old notes. Even with newly planned projects, someone else has frequently already specified what methods will be used. Budget constraints are always present. Too, data are frequently suspect. Operators were tired, a storm was in progress, hole conditions were poor, time was limited, repeat measurements were not made, data were omitted; the list can be endless. We will, during the course of our evaluation, be required to verify, normalize, and correct data. One should *always* be suspicious of existing data.

12.6 Research

The place to start any project of this type is in the office files, text books, published papers, in the library. One must find what has been done. Libraries and files are rich sources of information. They can save many hours of work and thousands of dollars. Texts have descriptions of the areas which you plan to evaluate. These descriptions involve studies done for government archives, instructing students, defending or instituting lawsuits, pre-construction studies, investment evaluations, and maps. All of these, especially the maps, can give valuable general or specific information.

Files of technical papers and technical symposia proceedings will have more detailed and in-depth information. And, don't forget, many organizations, besides the one you meet with once a month or once a year, are interested in this type of data. The State Geologist and State Engineer are good sources of preliminary and background information and advice. This can take may forms: cores, logs, texts, papers, drilling records, permit applications, and valuable experience.

Universities have computer files and search programs. They are also usually tied in with the computer systems of other libraries, including the public libraries. They can furnish titles, abstracts and copies of papers. They can pinpoint data sources. They charge a fee, but this is usually an inexpensive way to get information. They frequently are very slow, however.

Log and core libraries can usually give you a picture of the subsurface conditions before you go out to the field. The fact that the information may be petroleum oriented and/or old does not lessen its value for engineering or scientific purposes.

This "old" information can tell us what to expect. That, coupled with the requirements of the goal, will define what new data we will require. It will tell what area we might have to examine, how deep the target probably is, whether there is the possibility of faulting, the type of excavating machinery possibly needed, the possibilities of hole problems, and many other items. Include in this category the driller's experience.

Importantly, the background information allows us to specify the initial suite of measurements to be made on the project and the amount of time and cost needed to do them, budget core amounts, and plan for laboratory use.

12.7   New Information

If one is fortunate, he will be able to specify the entire suite of measurements to be made on a project. Unfortunately, we must often work with someone else's specification and with existing information. Too, important information is frequently missing.

12.8   Methods to be Used

On the other hand, a constant effort *must* be made to obtain enough and vital information within a reasonable budget. Usually we can gain much of the necessary preliminary information from the existing measurements. And frequently, fill-in measurements can be made on the existing cores and samples. It should also be remembered that the specifications will change continuously as the project progresses. The goal may remain constant and should not be confused with the specifications.

The evaluation methods to be used, will, of course, depend upon the project goal, the results desired, and the information available. The method will also depend upon our experience as we progress through the project. The "initial" set of measurements is exactly that; initial. Subsequent suites of measurements will almost certainly be different, depending upon the previous results.

The first step is to determine from the literature and surface studies the probable sequence of the formations and their ages. This is a stage which may have been completed before serious consideration of the exact site for the project.

The second step is to determine the probable target depth and/or goals of the project. If the project is a petroleum wildcat, the surface geology, literature and seismic studies will indicate the approximate depth and location of any promising horizons, such as structures or traps in sediments. If the project is engineering, the same preliminaries, plus and examination of the local history, will show the locations of faults, structures, and possible problems. Mineral and engineering projects may include previous airborne studies. Scientific studies will probably have to be handled on an individual basis. This type of wide area preparation is inexpensive and tremendously valuable.

The next step is to clear up any land acquisition, legal, and economic problems. Permissions must be obtained and access roads located and/or made.

Once the situation is known and stabile, the initial needed and desired measurements can be specified. It is better from a data standpoint, of course, to have too many measurements than not enough. There are, however, always the limiting factors of time and cost. In all types of projects, identification and sequence of the formations is of prime importance. This locates and verifies the preliminary studies. If the desired horizon cannot be identified, time and money can be wasted. Thus, the measurement suite must be specified with these things in mind, and not the "target" information exclusively.

The goals of the project, the target, will dictate the types of measurements to be made. We have already seen the need for surface studies. If the project is for petroleum, we also will need to know the porosity, permeability, saturation values, and rock types. This dictates, at least, resistivity, porosity, and core measurements. We are primarily (but not exclusively) interested in the formation fluids. A mineral or engineering project will have a similar set of requirements, but perhaps, in a different order of importance. We will be primarily (but not exclusively) interested in the formation rock materials. A scientific project will probably demand more specific and detailed types of measurements. Frequently, the list of different types of measurements for a scientific project is shorter and available time is greater. The reliability and accuracy of scientific measurements should be the best. This is not always the case, probably because the persons specifying the suite are most likely not used to working with geophysical contractors.

Other parameters than those initially identified may be needed. These may be fluid composition, clay content, structure, gas composition, and many others. The need for some of these may not show up until the initial measurements have been made and examined. Sometimes additional data are needed after the analysis is already underway. This possibility should be kept in mind during the planning stage. Each type of project will have its own set of possible or unexpected requirements, in addition to the foreseen ones.

There is also the probability that, at some point, no further measurements can be made. This means that *the most important measurements should be made first.* Some measurements cannot be made after other measurements. A temperature log should be logged going down into an undisturbed borehole. A gamma ray measurement should not be attempted closely after any neutron measurement. Bad hole or terrain conditions can stop a project at any point. There will always be some kind of restrictions, and an experienced operator is invaluable. There will always be some information that we will wish we had, but do not. Plan carefully and schedule the sequence and priorities.

12.9 Analysis Methods — Standard Measurements

Now! How will we prepare the information we have acquired? The following section is a collection of some of the analysis methods from earlier chapters, existing literature, and a few other methods, suggestions, and comments. The general procedures are much the same for petroleum work and for non-hydrocarbon investigations. Thus, the step-by-step path will be outlined with needed exceptions and differences for formation evaluation for different types of goals.

The first step in analyzing the data on hand is to make depth corrections and to correlate all of the information. This has a two-fold purpose. It establishes the probable correct depth sequence for each type of measurement. It also draws attention to missing, out-of-sequence, or erroneous information.

12.9.1 Normalizing Data

Errors occur when gathering and processing data. We are fortunate in that we usually have many data points from many different sources. We will find that some of the data points, odd ones, do not fit well with the others. The question then, is, are these odd points in error or are they anomalous, but correct?

We must first examine the conditions under which an odd point was gathered. Was it taken in an area where we would expect, or not be surprised at, a sharp change. If it was, it may be correct. If not, it is suspect. Can the odd point be reconciled with physical laws? Can it be explained by probability? If not, the point is in error. If it can, maybe it is correct.

Once we have decided that a point is in error, perhaps we can correct it or estimate a very probable value. *If we do correct it or estimate a value, the point must be labelled thus.*

Estimated corrections to points must be logical. For example, a contour line on a map usually changes smoothly. An odd point which changes out of sequence can be moved or changed to fit the sequence at that point, *if we are very certain that the deviation from the norm cannot be explained as valid.* A gamma ray or *SP* curve transition across a clean, sharp bed boundary must follow a sinusoidal path. There are phenomena which can distort this shape, but they would not affect only one or two points.

12.9.2   Depths

The primary geophysical borehole log should be used as the depth and bed thickness standard. In deep holes, the depth measurements must be corrected for cable stretch, tool drag, and (sometimes) temperature. This easily allows possible absolute depth accuracies of one part in ten thousand (±0.01%). The correlation of a repeat run can be ten times better than this.

The accuracy is not as good when mineral-type logging equipment is used. Actual hole depths, however, are seldom deep enough to result in serious errors in absolute depths. Repeat depth measurements are very good. Even in mineral-type logging, an accuracy of ±0.1% and a repeat within ±0.01% is quite feasible. Possible errors can be caused by a mud buildup on the measuring wheel, especially in very cold weather. The type of measuring equipment is also a factor.

Mineral-type logging cable is usually not stretch corrected. This is because the holes are shallow, formation temperatures are low, and the tools are light. Many of the mineral contractors can furnish cable length corrections, if desired. Some of the mineral logging contractors have installed automatic cable measuring devices which accurately measure the cable length, against a standard, each time the cable goes into the hole. These devices are much more accurate than the standard measure wheel. Petroleum logging cable is stretch, temperature and drag corrected. Petroleum, and even mineral-type depth measurements will almost always be more accurate than pipe measurements.

Pipe strapping for depth measurement on a deep hole has a poor accuracy. Pipe strapping is usually done with the pipe lying horizontally on the pipe rack. Even if all other factors (stretch, temperature) are neglected, if an error occurs of 0.25 in (0.63 cm) on each 33.3 ft (10.1 m) length of drill pipe, the accumulated error at a depth of 10,000 ft (3049 m) is 75 in (190 cm); more than six ft or almost two m (almost 1 part in 1000). When stretch, manufacturing tolerance, makeup gap, and temperature expansion are added, the errors become intolerable.

Present day practice is to run a natural gamma ray log with each run of other logs. Special care is taken with the accuracy of the depths of the primary combination log. The logs, the cores, cuttings of other runs are depth-matched to the primary log depths.

12.9.3   Cuttings Lag

Cuttings depths are not reliable even if the surface-arrival lag is carefully calculated. Core depths are more reliable. Even with cores, however, depth errors can be caused by missing sections and miss-counted pipe lengths. Occasionally, sections are inverted or incorrectly located when being put into the core box. These are not always easy to spot. Both cores and cuttings should be carefully depth corrected to the depths of the primary log. Rock materials should be identified on the log.

Lag is the time interval required for the cuttings to travel from the drill bit to the shale shaker. The lag, L_t, in minutes, in English units, is

$$L_{t,e} = \frac{0.105D\left[\left(id_{csg}\right)^2 - \left(od_{dp}\right)^2\right]}{R} \qquad (12.1a)$$

or, in metric units, is

$$L_{t,m} = \frac{78.54D\left[\left(id_{csg}\right)^2 - \left(od_{dp}\right)^2\right]}{R} \qquad (12.1b)$$

In terms of depth, L_D, it is

$$L_D = L_t R_t \qquad (12.1c)$$

where
   D = the drill bit depth, in feet or meters
   id_{csg} = the inside diameter of the casing, in inches or centimeters
   od_{dp} = the outside diameter of the drill pipe, in inches or centimeters
   R = the pumping rate in gal/min or l/min
   R_t = the drilling rate, in ft/min

Formation materials which were identified with the core and cuttings listings should be matched with the primary log response. The core and cuttings depths should always be adjusted to those of the primary log depths, because of the greater control and accuracy of the wireline log. Correlation of the log response with the core and cuttings materials identifications is vital to the process of formation evaluation. A common example, increasing shaliness in a sediment usually results in an increase of the natural gamma ray deflection. It will reduce the resistivity reading and increase the apparent neutron porosity. The density values will be only slightly changed.

Drilling time logs, too, should be matched to the responses of other logs and the cores. Shales and soft sands drill faster. Cementation slows drilling. Usually, the drilling time log can be easily correlated with the spontaneous potential log and/or the gamma ray log. The geophysical borehole logs may also be used to correct the physical depth of the seismic and VSP logs (since their displayed depth are time), and any other surface measurements. This should be done *before* any detailed analysis is attempted.

Once the logs, cores, cuttings log, drilling time log, mud log, VSP, seismic, and any other type measurements have been depth corrected, we can move to the next step.

12.10   Identifying the Formation Sequence

The core sequence, cuttings sequence, and most of the geophysical log tracings will reflect, within reasonable limits, the actual formation sequence of the zones and beds. The geological column diagram, which can be obtained from the State Geologist and from texts, should be compared with the measurements. Formations and zones should be identified from the geological column and marked on a separate copy of the primary log. It is handy to have in the file, a paper copy of the primary log and write the formation zones on it, at the proper depths, as each zone is identified. In sediments, the gamma ray-electric log is a good one to start with. Usually, anomalous sequence variations are noted on the geological column. Therefore, look for any variations which may apply to the particular area which you are investigating. Match the log response with the likely formation zone on the column. These curves (*SP*, resistivity, and gamma ray) all respond to sands, shales, and carbonates. In hard rock environments, the density and/or the acoustic log with the natural gamma ray are good ones to consider.

In non-hydrocarbon exploration, a common practice is to combine on a sheet, at the same depth scales, the core log, the cuttings log, the geophysical borehole log, and the geologist's remarks, for each borehole.

12.10.1   Characteristic Signatures and Values

Do the established sequence and the log-determined one agree? If they don't, *find out why!* There has to be a reason. If they do agree, go on. Once a logical sequence is established on the primary log, the other geophysical logs can be matched to it and their characteristic signatures noted.

Characteristic signatures are important. They should conform, to a great extent, to what we would expect for that particular formation material and situation. Obviously, they will not conform exactly. If they did, there would be no need to explore. Also, they may often be used over wide areas to identify a particular horizon. Anomalous features can be particularly valuable because they are easily recognized. These features may be a thin coal streak or an isolated carbonate layer. A uniform progressive change of deflection can also be used. Some of the measurement values may be used. The author has identified Central Texas and Wyoming shale layers over wide areas by their characteristic radiation levels. The resistivity signature in a petroliferous zone can tell much about the deposit and can be used quantitatively. Some features of this are discussed further in this chapter.

In hard rock environments it is especially important to recognize characteristic signatures and values. Frequently, this is the best way to identify or verify extreme dip environment, overthrust, inversions, fracturing, faulting, and repeat sections.

Once formation sequences and their anomalies are identified and verified, target zones can be more easily identified.

12.10.2 Target Zones

A target zone is the zone which may be mineral-bearing or hydrocarbon-bearing. In the other types of work, it may be a general area or volume. For a driller it may be depth or time. It may be small in area and volume, as for the drainage of an oil well, or it may be large, as for a mining location. In the case of an engineering project, the target may be a whole horizon, such as a particular competent zone in sediments. The target could even be the lack of some feature, such as fractures or faults. Environmental project targets may even be chemical or dynamic. A chemical target may be a salinity level of a specific salt. A dynamic target may be fluid flow channels or directions and rates. The target in scientific work may be unknown. We may be simply gathering data for the record, for normal occurrences, or unknown anomalies. We would expect the target to be in some specific formation (i.e., the Tertiary Formation). We would expect it to have some identifying characteristic, such as a high radiation level. This is one of the reasons for gathering data and a reason for formation evaluation. The target is the goal of our exploration and/or development.

Keeping these things in mind, we will examine and review some of the methods and uses to pinpoint these characteristics.

12.10.3 Visual Examination

There are many things we can learn from an initial, visual examination of the data for a project. The usual items to note are evident; the total depth, the casing depth, the bed thicknesses, the number and types of measurements. We have already looked at the problems of information quality, geological sequence, correlations, and depth adjustments. The next step is to carefully examine the data for features of interest.

The shapes and amplitudes of the curve deflections, the core record numbers and numerical dumps can give an immediate view of the formation features. Low readings of electric logs suggest shales or shaliness, especially in sediments. These shales should be located in the core sequence. They probably will not be evident in the cuttings. Higher readings in sediments suggest sands and/or porous carbonates. These should be evident in both the cores and cuttings records. The relative deflections of the spontaneous potential (*SP*), resistivities (R), and natural gamma ray (GR) values can be functions of the degree of shaliness of a sand or porous carbonate. Short spacing resistivity curves (except for the Normal resistivity curves) will locate bed boundaries accurately, as will the inflection points of the *SP* and GR. These features will also indicate thin beds and laminar structures. Repeated signatures may suggest the presence of certain kinds of faults. Missing zones may also suggest faulting. This last can also suggest lost core and/or an instrument fault.

A hard-rock environment may have been anticipated in the initial stages of the project and the measurement suite designed for it. This environment will be evident from the drilling time logs and, of course, from the records and the cores. Hard-rock environments are frequently (but not always) characterized by high electrical resistivities, featureless *SP* curves, high densities, low porosities, smooth to-gauge holes, competent cores, and long drilling times.

Some hard rock structures have systems of micro-fractures. These show up well on resistivity measurements as lower than expected readings and sometimes on the core records. Density readings can help identify rock types. Fault intersections show well on density, focused resistivity, and caliper curves. Massive sulfides have extremely low electrical resistivities when they exceed certain concentrations.

One should look at *all* of the data, picture what might cause a particular combination of features or responses, and satisfy himself that he can explain each feature. Each feature means something. Many of them are of vital importance.

When visually checking geophysical data keep the normal, expected limits for the formation zones in mind. If the data go outside those limits, they may be correct, but *find out why*. If possible, repeat the measurement. A real anomaly will usually repeat within reasonable limits. A false one probably won't.

12.11 Rock Type

Identifying the zone rock type is a shorter job, but more complex, than that of determining S_w. It has however, much wider application than saturation calculations. Any and all of the standard logs which we have discussed so far, can be used. Some of the methods, such as rock mechanics, chemical analysis, and historical records are also valuable for this purpose and for extensions.

Any sedimentary zone which has gamma ray emissions higher than 30 or 35 APIγ must be suspected as containing clay or shale. Any sand with an emission lower than 20 APIγ probably is shale-free. Carbonates are usually low emission (but not always). The gamma ray reading in a coal is often a function of its ash content.

Hard-rock environments are not as easy to evaluate with respect to shale or clay. Alteration can occur. And, the altered material may be radioactive, even if the parent or unaltered rock is not. This does not necessarily indicate the presence of clay or shale. The density system is excellent for determining hard rock types if the zone is low gamma radiation. The bulk densities of many igneous and metamorphic rocks can be found in literature. Neutron activation techniques are useful, also.

The gamma ray curve, in any of its forms, is excellent for determining bed boundaries. This is because the system employs essentially a point detector. The presence of the normal, inevitable statistical variation of radioactive events may tend to hide the bed boundary, especially if the counting rates are low. The bed boundary, however, is *always* at the inflection point of the recorded curve. For best results the counting rate should be as high as possible.

A long time-constant in an analog system will shift the inflection point, along with all other curve features, in the direction of the tool travel. This can be minimized by short time constants and slow logging rates. This problem is true with downhole, airborne and surface analog radioactivity measuring systems.

Digital radioactivity equipment is not as plagued by such logging motion errors as the analog systems are. This is because the averaging method of a digital system are symmetrical and quite different form the analog methods. Either type system is excellent for stratigraphic and depth determinations. The resolution of either analog or digital systems will benefit from combining repeat measurements.

The natural gamma ray curve is excellent for determining locations and qualities of shales and non-shales. It should, of course be used with other logs. Figure 6.3 in Chapter 6 shows some typical emission ranges for a number of sedimentary materials. Note that the clay minerals (shales, mudstones, silts) all have substantially higher gamma emissions than most of the non-clay minerals. Of course, rock which contains substantial amounts of uranium, thorium, bismuth, potassium, etc. do not conform to the "normal" range. Also, clays may be potassium compounds, which are radioactive. They may, on the other hand, be compounds of sodium and other non-radioactive elements. Shales show an even wider range of variation of radioactive characteristics.

The readings and estimates of the natural gamma ray curve must be confirmed and refined with other methods and/or curve types. Thus, the

SP-curve responds well to the sand/shale sequences and is good for determining bed boundaries. Since they are referred to the shale potential, they will read at or near zero potential. The resistivity curves generally read lower in shales and shaly zones than they do in clean zones. Neutron porosities are likely to be abnormally high, compared to the effective porosity, due to the common presence of water and chlorine compounds.

Resistivity curves are poor rock-type indicators, except for shales. This is because they are responding primarily to the usually high conductivities of the waters within the interconnected pore spaces of the rock and the bound water and usually low resistivities in the clays of the shales.

Considering their cost and ease of application, some of the cross-plotting methods are, by far, the best methods for determining the actual and/or general mineral type of the rock (the rock matrix, in petroleum work). Examples are shown in Figures 12.1 to 12.4. These cross-plots have been adapted from those in the contractor's chart books and other literature. Be sure to use the equations or the original drawing in the chart book for any actual work. *The redrawn curves in this text* may not be accurate. There are many more types of cross-plots and analytical charts available in the chart books and the literature than are shown here. Also, there are as many types as one cares to generate. One need not be limited to the published charts. You only need to plot two or more applicable parameters against each other. Strictly speaking, one parameter may even be time, depth or frequency of occurrence.

12.11.1   Potassium Minerals

Potassium is a light, monovalent, highly reactive metal. Its density is 0.862 g/cc. It is one of the alkali metals. This family also includes hydrogen, lithium, sodium, rubidium, cesium, and francium. It is never found naturally in pure form. Most of its simpler compounds are highly soluble in water. It occurs, however, in massive deposits of soluble evaporites which have been isolated from ground water solution. Many of the simple compounds decompose in hot water. Many of the common potassium minerals are listed as insoluble in the range of 0 to 100°C.

Potassium compounds, such as feldspar and mica, are common in rock materials. Potassium minerals are one of the more common constituents in all rock environments. Many of these weather and otherwise alter into the family of clay minerals. Potassium mineral alteration products are common constituents of shales. See Table 7.5 for physical properties of some potassium minerals.

Source rocks of potassium compounds are many. Some of the commonly found ones are K-feldspar, volcanic rock, mica, granite, granodiorite, diorite, and basalt. Other sources are the many potassium-containing evaporites, clays, and shales. The mineral may be transported in either particulate form or in solution.

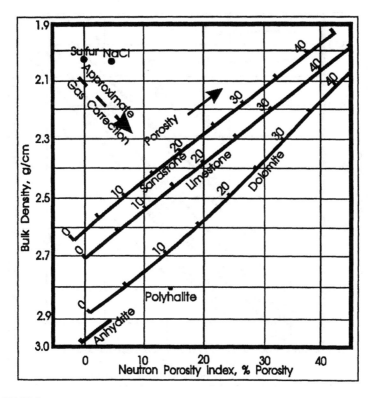

FIGURE 12.1
A porosity lithology crossplot using the bulk density and the SNP neutron porosity (After Schlumberger, 1989).

The radioactive isotope of potassium is potassium-40, $_{19}K^{40}$. It occurs as a constant proportion of naturally potassium (0.0119%). It decays by beta and neutrino emission to $_{20}Ca^{40}$ or by electron capture to $_{18}Ar^{40}$, with the emission of gamma rays. About 40% (average) of the gamma ray emission of sands is from the potassium compounds. This is shown in the series diagram, Figure 6.2.

Radioactivity logging systems are important in the exploration for and the development of evaporite minerals. Non-potassium evaporites often have low radioactivity. The minerals which are potassium compounds can be detected directly and quantitatively with the gamma ray logs. The grade may be determined in the same manner as with the uranium mineral grade determination. The calibration must, of course, be made for a potassium mineral, such as the equivalent potassium oxide or other simple potassium salt (eK_2O). The K channel (the $_{19}K^{40}$ curve) is excellent for this purpose. The KUT system may also be used with the density system and/or the neutron systems to identify the types and mixtures of evaporites. See Figure 7.10 for a log through a "potash" layer.

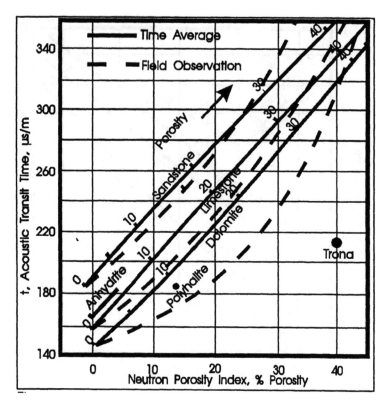

FIGURE 12.2
A porosity lithology crossplot using the acoustic travel time and the neutron porosity (After Schlumberger, 1989).

12.11.2   Shale and Clay Content

The shale and/or clay content of a zone are important parameters whenever fluids must be moved through a zone. This component bears directly upon the permeability of a medium and can greatly influence the porosity calculation. In most cases, the influence is a proportional factor of the volume of clay present. It is usually considered separately from rock identification because it is often handled quantitatively. Hydrocarbon projects are especially concerned with the clay/shale contents of zones because of the great influence the clay has upon the response of the measuring instruments.

Non-hydrocarbon projects are concerned with the clay/shale content of the formation zones for some of the same reasons as hydrocarbon projects. The clay content greatly affects the fluid flow or percolation properties of a permeable zone because it alters the effective porosity and the tortuosity of the zone. The clay content of a zone bears directly on fluid contamination problems. Construction of basins, lakes ponds, and canals must take

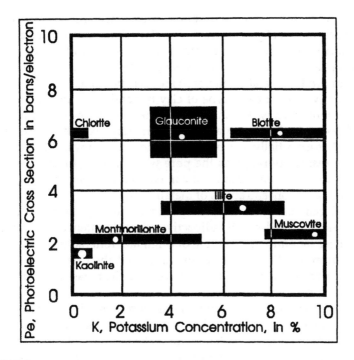

FIGURE 12.3
A crossplot of Pe against the K% concentration (After Schlumberger, 1989).

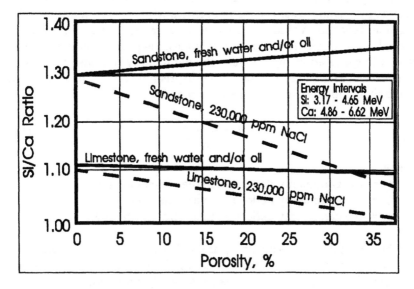

FIGURE 12.4
The measured porosity crossplotted against the Si/Ca ratio (Courtesy of Atlas Logging Services Division of Western Atlas International, Inc.).

the soil clay content into consideration. Thus, dewatering and *in situ* leaching projects are greatly concerned. The clay/shale content also affects the load bearing quality, the overburden rippability, and the overburden removal problems. It will affect the slope stability of a mine and its relatively immunity to weather factor in an open-pit mine.

The relative amount and type of clay mineral will affect the resistivity measurements because the shale usually has a very low electrical resistivity and occurs in the pore space and/or interlayered with sands and carbonates. It is essentially non-permeable and non-removable and has much the same resistivity range as the pore water. It also affects the neutron porosity measurement because it can contain a large amount of bound water and large concentrations of metallic ions.

Natural gamma ray measurements are affected by the shale/clay content because of the adsorbed radioactive mineral content of the clay. A portion of these ions and minerals are or originate from radioactive elements, such as uranium, thorium and potassium and their daughter elements. The solid portion of the clays themselves may be potassium compounds, which contain gamma emitters.

All of the natural gamma ray systems lend themselves to the quantitative determination of the clay/shale content of a rock zone. It is necessary to find a reading, real or probable, for a clean (non-shale) zone and for a pure clay or shale. These values can often be found in adjacent zones. The assumption must be made, of course, that the adjacent zone has the same characteristics as the zone of interest, with respect to the clay and to the sand or limestone. Many empirical methods are available and listed in contractor's handbooks. The major problem with these methods is finding suitable deflection values for a clean sand or carbonate and for a pure shale. If these can be found in nearby zones, the problem is minor. In this case, a linear relationship between the gamma ray response to the shale can be used. This can be reliably done surprisingly often. The relationship of the shale influence and the shale content, $I\gamma$, is

$$I_\gamma = \frac{\gamma - \gamma_s}{\gamma_{sh} - \gamma_s} \qquad (12.2)$$

where

γ = the gamma ray deflection in the zone of interest
γ_s = the gamma ray deflection in a clean sand
γ_{sh} = the gamma ray deflection in a solid, nearby shale

If, however, the clean sand and pure shale values cannot be found, there are cross-plotting methods for estimating their properties. The value of "N", (this is the same value of N as from the M_N cross-plot),

$$N = \frac{\phi_{N,f} - \phi_N}{\rho_b - \rho_f} \qquad (12.3)$$

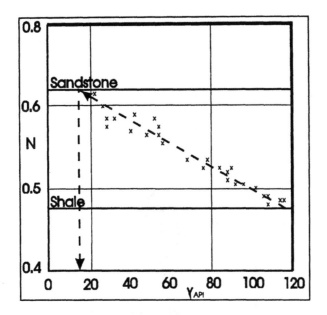

FIGURE 12.5
A crossplot of G_R vs. N, for shale and sand values (After Schlumberger, 1989).

can be plotted against the G_R or Th readings. A linear regression will show a probable trend of the sand or carbonate values. This can be extrapolated to the value of N for a pure quartz or carbonate for N. There, a value of the probable G_R or Th deflection for a clean zone will be indicated. The clean sand value of N will be near 0.64 to 0.68 or 0.59 to 0.62 for a limestone, depending upon the salinity of the borehole fluid. If the values of a clean sand (or limestone) are used, the value of N, for a clean sand, can be calculated. (Figure 12.5).

Another type of cross-plot for use in determining the clean sand (carbonate) and pure shale values makes use of the apparent density-derived porosity and apparent neutron porosity values. Sand (carbonate) points and shale points are plotted to establish the trends to the sand (carbonate) line and the shale/clay line. This is illustrated in Figure 12.6.

A plot, similar to the last one, can also be used to determine the percent volumes of the porosity and shale content. The assumption is made that the density and the neutron porosity responses are linear with respect to the shale volume. This is shown in Figure 12.7.

12.11.2.1 Clay Minerals

Refer to Table 7.1 for radioactive emission levels of clay and shale minerals. It is evident that most shales show high potassium and thorium contents, but lower uranium contents. The clay molecules in shales do not

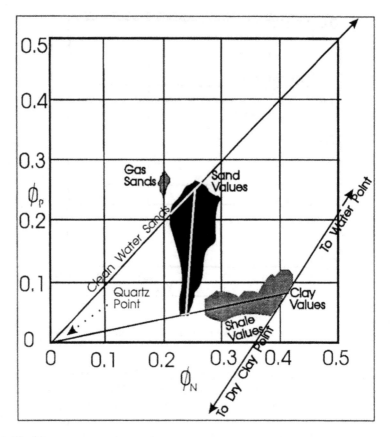

FIGURE 12.6
A ϕ_p vs. ϕ_N crossplot to determine sand and shale values (After Schlumberger, 1989).

always, however, contain radioactive elements. Tables are available in various handbooks showing clay compositions. The radioactivity of clays and shales is mostly due to the absorption and adsorption properties of the clays.

Figure 7.6 shows a log illustrating the high radioactivity of the Eagle Ford shale as due to the uranium content. It has, however, a low potassium and only a moderate thorium content. This log also compares the low potassium content of the Eagle Ford with the high potassium Del Rio shale. It is obvious that these two shales have quite different compositions.

Uranium content in clays and shales can be anomalous. It can vary widely, due to their histories. Kaolinite is low in potassium. Bentonite shows a high thorium content. Thus, care must be exercised when calculating a shale volume from a gamma ray log.

Shales in contact with uraniferous solutions can have the uranium migrate into them. Because of the low permeability of most shales, the radioactivity due to the uranium daughters will often linger long after the solution or mineral body has moved on down dip.

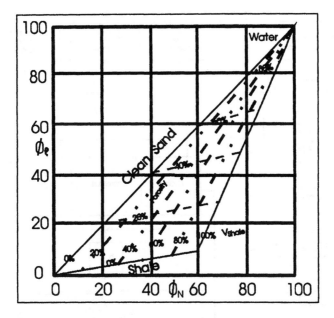

FIGURE 12.7
A crossplot to determine V_{SH} (After Schlumberger, 1989).

Organic-rich shales will show anomalous uranium because of their reducing environment. This will cause unusual amounts of uranium to be retained by the shale either during deposition or post-depositional. These shales should show a normal thorium amount and any amount of potassium, depending upon the type of clay in the shale. Organic shales usually have higher radioactivity than inorganic shales because of the reducing nature of the contained organic material. Most appear to be high in potassium and uranium. Some, such as the Kuperschiefer shale in Europe, are rich enough in uranium to be mined. Many have extensive fracture systems, inter-bedded silts, chert, and/or carbonate beds. Some are good oil reservoirs. The relationship between organic carbon and the Th/U ratio, in three black shales from West Virginia and Kentucky, can be seen in Figure 7.14.

Anomalous thorium readings with low or high potassium readings can indicate the clay type in a shale. Examples are shown in Figures 7.2 and 7.11. This latter probably also indicates exposure to migrating uranium-rich solutions at some time in the recent past. The thorium channel value of the three-channel gamma ray system, KUT log is better than the gross-count gamma ray (GCGR) curve for estimating the clay content of a shaly sand. This is because both the potassium content and the uranium content of a shale can vary widely, but the thorium content is relatively uniform. The concentration ratios of the three radioactive families, especially Th/K, can be used to trace a specific shale over areas as big as a project field. On a log of this type, the calculation of the volume of shale would be much

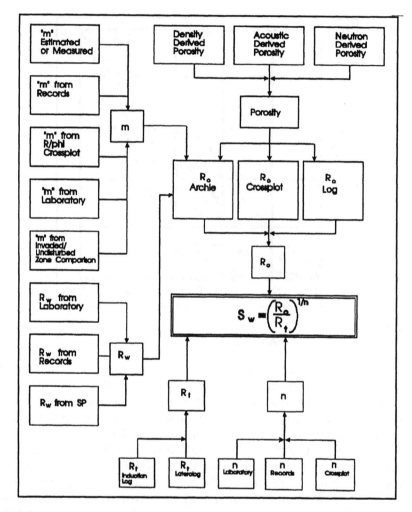

FIGURE 12.8
The major task of petroleum formation evaluation.

more realistic if the thorium curve would be used, instead of the total count, potassium, or uranium curve.

12.11.3   Carbonates

Carbonates, which are normally low in radioactivity, can contain authigenic feldspars and argillaceous materials. These increase their potassium content. Carbonates may also have their radioactivity level raised by precipitation of uranium from solution post-genetically. They may also cause ferric hydroxide-uranium mixtures to co-precipitate, because of the high pH of the carbonate. Conversely, radioactivity levels may be lowered by crystallization or re-crystallization of calcium carbonate to calcite and subsequent dolomitization.

12.12 Quantitative Analysis

Good data lends itself to quantitative or semi-quantitative analysis. Poor data can be confusing, misleading, and expensive.

12.12.1 Shale Identification

In sediments, the G_R can be used to estimate the amount of shale or clay V_{sh}, of a zone. V_{sh}, of course, has a bearing upon the permeability and producibility of a zone for petroleum, water flood, enhanced recovery, and *in situ* leaching projects. It bears upon the water and excavation problems for a mine and other engineering projects. Figure 6.13 and Equations 12.3 to 12.7c deal with the calculations for estimating V_{sh}. Briefly, the volume of shale, V_{sh}, is in Tertiary and younger sands, I_R is the gamma ray index and is

$$I_R = \frac{\gamma - \gamma_s}{\gamma_{sh} - \gamma_s} \tag{12.3}$$

$$V_{sh} = 0.083\left(2^{3.71 I_R} - 1\right) \tag{12.4a}$$

and in older sands (Larionov, 1969),

$$V_{sh} = 0.33\left(2^{2 I_R} - 1\right) \tag{12.4b}$$

Another suggested relationship is from Clavier, et al. 1977

$$V_{sh} = 1.7 - \left[3.38 - \left(I_R + 0.7\right)^2\right]^{1/2} \tag{12.4c}$$

and also, in Miocene and Pliocene in Louisiana (Stieber, 1970),

$$V_{sh} = \frac{I_R}{3.0 - 2.0 I_R} \tag{12.4d}$$

These are plotted in Figure 6.13.

where
γ = the natural gamma ray reading in the target sand
γ_s = a clean sand
γ_{sh} = a nearby, pure, solid, shale.

In the case of a shale containing *only* potassium shales, success has been reported using the relationship

$$V_{sh} = \frac{\rho\gamma - \rho_{sh}\gamma_{sh}}{\rho_{sh}\gamma_{sh} - \rho_s\gamma_s}$$

(12.4f)

The relationships of Equations 12.3 to 12.4d will work much better if the thorium (Th) curve of the KUT log is used instead of the total count G_R curve. The method works reasonably well in porous carbonate zones.

The value of V_{sh} should agree reasonably well with the assessments from the cores. Of course, the difference in sample size should be taken into account. The two methods will not agree if anomalous radioactivity is present. In this case the core values should be used and averaged over each interval chosen.

Recent studies and presentations tend to discourage the use of the Larionov relationships in favor of a linear or straight-line relationship. This author has had reasonable results using the linear relationship, but using the potassium curve (K curve) of the KUT system. The problem is, of course, that the shales usually have mixtures of several clay types. And, the composition and types of the clays in a shale and their relative amounts varies widely, as does the adsorbed mineral content of the clay.

The spontaneous potential (SP) curve can also be used to estimate the shale content of a sand. If the sand is laminated and the shaly layers are lower resistivity than R_t, then

$$V_{sh} = 1.0 - \frac{PSP}{SSP} = 1 - \alpha$$

(12.5)

where PSP is the pseudo-SP, the reading in the shaly sand and SSP is the static-SP, the reading in a clean, infinitely thick sand (Doll, 1949).

The resistivity curve amplitude is usually sharply reduced by the presence of shale in a zone. The exception is in very saline sands, such as are found in the Texas Gulf Coast. Two relationships which are used are

$$V_{sh} = \left(\frac{R_{sh}}{R_t}\right)^{1/b}$$

(12.6)

where

R_{sh} = the shale resistivity
R_t = the true formation resistivity
b = 1.0 if $R_{sh}/R_t = 0.5$ to 1.0
b = 2.0 if $R_{sh} \to R_t$

and

$$V_{SH} = \left\{ \frac{R_{sh}(R_{max} - R_t)}{R_t(R_{max} - R_{sh})} \right\}^{1/b}$$ (12.7)

where

$b = 1$, if $R_{sh}/R_t \geq 0.5$
$b = 2(1 - R_{sh}/R_t)$, if $R_{sh}/R_t < 0.5$

The neutron porosity system also responds to the shale/clay content because of the water contained in the clay. This approximation works best in gassy zones. It is poor in sands with small amounts of shale:

$$V_{sh} = \frac{\phi_N}{\phi_{sh}}$$ (12.8)

where

ϕ_N = the neutron-derived porosity
ϕ_{sh} = the porosity reading in a pure shale

The density, neutron porosity, and the acoustic P-wave travel time can be combined to estimate V_{SH}:

$$V_{sh} = \frac{\rho_b(\phi_{N,ma} - \phi_{N,f}) - \phi_N(\rho_{ma} - \rho_f) - \rho_f\phi_{N,ma} + \rho_{ma}}{(\rho_{sh} - \rho_f)(\phi_{N,ma} - \phi_{N,f}) - (\phi_{N,sh} - \phi_{N,f})(\rho_{ma} - \rho_f)}$$ (12.9a)

$$V_{sh} = \frac{\rho_b(t_{ma} - t_f) - \phi_N(\rho_{ma} - \rho_f) - \rho_f t_f t_{ma} + \rho_{ma} t_f}{(\rho_{sh} - \rho_f)(t_{ma} - t_f) - (t_{sh} - t_f)(\rho_{ma} - \rho_f)}$$ (12.9b)

$$V_{sh} = \frac{\phi_N(t_{ma} - t_f) - t(\phi_{N,ma} - \phi_{N,f}) - t_{ma} + \phi_{N,ma} t_f}{(t_{ma} - t_f)(\phi_{N,sh} - \phi_{N,f}) - (t_{sh} - t_f)(\phi_{N,ma} - \phi_{N,f})}$$ (12.9c)

where

ρ_b = the formation bulk density
ρ_{ma} = the formation rock matrix density
ρ_f = the formation fluid density
ρ_{sh} = the pure shale density
$\phi_{N,ma}$ = the neutron apparent matrix porosity
ϕ_N = the neutron apparent formation porosity
$\phi_{N,f}$ = the neutron apparent fluid porosity
$\phi_{N,sh}$ = the neutron apparent pure shale porosity
t_{ma} = the acoustic matrix travel time
t = the acoustic formation travel time
t_f = the acoustic formation fluid travel time
t_{sh} = the acoustic pure shale travel time

This author has not had a chance to check the relationships of Equation 12.9. They were presented in the 1985 *Dresser Atlas Log Interpretation Charts*. They are shown here in case they might prove useful.

12.12.2   Hard-Rock and Non-Hydrocarbon Environments

Hard-rock environments are not logged with wireline geophysical logs as frequently as are sedimentary environments. This is because hard rock holes are usually cored. The general feeling is that logs are not needed if you have cores. This assumption is certainly untrue. Both are needed. There are many differences in concept between the two. They do not replace each other, they complement each other.

Further, hard-rock logging equipment must be designed for the purpose of logging in hard rock environments. Petroleum-logging equipment is designed for large holes in sedimentary environments to look for fluids from large holes in sedimentary environments; for example, petroleum-resistivity logs are excellent from about 0.02 up to about 2000 ohms. This range covers the needed hydrocarbon deposit parameters quite adequately. Mineral-resistivity systems are available which operate well to 10,000 ohms and can be used to 30,000 ohms. Petroleum natural gamma ray equipment is designed to be linear to about 150 APIγ units (typically about 400 counts per second). Good linearity is not very important, here. The usual digital mineral system is linear to about the equivalent of 10,000 APIγ units (about 10,000 counts per second) and can be corrected for use to six times that counting rate. Hard rock interpretation methods usually assume that there is little or no mudcake. Most of the needed logging systems are available in equipment designed specifically for the expected environment and usage. These should be investigated instead of just using the petroleum methods. Remember, petroleum people are looking for different materials than are mineral, engineering, and oceanographic investigators.

Calibration methods and standards are greatly different for the petroleum and the non-hydrocarbon systems, also. It is wise to specify and use the type of equipment and methods designed for your type project.

In sediments, the gamma ray (GR) will normally read between 5 APIγ units and 150 APIγ units. If it goes below 5 APIγ units, it may suggest a coal or a massive rock zone. On the other hand, the low reading may be caused by a faulty detector or discriminator circuit. If it goes above about 150 APIγ units, it may indicate a uraniferous zone. With an analog circuit, it can also indicate a broken conductor in the tool or cable. With a digital system, it may indicate a faulty counting circuit.

Similar possibilities will exist for hard-rock environments. Each type of environment will have its own normal limits. Anomalies are often more sharply bounded and have greater contrasts than in sediments. Anomalies in low-radiation rock frequently indicate fracture or fault zones. Thick, less intense anomalies may be micro-fracture systems.

Spontaneous potential and resistivity systems are sensitive to redox state. While this is not important in the highly reduced zones normal in deep petroliferous environments, it can be valuable in shallow and in non-hydrocarbon applications.

Gamma ray (GR) measurements may detect oxidation alteration of the formation material in both sediments and hard rock environments. The alteration product of many hard rock materials is clay. Clay has the property of electrically attaching solution ions, which may include uranium ions. Thus, the altered material will frequently show medium- to high-grade radioactivity which may be appreciably higher than that of the parent rock. The probability also is that its resistivity will be lower than the parent rock.

Sulfides are reduced minerals, as opposed to oxidized minerals. They will tend to reduce any oxidizing materials, especially solutions with which they come in contact. These solutions frequently carry small amounts of oxidized (uranic) uranium. In general, uranic ions are much more soluble than the more reduced uranous ions. Thus, sulfide deposits will tend to deposit out uranium. We frequently see anomalous radioactivity associated with sulfide deposits. The uranium mineral can be anywhere from trace amounts to ore-grade mineral. There is some thought that this process is aided by the action of anaerobic bacteria.

Similar phenomena can be observed with other reducing materials. Hydrocarbon gas seepage through fault zones can cause radioactive mineral to deposit in the fault. This is used by some geologists in New Mexico as a petroleum exploration tool. Also in New Mexico are ore-grade silver deposits around gas seepage.

Resistivity systems have similar comments and normal excursion limits. The resistivities in sediments may run from 0.03 ohms in a high-temperature, high-salinity, high-porosity sand to 2000 ohms in a low temperature, low salinity, low porosity sand, especially if it contains gas or hydrocarbon.

In a hard rock environment, the resistivities will usually be high. A massive rock zone will have uncorrected values on the order of 10,000 ohms or higher, with much of the conductivity coming from the borehole fluid. There are exceptions and they are significant. Sharp, isolated, lower resistivity excursions could suggest open fractures or faults. A brecciated or an altered fault zone will show an appreciably lower resistivity than the main rock body. It will often have some repeatable character. Multiple thin fault zones usually have some recognizable signature. Thick, moderately lower resistivity zones may contain extensive micro-fracture systems.

Sulfide minerals generally have low electrical resistivities — high conductivities. We might expect them to influence the bulk resistivity of the containing rock. In low concentrations, however, they are disseminated. The crystals or particles may be surrounded by the high-resistivity rock material. Thus, the bulk resistivity may remain high even when the concentrations of sulfides approach 10% or 15%. At some concentration, depending upon the structure, the particles appear to contact each other and form continuous electrical paths. The resistivity then drops precipitously. Extremely low resistivity («0.1 ohmmeter) readings can then be]present. Often, the readings are lower than the range of the logging system. The same phenomenon may occur with formation carbon and virgin metals.

Coal minerals have typical resistivity trends, different from other sediments. Lignites have low resistivities (often less than 20 Ωm). This is because they have high ash (clay mineral) and moisture (water) contents. The ash content may be 10% or more by weight. Moisture contents may by 30%. Thus, lignite bulk resistivities are generally low, 20 Ωm and lower.

Bituminous coals have lower ash and moisture contents. The ash may be as low as 1 to 2%. Moisture can be very low. Volatile contents, which include hydrocarbons, are high. These coals have high resistivities, which can be a rough index of the quality of the coal. R_t may run from 100 Ωm to >10,000 Ωm.

Anthracite coals begin to show electrical conduction by the carbon. These coals have very low ash, moisture, and volatile content. Thus, their electrical resistivities generally are less than 10 Ωm and can be very low.

12.12.3   Quantitative Porosity and Saturation

Probably the best known quantitative applications of formation evaluation are for evaluating hydrocarbon occurrences. The best known part of this, the determination of water saturation, S_w, from which oil saturation, S_o. In addition to S_o we need to know many other items. Many of these are listed in Table 12.2.

The first item in the list of Table 12.2 is the geological sequence. This was discussed at length, earlier in this chapter. Identification of the target zone is an essential part of verifying the formation sequences.

TABLE 12.2

Data categories needed for petroleum
formation evaluation

1.	Geological sequence
2.	Formation water salinity
3.	Water/hydrocarbon saturation
4.	Rock type
5.	Shale/clay content
6.	Permeability
7.	Formation fluid types
8.	Formation fluid distribution
9.	Formation fluid pressure
10.	Formation rock strength and competence
11.	Extent of the zone
12.	Well potential
13.	Field potential

12.12.4 Saturation Determination

Determining the fluid saturations, will probably take the greatest amount of time and effort. It, however, is one of the most important information sets.

The procedures for determining saturation are covered in detail in Chapter 5. The goal of these procedures is to determine the fluid saturations, especially the hydrocarbon saturations, of the target zone or zones. The principle of this is diagrammed in Figure 12.8. We will see how this goal is reached.

The suite of measurements for saturation determinations is centered around the resistivity measurements. The other method which are used will depend upon the type of formation material, the borehole environment, and the type of project on which they are used.

To determine hydrocarbon saturation (S_h, S_o, S_g), we must first determine the water saturation, S_w. This is because our methods determine S_w rather than S_o or S_g. The value of S_w is

$$S_w = \left(\frac{R_o}{R_t} \right)^{\frac{1}{n}} \qquad (12.10)$$

where

  R_o = the formation resistivity when 100% water saturated
  R_t = the true formation resistivity
  n = the saturation exponent (usually assumed to be equal to 2)

Figure 5.15 shows the problem well. In order to find the items of Equation 12.10 we will almost certainly need to find the values for the

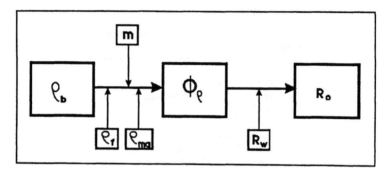

FIGURE 12.9
The flow diagram for R_o from ρ_b.

porosity, ϕ, a good value of the porosity exponent, m, a value for the formation water resistivity, R_W, and we must find, measure, calculate, or otherwise determine values for R_t, R_o, and n.

The values of R_t will come from the deep-resistivity log. Since the values of R_t reflects the presence of hydrocarbon and/or gas, their determinations must be the best possible. The corrections must be made as accurately as possible. The entire reserve calculation hinges on these values.

A value of R_o should theoretically be from the same point where the value of R_t is measured. The values of R_o are the those which R_t would have if the zone were 100% water saturated; *if no hydrocarbon or gas were present*. R_o, of course, can be a difficult value to obtain, especially at the point where there is hydrocarbon or gas. We can obtain approximate values of R_o in several ways. If the zone is uniform in character and if the hydrocarbon has migrated toward the top of the zone, we may be able to find a portion, near the bottom of the zone, which has nearly 100% S_w. An average value of R_o can frequently be found here. This type situation would result in a resistivity log which would have a signature similar to that in Figure 5.2.

The selection of an average or probable value of R_o is greatly aided if a curve of probable R_o values has been calculated from the porosity log. The usual parameter used for this is the density-derived porosity. Any of the porosity curves can be used, however. This computer-calculated curve is often plotted on the resistivity log.

If R_o is calculated from the density values using Archie's relationship (for example), several other values must be accurately determined. The flow diagram for this is shown in Figure 12.9. Although Figure 12.9 uses the density value, the reading from the acoustic curve or the neutron-porosity curve can also be used. As you can see from the diagram, the values of ρ_f, the formation fluid density, ρ_{ma}, the formation rock matrix density, m, the cementation exponent, and R_w, the formation water resistivity, must be accurately decided upon.

The value of ρ_{ma} can be measured accurately or closely estimated from the cores. Cores, however, are not always available. Cuttings logs can help. Lag and possible material loss and alteration in the transport of the cuttings to the surface and subsequent screening are a serious problem. An examination of the washed cuttings particles can often indicate the rock type. Thin sections can often help identify the rock material, allowing a very close estimate of the value of ρ_{ma}. Tables and handbooks are available and can be a great help in estimating ρ_{ma}. Also, refer to the tables in the chart books of the major logging contractors. This method is recommended.

Evidence from other logs and past experience can be quite reliable, if care is taken. Clean zone parameters are easier to estimate than are shaly zone ones. In clean zones, the three "porosity" logs should read the same porosity value, if the rock matrix value is correctly picked. In this case, the value of the borehole-corrected apparent porosity is likely to be more nearly correct.

The fluid density (ρ_f) and resistivity (R_w) are best measured from samples of the actual formation fluid. Many drillers are skilled at obtaining good formation water samples from a zone. This is not always practical, however, and may be expensive and time consuming. Fluid sample takers are available. Many core laboratories, State Geologists, State Engineers, and some logging companies maintain libraries of formation water compositions and resistivities for their areas.

The spontaneous potential curve has traditionally been used to calculate the value of R_w (actually, the value of R_{we}, the effective NaCl resistivity), and from that, ρ_f:

$$R_{w,e} = 10^{\left(\frac{E_{SSP}}{K_{SP}}\right)} R_{mf,e} \qquad (12.11)$$

where

E_{SSP} = the value of the Static Spontaneous Potential
K_{SP} = the SP temperature correction
$R_{w,e}$ = the effective NaCl value of the formation water resistivity
$R_{mf,e}$ = the effective NaCl value of the mud filtrate resistivity

The values of R_{mf} can be obtained from the shallow-resistivity values. They are also one of the "tornado" chart evaluations. $R_{mf,e}$ can be estimated from the charts shown in Figures 4.6 and 4.7. The value of K_{SP} can be calculated from the formation temperature or from the borehole temperature at the formation level. The equation, in degrees Fahrenheit, is

$$K_{SP} = bT + c \qquad (12.12)$$

where
> T = the zone temperature in degrees Fahrenheit or Celsius
> b = 0.133 mvs/°F or 0.24 mvs/°C
> c = 60.77 mvs, if °F are used or 65.03 mvs if °C are used

The value of E_{SSP} is to be obtained, with the appropriate corrections, from the *SP* log. The value of R_w can be estimated from the chart Figure 4.7. There are also other charts available for this conversion, from other contractors.

The relationship between R_w and NaCl salinity, at 75°F or 24°C, is

$$ppm = 10^x \tag{12.13a}$$

where

$$x = \frac{3.562 - \log(R_{w,75} - 0.0123)}{0.955} \tag{12.13b}$$

$$R_w \approx 0.0123 + \frac{3647.5}{(ppm\ eNaCl)^{0.955}} \tag{12.13c}$$

and the temperature correction relationship is

$$R_{w1}(T_1 + a) = R_{w2}(T_2 + a) \tag{12.13d}$$

where a 6.77, if degrees Fahrenheit are used or 21.5, if degrees Celsius are used. The value of ρ_f can be estimated from the value for R_w. The resistivity/salinity/temperature/chart, such as that of Figure 4.12, can be used to determine the probable eNaCl concentration. The chart in Figure 12.10 can be used to determine the density, ρ_f, as functions of temperature/pressure/salinity, assuming the total dissolved solids (TDS) are all NaCl. Alternatively, the density of the formation water can be calculated if the solids are identified:

$$\rho_f = \left[c_1\rho_1 + c_2\rho_2 + \ldots + c_n\rho_n + (10^6 - \Sigma c)\rho_w \right]10^{-6} \tag{12.14}$$

where
> c = concentration of a solid in ppm
> ρ_n = the bulk density of a solid
> ρ_w = the density of water at formation temperature

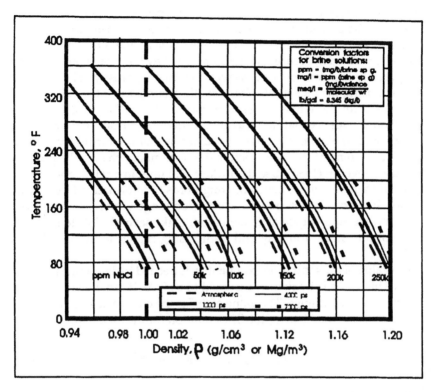

FIGURE 12.10
The relationship between NaCl solution density and temperature and pressure (Courtesy of Atlas Logging Services Division of Western Atlas International, Inc.).

If the salt composition of the formation fluid is known, the Dunlap method (shown in the *Schlumberger Log Interpretation Charts* and the *Atlas Wireline Chart Book*) can be used to determine the effective NaCl parts per million from the actual solute amounts. This can be used in the other direction, also.

The value of the exponent "m" (again, referring to Figure 5.14 and Table 5.2 in Chapter 5) may be best found by performing resistivity and porosity measurements upon cores from the interesting zone. To accomplish this, the core must be cleaned and dried. Then measure the porosity, ϕ. The core is then saturated with a solution of NaCl of a known resistivity. The chart Gen-3 in the *Schlumberger Log Interpretation Charts* book and chart 1-5 in the *Dresser Atlas Log Interpretation Charts* book will aid in determining the solution of NaCl. The bulk resistivity is then measured. At any particular porosity, Archie's relation will solve for "m":

$$m = \frac{\log R_w - \log R_o}{\log \phi} \qquad (12.15)$$

Since cores are not always available, the next best solution is to cross-plot R_t against the porosity or a function of porosity on a *Pickett* plot. The trend of the lowest values defines R_o. The slope of the R_o line is the value of "m". This is explained in Figure 5.15 in Chapter 8. Any *function* of porosity from the porosity logs may be used, as well as the core porosity values.

When using the Pickett plot it is usually assumed that the value of "m" is constant for the whole range of porosities, from 0% to 100%. There is substantial evidence that "m" is a function of the tortuosity and can change widely within a zone. We have already seen that the tortuosity changes because of some of the same factors that commonly change the porosity. See also Doveton (1986). Use the Pickett plot only in a relatively uniform zone and where there is substantiating porosity information to verify the values. In a zone which might have a wide range of porosities and/or a high and varying degree of cementation, make several plots to cover the range of porosities. You may find a surprising change in the actual value of "m".

The value of the saturation exponent, n, is needed to determine the value of S_w. This value is frequently assumed to be 2.00. It can, however, be different, and it can change. The value of "n" is the change of slope of the saturation lines of the Pickett plot. The equation for the Pickett plot trends is

$$\log R_t = \log R_w - m \log \phi - n \log S_w \qquad (12.16)$$

If this is solved for "n":

$$n = \frac{\log R_w - m \log \phi - \log R_t}{\log S_w} \qquad (12.16a)$$

The best way to determine the value of "n" is be laboratory measurement. Determine ϕ of the core samples, taken from known regions, by a gas method or any other nondestructive method. Saturate core samples with a known salinity of water. Measure the value of R_o and determine "m". Then partially de-saturate the samples by drainage or centrifuge, measure R_t, and determine S_w (i.e., by the change in weight and/or by the volume of water expelled). The value of "n" can be found with Equation 12.16a.

Now that all of the contributing factors have been determined, the actual values of S_w and S_h can be calculated from the log values:

$$S_w = \left(\frac{R_o}{R_t} \right)^{\frac{1}{n}} \qquad (12.17)$$

and

$$S_h = 1 - S_w \qquad (12.18)$$

Note that the calculation of saturation and porosity have extensive applications in non-hydrocarbon work. ϕ and S_w calculations can locate the top of the water table and define the transition distance precisely. They can predict exactly how much dewatering must be done for a mine.

Determining saturations allows a convenient method for estimating how much hydrocarbon might be produced. This is the moveable oil saturation and plot method, MOS/MOP.

12.12.5 Reserve Calculation

The calculated values for S_w, S_g, or S_o (S_h) must, of course, be applied to a reserve determination before an oilfield (or even a single well completion) or a mine can be considered. A method (Amyx et al., 1960) can be conveniently used for hydrocarbons:

$$N_h = \frac{K_h A h \phi S_h}{B_h} \qquad (12.19a)$$

where

N_h = the initial hydrocarbon reserve, in standard barrels, or gas reserve, in cubic feet or cubic meters

K_h = the conversion factor for oil in standard barrels per acre foot (7758) or per hectare meter (62878), or for gas in cubic feet per acre foot (43,560) or in cubic meters per hectare meter (40843) at formation pressure and temperature

A = the drainage area, in acres or hectares

h = the zone thickness in feet or meters

ϕ = the fractional porosity of the zone

S_h = the initial saturation of the zone with oil or gas, dimensionless

B_h = the initial oil formation volume factor, in barrels per standard barrel, or the initial gas formation volume factor, in cubic feet per standard cubic foot or in cubic meters per standard cubic foot

The formation volume factor treats the change in the oil or gas volume as it is transported from the formation conditions to standard (65°F, 18.3°C and 1 atm.) conditions.

The reserve procedure is essentially the same for mineral production. For mineral ores, the equation becomes

$$N_m = K_m A h G \qquad (12.19b)$$

where

   K_m = the conversion factor, in tons of ore per acre foot or hectare
          meter
   A = the area considered, in acres or hectares
   h = the average thickness of the mineralization, in feet or meters
   G = the average fractional grade of the mineral, dimensionless

Equations 12.19 determine the probable reserves in place. Not all of the hydrocarbon can be produced, however. Thus, the value of N_o, N_g, or N_m must be multiplied by the production efficiency factor, E, to find the probable potential production. Equations 12.19a and b will also give the amount of mineralization in place. E is always less than 1.

The efficiency factor, E, for hydrocarbon is simply the amount of hydrocarbon in place, minus the amount which will not be producible because of problems with economics, surface tension, coning, dilution, etc., and divided by the original amount in place. In a mineral project, this factor will involve the portion of the mineralization which it too low a grade or too thin a bed to consider economically, dilution by overburden and floor, working space (in an underground mine), and transportation costs. Both the hydrocarbon and mineral factors are, at least, largely economic.

12.13   Cross Plots

12.13.1   MN Cross-Plot

The M_N cross-plot is a Schlumberger method for eliminating the effects of porosity and constructing a lithology, only, cross-plot. The values of bulk density, neutron porosity, and acoustic P-wave travel time are combined in such a way as to minimize the effects of the formation fluids and porosity. The value of "M" is

$$M = 0.01 \frac{t_f - t}{\rho_b - \rho_f} \qquad (12.20a)$$

The value of "N" is

$$N = \frac{\phi_{N,f} = \phi_N}{\rho_b - \rho_f} \qquad (12.20b)$$

The values of "M" and "N" are plotted against each other. Each of several rock types have well defined positions or areas in which logged points

TABLE 12.3

Example values of "M", "N", "A", "K"

Material	M	N	A	K
Fresh Mud				
Anhydrite	0.71	0.52	1.88	1.35
Dolomite	0.78	0.49–0.51	1.84–1.91	1.46–1.53
Gypsum	1.08	0.38	2.45	2.61
Limestone	0.83	0.59	1.61	1.57
Quartz	0.81–0.84	0.62	1.47	1.23–1.27
Shales	0.55–0.65	0.5–0.6		
Salt Mud				
Anhydrite	0.72	0.53	1.98	1.39
Dolomite	0.80	0.52–0.54	1.74–1.91	1.47–1.53
Gypsum	1.62	0.45	2.65	2.69
Limestone	0.85	0.62	1.71	1.41
Quartz	0.83	0.68	1.56	1.27–1.31
Shales	0.55–0.65	0.5–0.6		

Compensated Neutron Porosity Log, only.

will fall. Table 12.3 lists the approximate "M" and "N" values for some formation rock materials. *Note that these values apply to the compensated neutron porosity logs, only.* See Figure 12.11.

12.13.2 The AK Cross-Plot

A similar cross-plot, published by Dresser Atlas (now Atlas Wireline Services) is the AK plot. This author has found it helpful to use both the M_N and the AK cross-plots, on occasion, to separate some of the materials, especial the carbonates. The value of "A" is

$$A = \frac{\rho_b - \rho_f}{\phi_{N,f} - \phi_N} \qquad (12.21a)$$

and "K" is

$$K = 0.01\frac{t_f - t}{\phi_{N,f} - \phi_N} \qquad (12.21b)$$

Values of "A" and "K" (for the compensated neutron porosity logs) are also listed in Table 12.2.

12.13.3 The MID Plot

Another popular method for determining rock type is the Mineral Identification (MID) plot of Schlumberger. The MID plot uses a calculated apparent

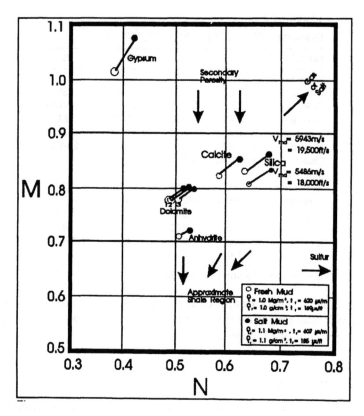

FIGURE 12.11
The M_N crossplot (After Schlumberger, 1988).

matrix travel time, $t_{ma,a}$, (note that Schlumberger does not use the apostrophe between the subscript "*ma*" (for "matrix") and "*a*" (for "apparent"). This is done in this text for clarity)

$$t_{ma,a} = \frac{t - \phi_{t,a}t_f}{1 - \phi_{t,a}}$$

(12.22a)

for the porosity calculated by the time average method and

$$t_{ma,a} = t - \frac{\phi_{t,a}t}{c}$$

(12.22b)

The apparent matrix density, $\rho_{ma,a}$, is also used:

$$\rho_{ma,a} = \frac{\rho_{b,a} - \rho_f\phi_{t,a}}{1 - \phi_{t,a}}$$

(12.22c)

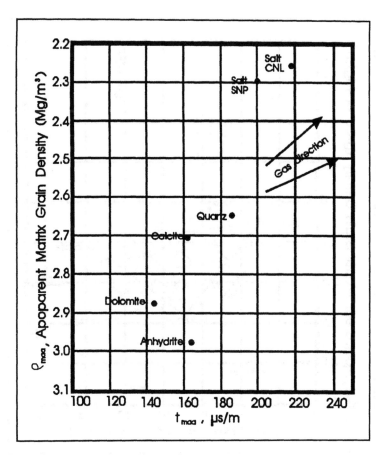

FIGURE 12.12
The MID Plot. (Schlumberger Well Services, Inc., 1988.)

where

t = the corrected log value of the acoustic P-wave transit time
t_f = the transit time of the pore fluid
$\phi_{t,a}$ = the apparent total porosity
c = a constant ($c \approx 0.68$)
$\rho_{b,a}$ = the corrected, measured log bulk density
ρ_f = the density of the pore fluid

Charts for determining the values of $t_{ma'a}$ and $\rho_{ma,a}$ from the logged values of t and p, respectively, as a function of porosity, will be found in Log Interpretation Charts by Schlumberger Well Services, Inc.

The value of $t_{ma'a}$ and $\rho_{ma,a}$ are then plotted on a MID cross-plot grid to determine the probable lithology (Figure 12.12 and 12.13).

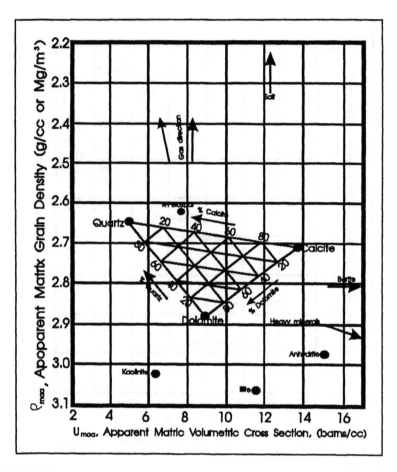

FIGURE 12.13
A variation of the Schlumberger MID plot (After Schlumberger, 1988).

A variation of the above MID plot just shown is one which plots $\rho_{ma,a}$ against the apparent volumetric cross section of the matrix, $U_{ma,a}$, from the Litho-Density* system,

$$U_{ma,a} = \frac{P_e \rho_e - \phi_{t,a} U_f}{1 - \phi_{t,a}} \qquad (12.23a)$$

where
U_f = the volumetric cross section of the pore fluid (b/electron)
P_e = the photoelectric cross section index, from the Lithodensity log (b)
ρ_e = the electron density of the formation material (b/electron)

$$\rho_e = \frac{\rho_b + 0.1883}{1.0704} \qquad (12.23b)$$

---

* Litho-Density is a registered trademark of Schlumberger.

TABLE 12.4

Some values of the photoelectric parameters.

	P_e, b/e	$\rho_{b,a}$	U, b/cm³
Quartz	1.810	2.64	4.780
Calcite	5.080	2.71	13.800
Dolomite	3.140	2.85	9.000
Anhydrite	5.050	2.98	14.900
Halite	4.650	2.04	9.680
Siderite	14.700	3.89	55.900
Pyrite	17.000	4.98	82.100
Barite	267.000	4.09	1065.000
Fresh Water	0.358	1.00	0.398
Salt Water 100k ppm NaCl	0.734	1.05	0.850
Salt Water 200k ppm NaCl	1.120	1.11	1.360
Oil (nCH_3)	0.119	$1.22\rho_{oil}$-0.118	$0.136\rho_{oil}$
Gas (CH_4)	0.095	$1.33\rho_{gas}$-0.188	$0.119\rho_{gas}$

After Schlumberger.

Table 12.4 lists some of the values of P_e, ρ_a, and U from the *Schlumberger Log Interpretation and Principles*.

The many types of cross-plots are limited only by the data available and the imagination of the analyst. The published ones include Th vs. K for clay types and volumes, Th vs. K for evaporite determination, EPT propagation time vs. $\rho_{ma,a}$ for rock type, P_e vs. K for clay type (Figure 12.8c), P_e vs. Th/K for clay type. There are many others.

Cross-plot of ρ_b, G_R, ϕ_{Na}, and R_t can be used in various combinations and ratios to analyze coals. A cross-plot of Th vs. U can suggest the makeup of a uranium deposit.

In many cases, identifying the location of the individual points can add a third dimension (depth or distribution) to the plot. Identifying the number of data points at each location can make the third dimension the frequency of mineral occurrence.

12.13.4 Simultaneous Equation Method

One of the original methods for analyzing complex lithologies is the use of simultaneous equations. A well known principle in mathematical solutions is the need for one equation for each unknown. These equations can be solved by substitution or by matrix methods. When programming for computer solutions, the matrix method is frequently the faster.

A limey sand of known porosity might be defined, for example, by

$$\rho_b = \phi\rho_f + (1-\phi)(V_s\rho_s + V_{ls}\rho_{ls}) \tag{12.24a}$$

where V indicates the fractional bulk volume of the indicated component, and

$$t = \phi t_f + (1-\phi)\left(V_s t_s + V_{ls} t_{ls}\right) \qquad (12.24b)$$

where t is the acoustic P-wave travel time. This pair of equations can now be solved for V_s and V_{ls}.

The simultaneous equation method can be extended to any number and type of relevant unknowns, as long as an independent equation can be written for each unknown item.

If it is assumed that the number of unknowns is limited and known, then one of the equations can define the situation:

$$1 = V_1 + V_2 + \ldots + V_n \qquad (12.24c)$$

Of course, if the actual number of components is greater than "n", an error is introduced. The error will be a factor of the number of unlisted unknowns, the difference of physical parameters from the norm, and the relative volumes of each.

12.13.5   Frequency Plots

Any of the statistical parameters can be plotted against time, depth, or other parameters to indicate the makeup, number, and distribution of the formation components.

The statistical plots can be analyzed by ordinary statistical methods to gain additional knowledge about the formation zones. A Gaussian distribution might normally be expected for a single component. A skewed distribution might indicate trace amounts of another component. A skewed distribution could also suggest a mechanical disturbance at some time in the past. Additional peaks could indicate additional components.

A hypothetical frequency plot of a sand with thin limey streaks is shown in Figure 12.14. This sand would have an effective average porosity of about 19%.

12.14   Permeability

Permeability is important in hydrocarbon projects because it determines the relative ease with which the can be produced. It can also allow an estimate of

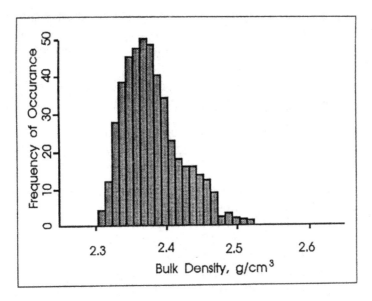

FIGURE 12.14
A hypothetical frequency plot.

the relative composition of the produced fluids. For an *in situ* leaching program or a petroleum water flood, the permeability must be known. A mine site must have water control, which will depend upon the permeability. Dam and other construction sites must be concerned with the flow of subsurface fluids, as well as their amounts, for construction integrity. A low-permeability engineering project site presents an entirely different problem from a contamination standpoint than does a high-permeability one.

Permeability is almost always an economic problem as well as a physical one. In a hydrocarbon-bearing horizon, regardless of the absolute reserve *in situ*, the lower the permeability the higher the production cost will be. Thus, there is usually a permeability cutoff below which the extraction cost is too high. In a mineral or engineering, and environmental cleanup project, the cost of the project typically increases as the permeability increases. This reflects the costs of the pumps, pumping, coffer dams, dikes, levees, permeability barriers, liners, and so on.

Wireline geophysical measurement of permeability is difficult and not well understood, as yet. The best methods to determine permeability are performed on core samples in the laboratory. Core analysis is well advanced and quite accurate. The two biggest problems are translating the laboratory results to subsurface conditions and the statistical error due to small sample size and sample variation.

Drillstem flow tests can give good information about the formation production permeability. These are commonly performed on oilwells before difficult production decisions are made. Somewhat cruder flow tests, but equally valuable, are routinely performed on water wells. These last, have the additional advantage of cleaning the producing sand of invading drilling fluids.

Permeability, k, is discussed in *Introduction to Geophysical Formation Evaluation*. In general form, k is defined by Darcy's law (Amyx, et al., 1960):

$$v_s = -\frac{k}{\mu}\left(\frac{dp}{ds} - \frac{\rho g}{1.0133} \frac{dz}{ds} 10^{-6} \right) \tag{12.25}$$

where

V_s	=	the volume flux across a unit area per unit time, along "s", in dynes/cm² atmospheres
k	=	the permeability of the sample, in darcys
μ	=	the viscosity of the fluid, in centipoise
dP/ds	=	the pressure gradient along "s", in atmospheres/cm
p	=	the fluid density, in g/cm³
g	=	the acceleration due to gravity, 980.665 cm/sec²
dz/ds	=	the vertical coordinate per unit length, cm/cm

In the steady state, horizontal flow, this reduces to

$$V_x = \frac{k}{\mu}\frac{dp}{dx} \tag{12.25a}$$

or

$$k = -Q\frac{\mu}{A}\frac{dx}{dp} \tag{12.25b}$$

where Q = the cross sectional area.

If the fluid is (or contains) gas, the compressibility of the fluid must be considered. Also, the Klinkenberg gas slippage factor, b, (which considers the effects when the pore diameter begins to approach the mean free path of the gas molecules) must be taken into account, because of the frequent small size of the pore spaces. This is especially important since, in the laboratory, the permeability is usually measured with a gas.

Thus, the permeability of a medium to a single-liquid phase which completely fills the pore space, k_L, is

$$k_L = \frac{k_g}{1 + \left(\dfrac{b}{P_{mean}} \right)} \tag{12.26}$$

where

k_g = permeability of a medium to a gas completely filling the pore space

b = the Klinkenberg constant for a given gas in a given medium

p_{mean} = the mean system pressure

Since most sedimentary beds are anisotropic, the vertical permeability, k_v, will be different from the horizontal permeability, k_h. This factor should be considered in both vertical and horizontal wells and is particularly important in horizontal wells.

As you can see from this, the determination of "k" is complex. It is convenient to estimate it with wireline methods. It should always be remembered, however, that the wireline permeability methods are empirical approximations and should be checked against core methods.

There has been progress, in recent years, in measuring "k" with wireline acoustic methods. The Stoneley wave velocity appears to be influenced by the permeability of the formation material. These methods, in general, are still in the experimental stage.

12.14.1   Permeability Estimated from the Resistivity Gradient

It is possible to estimate a probable average value of "k" from the gradient of the deep resistivity curve, if certain factors exist and certain events have occurred in the past (Tixier, 1949).

We must assume that our hypothetical permeable zone is relatively uniform in porosity, permeability, and composition. If a hydrocarbon exists in the zone, it may, because its density is lower than that of the formation water, have migrated upward in the zone. Then the residual saturation, below the main body of hydrocarbon, is a function of the time which has elapsed since the beginning of the process. Thus, it is also a function of depth, if the migration process is not complete. It also appears to be approximately linear with depth (*Schlumberger Principles and Applications*, 1989). This situation will appear on the deep-resistivity log (See Figure 5.2). The mean value of k is:

$$k = c\left(\frac{2.3a}{\rho_w - \rho_H}\right)^2 \tag{12.27a}$$

where

$$a = \frac{1}{R_o}\frac{R_{t_2} - R_{t_1}}{D_2 - D_1} \tag{12.27b}$$

and where

c = a constant, ≈ 20

R_{tn} = the value of R_t at depths 1 and 2, respectively

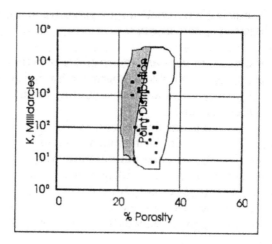

FIGURE 12.15
Core porosities and permeabilities from a California Well (After Schlumberger, 1989).

D_n = the depth at points 1 and 2, respectively
R_o = the value of R_t in the 100% S_w zone
P_w = the density of the formation water
P_h = the density of the hydrocarbon or gas

12.14.2   Permeability as a Function of Mineral content

Permeability appears to be a function of both porosity and tortuosity. It also is a function of grain size (*Schlumberger Principles and Applications,* 1989). Thus, it is reasonable to expect that it might be affected by mineral deposition within a zone. Mineral deposition can change all three of these factors. Figure 12.15 shows a permeability change of more than a factor of 4 magnitudes for a limited change of porosity (25% to 38%), which can be accounted for by post depositional mineralization.

The interaction between porosity and tortuosity is an interrelation which should be taken into account. It is normally not considered, however. At the very least, the value of "m" should be checked over the full range of zone porosities. An estimate or determination of tortuosity, with corrections applied to "m" is better.

12.15   Formation Fluids

12.15.1   Fluid Types

It is obvious that the formation fluid types should be an important consideration in petroleum work. It is also a factor, but lesser, in non-hydrocarbon investigations. The salinity of the formation water, the amounts and types of gases in the zone, the presence of even traces of hydrocarbons can change the situation on an oil effort, of a uranium project, deriving a coal mine plan, in environmental considerations, and many other situations.

The types and amounts of formation fluids can be well measured with wireline geophysical logs. Cores and cuttings are not reliable sources of formation fluids because of the washing by borehole fluids and the extreme temperature and pressure changes during recovery. Drill-stem tests, production tests, and formation tester samples are excellent and should be considered. They are quite reliable, but usually expensive. Surface measurement methods always have a high degree of uncertainty with respect to depth. This is because their depth component is historical, time, or electrical field, not actual physical depth.

Resistivity logs respond to the presence of water in the formation pore spaces and clays. Hydrocarbons and gases have high resistivities which are on the same order of magnitude as many of the common rock materials. Thus, rocks and hydrocarbons are indistinguishable from each other by resistivity measurements alone. Independent measurements of some factor of porosity are mandatory.

In a few locations, such as the coastal regions and the upper Amazon of Brazil and the shallow Rocky Mountain area of the U.S., the total dissolved solid (TDS) content of the formation waters is sometimes so low that the waters have extremely high electrical resistivities. In these locations it sometimes becomes difficult to distinguish between hydrocarbons and water with the standard logs alone. Possible methods are available, but are beyond the scope of this text (see *Non-Hydrocarbon Methods of Geophysical Formation Evaluation*).

The presence of gases is usually indicated by their high resistivities, acoustic signal distortions, and low densities. They also have low hydrogen densities, thus low apparent neutron porosities. In a liquid-phase, gases have about the same densities as the normally liquid hydrocarbons. In the gas phase, they follow the gas laws of Charles, Boyles, Klinkenberg, and others.

Liquified gases and liquid hydrocarbons both have lower densities than water. These densities of the liquid hydrocarbons are close enough to that of the formation water that a simple density measurement is not diagnostic by itself when it is diluted by the rock density component. Also, their hydrogen densities are near that of water, ruling out the use of the neutron porosity systems as saturation tools. Hydrocarbons are low enough density that they do float upward in water, however. Thus the characteristic signature is frequently recognizable on the resistivity curves.

The neutron porosity measurements respond well to hydrocarbons and gases, because of their sensitivity to the hydrogen density in the fluids. Liquid hydrocarbons appear much the same as water on the neutron ‘porosity curve. There is, however, a difference on the resistivity curve. In any gas, the hydrogen density is low. The neutron porosity curve shows this like a low porosity. The density porosity algorithm, however, assumes that the pore space is filled with water. Since the gas-phase of any fluid is usually lower density than the liquid-phase, the algorithm used in logging instruments interprets this as higher than true porosity. Thus, a comparison of the neutron and density porosity curves can locate the presence of gas phase pore fluids. This may be used to locate the respective depths for perforations and production purposes. It is also a good way to locate the top and transition zone of the water table. In a hard rock environment, it is one of the few ways to locate gas-filled isolated porosity. The petroleum industry uses this phenomenon as a "gas log." This is a neutron, density porosity log, usually combined with the resistivity curves. If the matrix parameters are set correctly, the three systems will indicate the same porosity in a water-filled environment. If the pore liquid contains liquid-phase hydrocarbon, the deep resistivity curve will read a lower porosity (higher resistivity) than the density and neutron (which will be nearly the same). If the pore space contains gas-phase gases, the resistivity reading will be high; the density porosity will be too high, but the neutron porosity will be too low. If the zone is shaly, the resistivity will be low, the neutron porosity will be high, and the density porosity will be near the true effective porosity. There are many uses for this procedure besides in the petroleum industry.

12.15.2   Fluid Distribution

Formation fluid distribution is closely related to the fluid types, the previous topic. Fluid distribution is a step beyond fluid types, however. In a single well or borehole, the two may be accomplished simultaneously. Distribution involves the horizontal extent and dynamic characteristics of the fluids, as well as the location at a single borehole. It also can involve the distribution of the phases in multiple phase fluids.

We normally expect certain conditions to exist. We expect gas to be above the oil in a hydrocarbon zone. We expect both to be above the water. We normally expect a porous zone to contain water and the water to contain dissolved solids; salts, especially NaCl. We further expect the total dissolved solids (TDS) to be nearly constant, both in amount and type, over a wide area. While these things are usually the way we expect them to be, this leads to simplifications and possible trouble, if we are not careful.

Gases are not always present in the formation in their gas-phase. Hydrocarbon gases, especially, but also non-hydrocarbon gases, are soluble in liquid hydrocarbons. Many are soluble to some extent, in water. If any of these things occur, the gas must be treated as a liquid. It will, of course, affect the physical characteristics of the liquid. These gases can also be dispersed in the drilling fluid and greatly alter its character. It should also be remembered that when a gas is in certain ranges of pressure and temperature, it can exhibit characteristics of a mixture of liquid and gas.

The horizontal distribution of fluid type can change, also. Common causes are anticlines and synclines, faults, pinchouts, changing shaliness, change of mineral deposition, and changes in porosity and permeability. If the subsurface fluid flow is not uniform and there are substantial inflows, the water can have salinity variations. In the Rocky Mountain region of the United States the snow melt each spring alters the salinity of the ground water to substantial depths. Salinity changes must also be expected in the vicinity of salt domes and other deposits.

The formation water salinity and resistivity can both be used to detect such dynamic features as salt plume down-dip of a dissolving salt dome and injected salinity tracers. For example, the *SP* potential, E_{SSP} is

$$E_{SSP} = -K_{SP} \log_{10}\left(\frac{R_{mfe}}{R_{we}}\right) \tag{12.28}$$

or

$$R_{we} = R_{mfe} 10\left(\frac{E_{SSP}}{K_{SP}}\right) \tag{12.29}$$

If there shale contamination is compensated for, careful measurement of E_{SP} and R_{mf} can locate salinity plumes by the lower resistivity of R_{we}, especially when plotted on a map.

The resistivity systems can also be used to locate plumes and check the SP. R_t, in the absence of any hydrocarbon and/or gas,

$$R_t \equiv R_o \tag{12.30}$$

In these circumstances, any of the deep resistivity measurements will measure R_o, when an independent porosity measurement is available:

$$R_o = R_w \phi^{-m} \tag{12.31a}$$

or

$$R_w = R_o \phi^m \tag{12.31b}$$

readily be distinguished from the changes of the salinity plumes. The plumes themselves will have a characteristic flowing shape, normally down the present dip.

Chemical differences in formation waters and other fluids are best measured by using laboratory techniques. This usually requires obtaining a sample of the fluid. This can easily be a difficult and expensive process. There are many laboratory methods, but will not be explained here.

Formation water samples can be taken by setting packers above and below the target zone. Fluids are then pumped until any contamination by borehole fluid invasion is eliminated. Samples should be collected in clean bottles and sealed immediately. The well number, date, depth, and length of time of the flow should be noted. The fluid temperature, during the discharge, can be close to the actual formation temperature (unless there are substantial quantities of gas dissolved in the water). It is worth while to measure and record this. Wireline fluid sample-takers are available. The are large diameter. They are also better for taking samples of the invading fluid than of the formation water.

Once laboratory values are obtained they can be plotted on a map, with the depth noted. It may be necessary to reduce all values to those at a standard temperature, so they can be compared. In the American oil fields this temperature is usually 65°F (18.3°C). Often, however, a standard temperature of 75°F (24°C) is used. *Be sure to check this*. This reduction has the further advantage of listing the values at a temperature used for mathematical calculations.

Analysis of formation water samples can determine the acid/base state (pH) (from which redox state (Eh) may be inferred), dissolved solid content and type, ion type and amount (if different from the TDS content), trace element content, dissolved gases (if the sample is analyzed soon after capture), radioactivity amount and type (from which Eh might also be inferred), and organic content.

This process, just described, can be used for studies of ground water flow, salt dome and contamination dynamics, tracer studies, cementing problems, and *in situ* leaching or water flood project planning.

12.15.3   Fluid Pressure

Formation fluid pressure is measured in a rough way when a borehole is drilled. The driller or mud engineer adjusts the weight of the drilling fluid to slightly over-balance and contain the formation fluid pressure. If the borehole pressure is too low, the formation fluids will enter the hole. The results of this can be quite destructive and expensive. Often, hole collapse and/or stuck drillpipe will occur. At the least, the drilling fluid will be diluted, destroying its properties. If the borehole fluid pressure is too high, lost circulation can result. Excess weighting materials in the mud can be quite costly. Controlling this problem is part of the skill of a good driller.

Formation fluid entering the borehole or mud filtrate leaving the borehole in small amounts will add a component to the spontaneous potential value. The equation for this component is

$$E_H = \frac{\xi R \zeta}{4\pi\mu} \Delta P \qquad (12.32)$$

where

    ζ = the dielectric constant of the moving fluid
    R = the resistivity of the moving water
    ς = the zeta potential, which is the potential due to the existence of fixed and moving layers of water
    μ = the viscosity of the moving water
    ΔP = the pressure drop across the interface

ΔP and ζ can be either polarity. Therefore, E_H can be either positive or negative.

This electrical phenomenon, seen by the spontaneous potential, has several names. It is the electrofiltration component of the normal *SP* curve. It is known in the oilfield as the "streaming potential." Its scientific name is the Helmholtz potential. At one time, about the beginning of the 20th century, it was thought to constitute the entire *SP* potential. We have since learned that the electrofiltration potential is only one of four or more components of the normal *SP* potential. The electrofiltration potential has been used to locate hot water entry in hard-rock geothermal wells.

There are other problems associated with formation fluid pressure. The weight of the formation materials above a zone — the overburden — is borne by the zone materials. The rock matrix of a zone essentially bears the weight of the rock material above it. The average bulk density of the sedimentary rock material is probably about 2.4 g/cm³. Thus, at 1000 ft (305 m) depth, the overburden pressure, due to the rock material is about 72 atmospheres.

The fluids above the zone are supported almost independently by the zone fluids. In average saline conditions, the fluid density is probably about 1.1g/cm³. Thus, the formation fluid pressure or head, at 1000 ft, is about 27 atm.

The rock "pressure" will be different in different directions — it is anisotropic. This is because of the structure of the rock. It will be greatest, however, vertically downward. The fluid pressure at any point will be the same in every direction. It will be a function of the average density depth and of the fluids above the point.

As shales are increasingly buried over time, the water between the clay platlets will be forced out (the shale will be compacted), but the shale solids will continue to support the overburden.

12.15.4   Overpressure

In some cases, however, the water is trapped within the shale. In this case, the water will eventually begin to support the rock material in addition to the fluid head. The rock overburden force will act in all directions, in this case. The shale then is said to be "overpressured". This can be a dangerous situation since the borehole pressure is adjusted to balance the *normal* fluid pressure.

Overpressure can be detected by the resulting changes in acoustic velocity, resistivity, and bulk density. The acoustic velocity is especially sensitive to this situation because the excess of water will cause the acoustic properties to skew toward the properties of the water, rather than normal shale properties.

Water/oil, gas/oil and gas/water interfaces can be detected from the log. If the measurement average gradient value is subtracted from the log reading, the difference will more easily point out the interface. The density and acoustic values may both be used for this. The resistivity curve is used for the water/hydrocarbon interface detection.

12.16　Structural Changes

Since the cores and logs, and even the cuttings, can be nearly as detailed or as general as we need them to be, these data can easily be used for determining the subsurface structure in any detail. The determination of the local stratigraphic or geologic column has already been discussed. Using the published geologic column as a reference, a local sequence can be determined.

The high-resolution focusing electrode resistivity systems are excellent for finding the partings in coals. This same technique can easily locate thin shale stringers which may drastically change the permeability. The density system is a good one to use to detect compositional changes in either sediments or in hard rock environments. Neutron activation systems are also good for this purpose.

A comparison of the density log, focused resistivity log, spectral gamma ray, and the acoustic log can give a detailed picture of a fracture and/or a fault system. The gamma ray, for example, may respond to both open and sealed fractures. The resistivity curve will respond to the open and shale-filled fractures, the density curve will show the differences between the open, shaly, and sealed fractures. The micro-resistivity and acoustic curves will give a picture of the relative sizes of the fractures. In hard-rock environments, the density and resistivity curves often respond well to the presence of faults and fractures.

Correlation of features from one borehole to another can show regional dip features, after converting depths to elevations. Close spaced gamma ray measurements can detail an ore body and, if a stabile calibrated gamma ray is used, allow grade calculation of some mineral types. Figure 12.16 shows a typical sedimentary uranium roll front deposit. Outlined by correlating five gamma ray logs through the mineralized zone.

Structural changes can often be detected by mapping the depths, or better, the elevations, of single features, such as the top of a zone. Figure 12.17 shows a hypothetical field which was drilled on an anticline and detected

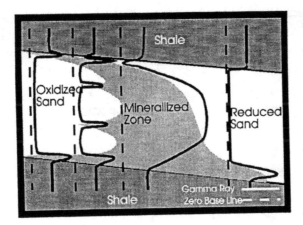

FIGURE 12.16
A hypothetical roll front and the probable gamma ray logs.

by mapping the elevation to the top of the same zone or formation for each hole. A regional dip was deliberately not included in the elevations. But, even with a regional dip, this method can show the probable locations of subsurface features. Figure 12.18 shows an area and its regional dip. There is probably a fault line running NE-SW.

12.17 Rock Formation Strength and Competence

Rock strength determination is a rapidly growing field at this time. Its importance is being realized and measurement and analysis techniques are being developed. Since many of the techniques involve methods which we have not yet discussed, the bulk of this category will deferred until a later volume. There are, however, several things relating to rock strength which can be gleaned from the standard methods which we have examined so far.

A fracture system in rock is usually filled with water. Even in a very porous rock this can often be detected. In a low or zero porosity rock, the contrast can be great between the rock and the water-filled fracture. The bulk density will be lowered, the bulk resistivity will be lowered, the acoustic velocity will be slowed, and the neutron porosity will be increased. Usually the mechanical integrity of the rock is affected. Surface seismic and vertical seismic profiling (VSP) are quite sensitive to faults.

The amount of the changes due to fractures will depend upon the relative volume of the fracture compared to that of the rock, within the measurement volume. To be detected by a resistivity system, the fracture volumes must be water- or clay-filled and interconnected. The signature of

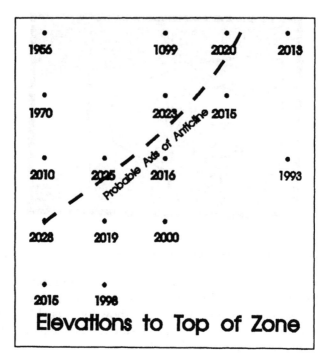

FIGURE 12.17
Use of log information to determine structure.

the fracture system will, of course, depend upon the character of the system and upon the type and resolution of the measurement. Large, widely spaced fractures may show up individually on any of them. A micro-fracture system in a granite or any other massive rock may be a smooth change of the measurement value where there should be none.

Fracture systems can also carrying flowing or moving fluids. Thus, a gamma ray log may detect the deposits from the uraniferous solutions that moved through them in the past. In some cases clays may have partially or wholly filled the fracture. In this case the radioactivity will contain energies characteristic of thorium decay products and, perhaps, potassium. These environments are detectable by the standard methods. Fluids entering or leaving the borehole by fractures will affect the spontaneous potential (*SP*) values. The *SP* will show unexpected changes where there should be predictable changes or none. This technique has been used by the Sandia Laboratories in geothermal wells.

Fracture surfaces and fill materials can reflect and/or attenuate acoustic waves. If the fractures are large (on the order of the acoustic wave length) and acoustic impedance contrast exists, the resulting interference patterns on a full-wave recording will be complex. Often they will show a typical "chevron" pattern.

The density log is a good indicator of competence in sediments. The higher the density, the more competent the zone is likely to be. This is frequently

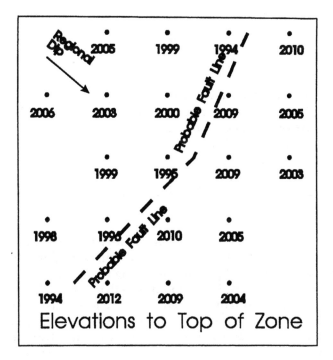

FIGURE 12.18
A probable fault, detected with logs.

used to estimate the probable "rippability" and stability of mine site over-burdens. The resistivity curve is useful in sediments because it tends to respond to cementation. Low porosity readings from any of the "porosity" systems are indicators of the same thing. Repeating or missing sequences sometimes indicates the presence of faults.

12.18 Zone Extent

The extent of a zone can best be done by drilling and mapping. The cores and logs must be carefully noted, correlated, and normalized. This process usually requires a multiplicity of drill holes. Any surface methods, such as resistivity, seismic, or induced polarization can greatly aid the task and must be carefully examined and integrated in the mapping.

12.19 Redox Effects

The redox potential is one of the components of the spontaneous potential. It is described by

$$E_{ox} = \frac{R_g T_k}{ZF} \ln \frac{kC_{ox}}{C_r} \qquad (12.33)$$

where
R_g = the universal gas constant
T_k = the absolute temperature, in degrees K
Z = the valence of the ion
F = Faraday's number
k = the reaction constant
C_{ox} = the concentration of the more oxidized component
C_r = the concentration of the less oxidized (more reduced) component

Note that because of the concentration term, the potential can be either positive or negative. This component is seldom noticed in hydrocarbon logging because the usual petroleum-bearing formation is deep and in a therefore more likely to be in a reduced state than are the shallower ones.

Redox (reduction/oxidation) phenomena are associated with many of the environments and operations with which we work. Many of the mineral deposits exist in particular redox states. They may have been transported and/or deposited by redox mechanisms. Hydrocarbon are excellent reducing materials; that is, they exist in highly reduced states. Atmospheric and surface redox-active materials have influences well below the surface. Fluids and solutions migrate from one redox condition to another. All of these actions are electrochemical and have and/or cause events we can exploit.

Oxidation is the loss of one or more electrons. It may or may not be the acquisition of an atom or ion of oxygen (O^{++}). Reduction is the opposite process. It is the gain of and electron or an atom or ion, or radical with a negative charge. We increase (oxidation) or reduce (reduction) the positive valence of the item gaining or losing the electron. It is an electrical phenomenon. The process of oxidation or reduction always changes the physical characteristics of a material. The differences, due to redox state, in physical properties of cuprous oxide (Cu_2O, cuprite) and cupric oxide (CuO, tenite) are compared in Table 12.5.

Many minerals are more soluble in water in one redox state than in the other. Usually, they are more soluble when oxidized. Thus, some oxidized metallic compounds are easily transported in solution. They are readily precipitated out of solution if the solution becomes more reduced. Many of our commercially valuable mineral deposits are in a reduced state, after being transported in solution and then encountering a highly reducing environment, such as sulfides or organic trash. Seepage of hydrocarbons frequently precipitates metallic compounds out of solution in a "halo" around the seepage. Redox state and pH often show sharp changes across the altered/unaltered interface of the surface oxidation and/or the water

TABLE 12.5

An example of the change of physical characteristics with a redox change

Material	Redox State	Molecular Weight	Color	Crystalline Form	Refractive Index	Density g/cm^3	Water Solubility	Other Solubility
Cuprous Oxide	Reduced	143.08	Red	Cubic Octahedral	2.705	6.0	Insol.	Sol. Acids
Cupric Oxide	Oxidized	79.54	Black	Isoclinic	2.63	6.3–6.49	Insol.	Insol.

TABLE 12.6

Approximate radioactvity contents of various minerals

Minerals	Potassium (%)	Uranium (ppm)	Thorium (ppm)
Accessory Minerals			
Allenite		30–700	500–5000
Apatite		5–150	20–150
Epidote		20–50	50–500
Monzanite		500–3000	$2.520(\times10^4)$
Sphene		100–700	100–600
Xenotime		$500–3.4\times10^4$	low
Zircon		300–3000	100–2500
Andesite (average)	1.7	0.8	1.9
Andesite, Oregon	2.9	2.0	2.0
Basalt			
Alkali basalt	0.61	0.99	4.6
Plateau basalt	0.61	0.53	1.96
Alkali olivine basalt	<1.4	<1.4	3.9
Tholeiites (orogene)	<0.6	<0.25	<0.05
(nonorogene)	<1.3	<0.05	<2.0
Basalt, Oregon	1.7	1.7	6.8
Carbonates			
Range	0.0–2.0	0.1–9.0	0.1–7.0
Average	(0.3)	2.2	1.7
Calcite, chalk, limestone, dolomite (all pure)	<0.1	<1.0	<0.5
Dolomite (clean, West Texas)	0.1–0.3	1.5–10	<2.0
Limestone (clean), Florida	<0.4	2.0	1.5
Texas, Cretaceous trend	<0.3	1.1–15	<2.0
Hunton Lime, Oklahoma	<0.2	<1.0	<1.5
West Texas	<0.3	<1.5	<1.5
Clay Minerals			
Bauxite		3–30	10–130
Glauconite	5.08–5.30		
Bentonite	<0.5	1–20	6–50
Montmorillinite	0.16	2–5	14–24

table. Basalt, feldspars, and other materials oxidize (alter) to various clay minerals. The physical changes accompanying a change of redox condition can be detected with our geophysical instruments and exploited to aid exploration and development. The electrical potentials, electrical resistivities, densities, and acoustic travel times *all* may change in a redox reaction.

The first redox state indicator to look for is the color. Examine the cores and cuttings closely. Bright colors usually indicate an oxidized state. Dull, dark colors usually indicate a reduced state. Every sedimentary uranium geologist knows white, yellow, red or orange sand and shale colors mean that the borehole is in the barren, oxidized interior of the deposit. The hole is still behind the main mineral body and he must look further down dip, until the colors begin to change.

Since the redox process is electrochemical, the various materials are accompanied by characteristic electrical potentials. This will affect the electric logs, especially the SP. While we do not read absolute potentials with the SP, we can see changes and relative potentials with standard equipment and a lead reference electrode in the mudpit. As the measure electrode approaches an oxidized zone, from a reduced one, the whole *SP* curve will shift to the left or negative direction. Of course, approaching a less oxidized (reduced) zone will result in a positive shift. This is especially evident in the shales above and below the zone.

The amount of the redox shift will depend upon the resistivity of the formation and the borehole (it is more evident in higher resistivities) and upon the amount of redox contrast. It may only be 2 to 3 millivolts or it may be 20 to 30 millivolts. The shift is also evident when approaching the surface oxidized zone in an open hole. The author has seen a shift of –90 millivolts at this zone.

The porosity logs can also be affected. Oxidation can result in a higher clay content with its higher bound water amount and longer acoustic travel time. The density log is not as sensitive, in most cases, unless the water table is being sought. One location uses the density system to locate the water table because gold deposits tend to be there, in that property.

12.20 Enhancement

Signal detection and (especially) transmission are not perfect. Any measurement will be distorted to some extent upon detection and transmission. Good engineering practice minimizes these problems, especially in the range of interest to a particular measurement. Substantial effort goes into designing detectors and transmission systems which will deliver signals which faithfully represent the measured events. The advent of digital techniques has greatly improved this aspect.

Even with the best design not all of the signal distortion is removed. For example, the gamma ray emission is a step function at a bed boundary. That is, the emission in a sand is low, while the adjacent shale emission is high. Across a sharp bed boundary the emission will change in a step fashion. But, because of the nature of gamma rays, their penetrating abilities, the detector, in our present systems, will always respond to the shale while still some distance in the sand (about 0.5 m) before actually arriving at the boundary. Likewise, it will still respond to the sand while a similar distance into the shale. This is true even if we disregard such design problems as detector response and finite length. The resulting response curve approximates a sinusoid across the boundary, with the curve inflection point at the bed boundary. The longitudinal distance of the distortion is a function of the geometry of the measured volume; its shape and distribution. This type of distortion is inherent in all of our present measurement systems, analog or digital.

The bed boundary problem can be reduced by increasing the resolution of the system, usually at the expense of increasing the clutter of greater detail. This is done by reducing the longitudinal extent of the measurement, preferably while retaining or controlling the lateral extent. There are, of course, limiting factors. Both surface and downhole measurements use these methods. Thus, we see focusing electrode resistivity systems and manipulation of induction logging fields.

The measurement volume geometries of induction logging systems are not as easily controlled as the focusing electrode systems. Induction logs are valuable, however, because of the ease with which lateral response can be controlled. Thus, borehole and near-borehole response can be minimized. The result is medium and deep lateral measurements requiring minimum correction.

The common combination resistivity measurement then is a deep and medium lateral depth investigation induction measurements and a shallow investigation, high resolution focusing electrode resistivity measurement.

There always will be, however, some bed boundary distortion in any present measurement curve, neutron, gamma ray, induction, focusing electrode, density, acoustic. Therefore, enhancement techniques have been developed to correct the response to more closely represent the probable true shape and character of the beds.

If the bed boundary response can be described (i.e., mathematically or electronically) a circuit or computer algorithm can apply the inverse distortion to the signal. This will make the signal more truly represent the true bed response. These are the enhancement techniques.

Enhancement techniques were first used to improve the photos and other measurements made by the Air Force and the NASA satellites. They were successfully applied to the gamma ray curve for uranium exploration and development (Czubek, 1978, 1983) as "deconvolution" techniques.

A simpler technique for gamma ray enhancement was developed by the Department of Energy (USDOE) as the Gamlog program. This was routinely run on both field and finished uranium logs for many years. Techniques which are apparently similar to the Gamlog program have been applied recently to induction log curves. These vary in complexity and ease of use. This program is discussed in Chapter 6.

Schlumberger Well Services, Inc. applies an enhancement to their Phasor Induction logs. It makes use of the quadrature component of the induced signal to correct the response of the conductivity component. The result is noticeably better in both resistivity reading, curve shape, and resolution than the standard dual induction system. This is discussed in Chapter 3.

BPB Instruments uses their Vectar (Vertical Enhancement by Contribution and Transformation of Associated Responses) to enhance their logs. "The vertical resolution of deep or compensated measurements, which usually have poor resolution, is enhanced by superimposing the resolution of a shallower and similar measurement onto the original curve " (BPB Instruments Vectar Services brochure). This is also discussed in Chapter 3,

The Phasor method is automatically a part of the Phasor Induction logs. The Vectar method can be applied to any induction log, density, neutron, and acoustic log.

The enhancement techniques can be quite valuable. They make the curve signatures much more realistic, more accurate, and often easier to read. Their value must be balanced, of course, against the time and cost of using them. Since the Phasor, Vectar and GAMLOG methods can be utilized in the field, they should probably be preferred over non-enhanced standard presentations. In all cases, the unprocessed record tapes should be retained.

12.21 Combinations

Many neglected combinations of measurements could be profitably used. The gold prospect mentioned in the previous paragraph might gain resolution if a neutron porosity system were used in combination with the density.

The author has used a combination of surface resistivity and downhole electric logs to get a three dimensional picture of a uranium prospect where the deposits tended to be at the salinity interface of marine and continental ground waters. A combination of surface induced polarization (IP) and resistivity with downhole electric and density logs was used to great effect on a silver/copper prospect.

12.22 Other Methods

There are many formation evaluation methods which have not been mentioned in this chapter nor in this text. Some are so common that they can easily be found. Many are listed in the Bibliography. Refer also to the contractor's literature, technical society papers and proceedings, where much of this information come from. Also, be sure to examine the books by Doveton, Dewan, Helander, Tittman, and Ellis.

Glossary

Symbols, Abbreviations, Subscripts, and Superscripts

A	Absolute, Area, Atomic weight (mass)
a	Apparent (partially or wholly uncorrected), Tortuosity factor
Ag	Silver
α	Alpha particle, proton
Å	Ångstrom unit
AMU, amu	Atomic mass unit
~	Approximately
$\approx$, $\simeq$	Approximately equal to
atm	Atmosphere of pressure
Au	Gold
API	American Petroleum Institute, Unit approved by the API
APIg	An arbitrary unit of gamma radiation; defined as 1/200th of the radiation from a typical Midcontinent shale.
APIn	An arbitrary unit of neutron intensity; defined as 1/1000th of the neutron emission in the Indian limestone of the API Neutron Model at the University of Houston
[AB]	Box AB, the distance between the two electrodes shown within the box. Any two electrode designations may be used.
B	Bulk Modulus
b	Bulk
Ba	Barium
β	Beta particle, Electron, ratio v:c
Bi	Bismuth
BIB	Bibliography
BHT	Bottom hole temperature
C	Carbon, Centigrade, Celsius, Conductivity (electrical)
Ca	Calcium
c	Electrical conductance, core
Cal	Caliper
cc	Cubic centimeter
CEC, C.E.C.	Cation exchange capacity
Ci	Curie
Cl	Chlorine
cl	Clay

cm	Centimeter
corr	Corrected
Csg, csg	Casing
D	Depth, Detector, Bit size (diameter)
d	Diameter, Darcy, Deep, Mathematical differential, dry, Diminution, deci-, ($\times 10^{-1}$), Rank difference
δ	Skin Depth
Δ	Difference
diff	Difference
dol	Dolomite
E	voltage, Volts, Voltage source, Young's Modulus
e	Effective, Electron
Eh	The potential of a half-cell, referred to a standard hydrogen half-cell, at any temperature
EMU, emu	Electromagnetic unit
est	Estimate
ESU, esu	Electrostatic unit
eV	Electron volt
F	Fahrenheit
FE	Formation evaluation
f, fl	Fluid, Flow line
ft	Foot
f, form	Formation
F_r	Formation resistivity factor
G	Geometrical factor
g	Gamma ray, Gram, Gas, giga-, ($\times 10^9$)
γ	Gamma ray photo, gamma ray
H	Hydrogen
h	Hydrocarbon, Borehole, Planck's constant
He	Helium
H_w	Hydrogen index
I	Electrical current
i	Invaded, Invasion, Invaded zone
IL	Induction log
I_r	Saturation index, Invaded,
ir	Irreducible
K	Potassium, Relative permeability
k	Permeability, kilo- ($\times 10^{-3}$)
keV	kilo (thousands of) electron volts
λ	Lambda, Wave length, Disintegration constant, Probability constant
L,I	Length, Linear distance, Lag
LL	Laterolog, focusing electrode
Log, log	Mathematical logarithm (base 10)
Ln, ln	Mathematical logarithm (base e)
ls	Limestone

M	Mass, Total rock cementation exponent, mega-, ($\times 10^6$)
m	Cementation exponent, meter, medium, mud, milli-, ($\times 10^{-3}$)
mc	Mudcake subscript
mf	Mud filtrate subscript
Mg	Magnesium
MeV	Mega (millions of) electron volts
ML	MicroLog
MLL	Microlaterolog, microfocussing electrode
MOP	Moveable oil plot
MOS	Moveable oil saturation
MSFL	Microspherically focussed log
μ	Micro, micron, viscosity, Shear Modulus
–	Negative, Minus (mathematical)
MWD	Measurement while drilling
N	Neutron
n	Saturation exponent, neutron, number (mathematical)
Na	Sodium
ν	Frequency
O	Oxygen
o	100% water saturated, oil, degree (superscript), initial (subscript)
Ω	Ohm
Ωm	Ohmmeter(s), Ohms meter(s)^2 per meter
P	Pressure, Phasor
p	Pipe
Pa	Protactinium
Pe	Photoelectric index
Pb	Lead
Φ	Angle of electron scatter
φ	Porosity
pe	Photoelectric
+	Positive, Plus (mathematical)
ppb	Parts per billion
ppm	Parts per million
%	Percent, 1/100
∝	Proportional to
psi	Pounds per square inch
PSP	PseudoSP (affected by clay mineral)
φ	Porosity
π	Pi, 1.414...
Q,q	Quantity, Volume per unit time
qz, qtz	Quartz
R	Electrical resistivity
r	Electrical resistance, Radium, Radial distance, Residual, Resistivity
Ra	Radium

Ref	Reference; References
ρ	Density
Rn	Radon
S	Saturation, Sulfur, Source, Shear
s	Saturated, Surrounding, Sand, Second
sh	Shale
Si	Silicon
σ	Micro cross section, Wave Number
Σ	Sum, Mathematical summation, Thermal neutron macro cross section
SP	Spontaneous (Self) potential
SSP, ssp	Static SP value
St	Stoneley
SWC	Schlumberger Well Services, Inc.
Sym, sym	Symbol
T	Time, Period
t	True, Unit (Interval) travel time, Circulation time, total, tortuosity
τ	Photoelectric cross section
TD	Total depth
Th	Thorium
Θ	Photon scattering angle
θ	Angle
Th	Thallium
θ	Angle
U	Uranium, Volumetric photoelectric index
V	Volume
v	Velocity
w	Water, wet
W	Weight, Work, Energy
WA	Western Atlas (Atlas Wireline) Logging Services
WC	Water cut
WGR	Water:gas ratio
WOR	Water:oil ratio
xo	Pertaining to the invaded zone
Z	Atomic number
Z/A	Atomic number: atomic mass ratio
/	Per, Divided by

Bibliography

Albright, J.N., Pearson, C.F., Acoustic transmissions as a tool for hydraulic fracture location, *JPE*, Dallas, August 1982.

Alger, Robert P., Harrison, Charles W., Improved fresh water assessment in sand aquifers utilizing geophysical well logs, *The Log Analyst*, Vol. 30, No. 1, SPWLA, Houston, Jan-Feb 1989.

Allaud, Louis and Martin, Maurice, Schlumberger, *The History of a Technique*, John Wiley and Sons, 1977.

Allen, L.S., Tittle, C.W., Mills, C.W., Caldwell, R.L., Dual-spaced neutron logging for porosity, *Geophysics*, Vol. 32, no. 1, pp60-68, 1967.

Amyx, J.W., Bass, D.M., Whiting, R.L., *Petroleum Reservoir Engineering*, McGraw-Hill, Inc., New York, 1960.

Archie, Gerald E., 1942, The Electrical Resistivity Log as an Aid in Determining Some Reservoir Characteristics, *A.I.M.E. Transactions*, v. 146, pp.54-61.

Bigelow, Edward L., *Fundamentals of Diplog Analysis*, Dresser Atlas, Dresser Industries, Houston, 1987.

Boatman, E. M., An Experiment of Some Relative Permeability — Relative Electrical Conductivity Relationships, Unpublished Master's Thesis, Dept. of Pet. Eng., Uni. of Texas, Austin, Texas, June, 1961.

Bothe, W. and Becker, Z., *Zeitung Physik*, Vol. 66, 1930.

Brown, A., and Hussein, S., Permeability from Well Logs, Shaybah Field, Saudi Arabia, Transactions of the 18th S.P.W.L.A. Symposium, 1977.

Burke, J.A., Schmidt, A.W. and Campbell, R.L.,Jr., The litho porosity cross plot, SPWLA, *The Log Analyst*, Vol. X, No. 6, Houston, Dec. 1969.

Cassel, Bruce, Vertical Seismic Profiles — An Introduction, Western Geophysical Company, Middlesex, U.K., 1984.

Calhoun, John C., Jr., Fundamentals of Reservoir Engineering, University of Oklahoma Press, Norman, 1955.

Campbell, Wm. M. and Martin, J.L., Displacement Logging — A New Exploratory Tool, *Journal of Petroleum Technology*, Dec. 1955, pp 233-239.

Carmen, P. C., *Flow of Gases Through Porous Media*, Academic Press, Inc., New York, N. Y., 1956.

Chadwick, J., *Nature*, 1932.

Chapellier, Dominique, *Diagraphies Appliquees a l'Hydrologie*, Lavoisier TEC & DOC, Paris, 1987.

Chombart, L.G., Well Logs in Carbonate Reservoirs, *Geophysics*, Vol. XXV, No.4, Tulsa, 1960.

Clark, Isobel, *Practical Geostatistics*, Applied Science Publishers, London, 1979.

Clavier, C., Coates, G., Dumanoir, J.L., The Theoretical Basis for the "Dual Water" Model for the Interpretation of Shaly Sands, SPE paper 6859, Denver, Colorado, October 1977.

Clavier, C. and Rust, D.H., MID Plot: A New Lithology Technique, SPWLA, *The Log Analyst*, Vol. XVII, No. 6, Houston, Nov-Dec 1976.

Cork, James M., *Radioactivity and Nuclear Physics*, D. Van Nostrand Co., Inc., Princeton, N.J., 1957.

Crain, E.R., *The Log Analysis Handbook*, PennWell Publishing Co., Tulsa, 1986.

Czubek, Jan, Natural Selective Gamma Ray Logging. A New Log of Direct Uranium Determination, *Nukleonika*, Vol. 13, 1968.

Czubek, J.A., Lenda, A., *G-Function in Gamma Ray Transport Problems*, Report No. 1042/Pl, Institute of Nuclear Physics, Kraków, 1978.

Czubek, J.A., Advances in Gamma Ray Logging, *Nuclear Geophysics*, Oxford, 1983.

Dakhnov, V.N., (Keller, G.V., ed.), Geophysical Well Logging, *Colorado School of Mines Quarterly*, Vol. 57, No.2, Golden, Colorado, 1962.

Darcy, H., *Les Fontaines Publiques de la Ville de Dijon*, Victor Dalmont, Paris, France, 1856.

Degolyer, E., Notes on the Early History of Applied Geophysics in the Petroleum Industry, *The Journal of the Society of Petroleum Geophysicists*, Division of Geophysics, A.A.P.G., Vol. VI, No. 1, July 1935.

Dewan, J.T., *Essentials of Modern Open Hole Log Interpretation*, PennWell Books, Tulsa, 1983.

Desbrandes, Robert, *Encyclopedia of Well Logging*, Gulf Publishing Co., Houston, 1985.

Doll, Henri G., The SP Log in Shaly Sands, AIME Paper T.P. 2912, 1949.

Doveton, John H., *Log Analysis of Subsurface Geology*, John H. Wiley and Sons, 1986

Dresser Atlas, *Well Logging and Interpretation Techniques*, Dresser Industries, Houston, 1982.

Dresser Atlas, Log Interpretation Charts, 1985

Eberline, H.C. and Shreve, J.D. Jr., Natural Resources Exploration: U.S.A.E.C. *Californiun-252 Progress*, Vols. 7-12, 1971.

Eisler, P.L., Huppert, P., Wylie, A.W., Logging of Copper in Simulated Boreholes by Gamma Spectroscopy, 1. Activation of Copper by Fast Neutrons, *Geoexploration*, 9 (1971) pp. 181-194.

Ellis, D.V., *Well Logging for Earth Scientists*, Elsevier Science Pub., New York, 1987.

Englehart, W. V., and Pitter, H., Über die Zusmamenhangen Zwischen Porositat, Permeabilitat, und Korgrobe bei Sanden und Sandstein, *Heidel. Beitr. Petrogr.*, 2, 1951.

Fertl, W.H., *Abnormal Formation Pressures*, Elsevier Scientific Publishing Company, Amsterdam/New york (1976)

Fertl, W.H., Status of Shaly Sand Evaluation, CWLS 4th Formation Evaluation Symposium, paper 1, Calgary, Canada, May 1972.

Fertl, W.H., 1979, Gamma Ray Spectral Data Assists in Complex Formation Evaluation, Transactions, 6th S.P.W.L.A. European Formation Evaluation Symposium, London, England.

Fertl, W.H., and Vercellino, W. C., Predict Water Cut from Well Logs, *Oil and Gas Journal*, June 19, 1978.

Fertl, W.H., Wichmann, P.A., How to Determine Static BHT from Well Log Data, *World Oil*, January 1977.

Frasier, D.C., Keevil, N.B., Jr. and Ward, S.H., 1964, Conductivity Spectra of Rocks from the Craigmont ORE Environment, *Geophysics*, V. 29, no. 5, pp. 832-847.

Frost, E., Fertl, W.H., October 1979, Integrated Core and Log Analysis Concepts in Shaly Clastic Reservoirs, CWLS Meeting, Calgary, Canada.

Garrels, Robert M. and Christ, Charles L., 1965, *Solutions, Minerals, and Equilibria*, Freeman, Cooper & Company, San Fransisco, CA.

Gearhart Industries, *Formation Evaluation Data Handbook*, Fort Worth, Texas, 1982

Geyer, R.L. & Myung, J.I., The 3-D Velocity Log; A Tool for In-Situ Determination of the Elastic Moduli of Rocks, Seismograph Service Corporation, Tulsa, 1970.

Givens, W.W., Mills, W.R., Dennis, C.L.., Caldwell, R.L., Uranium Assay Logging, Using a Pulsed 14-MEV Neutron Source and Detection of Delayed Neutrons, *Geophysics*, Vol. 41, No. 3, Tulsa, June 1976.

Glossary of Terms and Expressions Used in Well Logging; 2nd Edition, Society of Professional Well Log Analysts, Houston, 1984.

Goldman, L., Gold Detection In-Situ (personal communication), 1975

Gondouin, M., Tixier, M.P., and Simard, G.L., An Experimental Study on the Influence of the Chemical Composition of Electrolytes on the SP Curve, *Journal of Petroleum Technology*, Feb. 1957.

Green, William, R., *Computer-Aided Data Analysis*, John Wiley & Sons, New York, 1985.

Guy, J.O., Smith, W.D.M., Youmans, A.H., The Sidewall Acoustic Neutron Log, Conference paper: S.P.W.L.A. Twelfth Annual Logging Symposium, May 2–5, 1971.

Hallenburg, J.K., *HP41C Formation Evaluation Programs*, PennWell Books, Tulsa, 1984.

Hallenburg, J.K., Logcomp, *Petroleum Formation Evaluation Programs*, PennWell Books, Tulsa, 1985.

Hallenburg, J.K., *Introduction to Geophysical Formation Evaluation Methods*, Lewis Publishers, Boca Raton, FL, 1997.

Hallenburg, J.K., *Nonhydrocarbon Methods in Geophysical Formation Evaluation*, Lewis Publishers, Boca Raton, FL, 1997.

Helander, D.P., *Fundamentals of Formation Evaluation*, OGCI Publications, Tulsa, 1983.

Hertzog, R.C., Laboratory and Field Evaluation of an Inelastic Neutron Scattering and Capture Gamma Ray Spectrometry Tool, *Society of Petroleum Engineers Journal*, Oct. 1980,pp. 327-340.

Holt, Owen R., Relating Diplogs to Practical Geology, Dresser Atlas, Dresser Industries, Houston, 1980.

Jones, P. J., Production Engineering and Reservoir Mechanics (Oil, Condensate, and Natural Gas), *Oil and Gas Journal*, 1945.

Keller, G.V. and Frischknecht, F.C., Electrical Methods in Geophysical Prospecting, Pergamon Press, Oxford, 1966.

Keller, G.V., Electrical Prospecting for Oil, *Colorado School of Mines Quarterly*, Vol. 63, No. 2, April 1968.

Koerperich, E.A., Shear Wave Velocities Determined from Long and Short Spaced Borehole Acoustic Devices, *JPE*, October 1980, Dallas.

Koerperich, E.A., Investigation of Acoustic Boundary Waves and Interfering Patterns as Techniques for Detecting Fractures, *JPE*, Dallas, August 1978.

Lapp, R.E., and Andrews, H.L., Nuclear Radiation Physics, Prentice-Hall, Inc., New York, 1949.

Larinov V.V., Borehole Radiometry, Nedra, Moscow 1969

Lawson, B.L., Cook, C.F., Owen, J.D., A Theoretical and Laboratory Evaluation of Carbon Logging: IV. Laboratory Evaluation, SPE Paper 2960, A.I.M.E., 1970

Lawrence, Tony D., Continuous Carbon/Oxygen Log Interpretation Techniques, SPE Paper 8366, A.I.M.E., 1979.

Leburton, F., Sarda, Q., Trocqueme, F., and Morlier, P., Logging in Porous Media to Evaluate the Influence of their Permeability on Acoustic Waveforms, Paper Q, *S.P.W.L.A. Symposium Transactions*, Houston, 1978

LeRoy, L.W., and LeRoy, D.D., Subsurface Geology, *Colorado School of Mines*, Golden, Colorado, 1977

Lynch, Edward J., Formation Evaluation, Harper & Row, New York, 1962

Meehan, D.N. and Vogel, E.L., *HP41C Reservoir Engineering Manual*, PennWell Books, Tulsa, 1982.

Meyers, John E., Jr. High Temperature Helium-3 Detectors, Conference Paper, Institute of Electrical and Electronic Engineers, Boston, Mass., October 19-21, 1966.

Morris, R. L., and Biggs, W. P., Using Log-Derived Values of Water Saturation and Porosity, Transactions of the 1967 *S.P.W.L.A. Symposium Transactions*, 1967.

Myung, J.I. & Helander, D.P., Correlation of Elastic Moduli Dynamically Measured by *In-situ* and Laboratory Techniques, 13th Annual Logging Symposium, *S.P.W.L.A. Symposium Transactions*, Tulsa, 1972.

Myung, J.L.,& Henthorne, J., *Elastic Property Evaluation of Roof Rocks with 3-D Velocity Logs*, Solution Mining Research Institute, Atlanta, 1971.

NL Baroid/NL Industries, Inc., *Manual of Drilling Fluids Technology, the History and Functions of Drilling Mud*, Vol 1, Section 1, 1979.

Oliver, D.W., Frost, E., Fertl, W.H., *Carbon/Oxygen Log*, Dresser Atlas, Dresser Industries, Houston, 1981.

Overton, H.L., Lipson, L.B., A Correlation of Electrical Properties of Drilling Fluids with Solid Content, *A.I.M.E.*, 213:333-336, 1958.

Pickett, R. and Kwon, B.S., A New Pore Structure Model and Pore Structure Inter-relationships, Paper P, *S.P.W.L.A. Symposium Transactions*, Houston, 1975.?

Pirson, S.J., 1935, Effect of Anisotropy on Apparent Resistivity Curves, *Bull. A.A.P.G.*, v. 19, no. 1, pp37-57.

Raiga-Clemenceau, J., A Quicker Approach of the Water Saturation in Shaly Sands, Paper E, *S.P.W.L.A. Symposium Transactions*, Houston, 1976.

Raymer, L.L. and Biggs, W., Matrix Characteristics Defined by Porosity Computations, Schlumberger Well Services, c1970.

Raymer, L.L. and Hunt, E.R., An Improved Sonic Transit Time — To Porosity Transform, 21st *S.P.W.L.A. Symposium Transactions*, Houston, 1980.

Roy, A. and Apparao, A., Depth of Investigation In direct Current Methods, *Geophysics*, Vol. 36, No. 5, Oct. 1971, pp943-959.

Recommended Practice for Determining Permeability of Porous Media, American Petroleum Institute, APR RP No. 27, Sept. 1952.

Schlumberger, A.G., The Schlumberger Adventure, Arco Publishing, Inc., New York, 1982.

Schlumberger Educational Services, Log Interpretation Principles/Applications, Houston, 1987.

Schlumberger Well Services, Inc., Log Interpretation Charts, 1986.

Sentfle, F.E., Hoyte, A.F., Mineral Exploration and Soil Analysis Using *in situ* Neutron Activation, Nuclear Instruments and Methods, Vol. 42, pp 93-103.

Sharma, P.V., Geophysical Methods in Geology, Elsevier, Amsterdam, 1986.

Sheriff, R.E., Encyclopedic Dictionary of Exploration Geophysics, Society of Exploration Geophysics, Tulsa, 1973.

Society of Professional Well Log Analysts, The Art of Ancient Log Analysis, 1979.

Steiber, S.J., Pulsed Neutron Capture Log Evaluation in the Gulf Coast, *SPE Paper 2961*, Texas, 1970.

Steinman, D.K., John, J., *Californium-252-Based Logging System for In-Situ Assay of Uranium Ore*, IRT Corporation, San Diego, Cal., undated (c1977).

Timur, A., An Investigation of Permeability, Porosity, and Residual Water Saturation Relationships for Sandstone Reservoirs, *The Log Analyst*, July–August, 1968.

Tittman, Jay, Geophysical Well Logging, Academic Press, Inc., Orlando, Florida, 1986.

Tixier, M.P., June 16, 1949, Evaluation of Permafrost from Electric Log Gradients, *Oil and Gas Journal.*

Vennard, John K., Elementary Fluid Mechanics, John Wiley & Sons, New Yorl, 1961.

Weast, Robert C. (ed), *CRC Handbook of Chemistry and Physics*, 61st Ed., CRC Press, Inc., Boca Raton, Florida, 1981.

Wichmann, P.A., Hopkinson, E.C., McWhirter, V.C., The Carbon/Oxygen Log Measurement, S.P.W.L.A., Dresser-Atlas publication, 1977.

Winn, R.H., A Report on the Displacement Log, *Journal of Petroleum Technology*, Feb. 1958.

Winsauer, W.O., Shearin, H.M.,Jr., Masson, P.H., and Williams, H., 1952, Resistivity of Brine Saturated Sands in Relation to Pore Geometry, *American Association of Petroleum Geology Bulletin*, V 36, no. 2, pp253-277.

Woolson, W.A. and Gritzner, M.L., 1977, Borehole Model Calculations for Direct Uranium Measurement with Neutrons, The U.S. Energy Research and Development Administration, Grand Junction, CO 81501.

Woolson, W.A. and Gritzner, M.L., 1978, Evaluation Models of Active Neutron Logging Tools for Direct Uranium Measurement, The U.S. Department of Energy, Grand Junction, CO 81501.

Wyllie, M.R.J., Gregory, A.R., and Gardner, G.H.F., Elastic Waves in Heterogeneous and Porous Media, *Geophysics*, Vol. 21, No. 1, Tulsa, 1956.

Youmans, A.H., Wilson, J.C., Lebreton, B.F., Oshry, H.I., Lane-Wells Company paper, date unknown.

Zemenek, J., Low-Resistivity Hydrocarbon-Bearing Sand Reservoirs, SPE Paper 15713, Dallas, 1987.

Index

9 780367 579401